Edited by
Egbert Dittrich

The Sustainable Laboratory Handbook

Related Titles

DiBerardinis, L.J., Baum, J.S., First, M.W., Gatwood, G.T., Seth, A.K.

Guidelines for Laboratory Design

Health, Safety, and Environmental Considerations

4th Edition
2013
Print ISBN: 978-0-470-50552-6; also available in electronic formats

Wolfensohn, S., Lloyd, M.

Handbook of Laboratory Animal Management and Welfare

4th Edition
2012
Print ISBN: 978-0-470-65549-8; also available in electronic formats

Zhou, M. (ed.)

Regulated Bioanalytical Laboratories

Technical and Regulatory Aspects from Global Perspectives

2011
Print ISBN: 978-0-470-47659-8; also available in electronic formats

McMaster, M.C.

Buying and Selling Laboratory Instruments

A Practical Consulting Guide

2010
Print ISBN: 978-0-470-40401-0; also available in electronic formats

Edited by Egbert Dittrich

The Sustainable Laboratory Handbook

Design, Equipment, Operation

WILEY-VCH
Verlag GmbH & Co. KGaA

Editor

Egbert Dittrich
EGNATON e. V.
Mühltalstr. 61
Germany

Cover
The background image was kindly provided by WALDNER Laboreinrichtungen GmbH & Co. KG
Test tube. Source: istockphoto.com © pilipipa

FSC
www.fsc.org
MIX
Paper from responsible sources
FSC® C013604

Library of Congress Card No.: applied for

British Library Cataloguing-in-Publication Data
A catalogue record for this book is available from the British Library.

Bibliographic information published by the Deutsche Nationalbibliothek
The Deutsche Nationalbibliothek lists this publication in the Deutsche Nationalbibliografie; detailed bibliographic data are available on the Internet at <http://dnb.d-nb.de>.

Print ISBN: 978-3-527-33567-1
ePDF ISBN: 978-3-527-67954-6
ePub ISBN: 978-3-527-67955-3
Mobi ISBN: 978-3-527-67956-0
oBook ISBN: 978-3-527-33709-5

Typesetting Laserwords Private Limited, Chennai, India

Contents

List of Contributors

Markus Bauch
Industriepark Höchst
65926 Frankfurt
Germany

Thomas Brock
BGRCI
Kurfürstenanlage 62
69115 Heidelberg
Germany

Egbert Dittrich
Labdicon-Dittrich Consulting
Mühltalstr. 61
64625 Bensheim
Germany

Peter Dockx
Van Looy
Nordersingle 19
B2140 Antwerp
Belgium

Eberhard Dürr
Bindergasse 6
72131 Ofterdingen
Germany

Tobias Ell
Carpus+Partner AG
Wilhelmstraße 22
89073 Ulm
Germany

Oliver Engel
Industriepark Höchst
Siemens AG
65929 Frankfurt
Germany

Thomas Gasdorf
Broen-Lab
Drejervaegt 2
5610 Assens
Denmark

Walter Glück
Thermo Fisher Scientific
Robert-Bosch-Strasse 1
63505 Langenselbold
Germany

Markus Hammes
hammeskrause architekten
Krefelder Str. 32
70376 Stuttgart
Germany

Christoph Heinekamp
Dr. Heinekamp Labor- und
Institutsplanung GmbH
Gaußstr. 12
85757 Karlsfeld
Germany

Stefan Krause
Industriepark Höchst
Sanofi Aventis
65926 Frankfurt
Germany

Andreas Kühne
Bauakademie
Alexanderplatz 9
10178 Berlin
Germany

Thomas Lischke
Carpus+Partner AG
Forckenbeckstraße 61
52074 Aachen
Germany

Hansjürg Lüdi
H. Lüdi + Co. AG
Moosäckerstr. 86
CH8105 Regensdorf
Switzerland

Helmut Martens
MartensLabConsult
Schatzbach 31
84364 Bad Birnbach
Germany

Ines Merten
Dr. Heinekamp Labor- und
Institutsplanung
Schützenmattstr. 27
4051 Basel
Switzerland

Bernhard Mohr
Friedrich-Alexander-Universität
Erlangen-Nürnberg
Departement of Chemical and
Biological Engineering
Cauerstr. 16
91058 Erlangen
Germany

Ina-Maria Müller-Stahn
Dr. Heinekamp Labor- und
Institutsplanung GmbH
Gaußstr. 12
85757 Karsfeld
Germany

Ali-Yetkin Özcan
Industriepark Höchst
Infraserv
65926 Frankfurt
Germany

Jörg Petri
Bayer Pharma AG
13342 Berlin
Germany

Maike Ring
Carpus+Partner AG
Forckenbeckstraße 61
52074 Aachen
Germany

Fritz Runge
Bauakademie
Alexanderstr. 9
10178 Berlin
Germany

Roland Rydzewski
Simtecto GmbH
Johannes-Gutenbergstr. 1
87224 Ottobeuren
Germany

Berthold Schiemenz
Industriepark Höchst
Sanofi Aventis
65926 Frankfurt
Germany

Bernd Schubert
Tintschl BioEnergie und
Strömungstechnik AG
Goerdelerstr. 21
91058 Erlangen
Germany

Gordon P. Sharp
Aircuity Inc.
Newton
MA 2458
USA

Michael Staniszewski
Dr. Heinekamp Labor- und
Institutsplanung
Gaußstr. 12
85757 Karlsfeld
Germany

Norbert Teufelhart
GLS Gesellschaft für
Laborsicherheit mbH
Gaußstr. 12
85757 Karlsfeld
Germany

Christian Völk
Düperthal Sicherheitstechnik
Frankenstr.
63791 Karlstein
Germany

Franz Wermelinger
H. Lüdi + Co. AG
CH8105 Regensdorf
Switzerland

Burkhard Winter
Zum Fuellerwald 4
66664 Merzig
Germany

Hermann Zeltner
Dr. Heinekamp Labor- und
Institutsplanung GmbH
Gaußstr. 12
85757 Karlsfeld
Germany

Preface

The laboratory research world permanently moves forward; there is no stagnancy because it is inherent within research in general to achieve new horizons. Hence, buildings and their technical engineering services, laboratory equipment and plug-in units, and even architectural concepts are subjects of change.

Research tasks are complex and need interdisciplinary teams comprising chemists, biologists, pharmacists, and other experts from the entire life science, collaborating within teams under an increasing time pressure. The essential framework is surely communication gained mainly by architecture. But permanent communication between decision makers and planners is a precondition for a successful planning process.

This book shall perform pathways beside the state of the art on the search for optimally designed and efficiently operationable laboratories. The results are particular contradictions and different approaches from single authors whose equitable unassessed reference for me as publisher is important.

We are unable to perform general concepts for planning and management, but only can give hints in terms of a large number of scenarios and strategies that have to be carefully verified by operators, planners, and users regarding their applicability for each single laboratory situation. The reader should not be surprised if the content does not raise instantaneous solutions. These may be found by an iterative approach only and may be changed until the last day in order to strive for the best concept.

Laboratory equipment, planning, and technology and research are interdisciplinary and unfortunately did not generate any apprenticeship or studies for the time being. Thus, there is neither an accepted curriculum nor generally accepted basic knowledge as a precondition. We further note that to date experts have been recruited from relevant disciplines and educated oneself in a practice-orientated way. This book wants to contribute a wider basic expertise to all of those working with different disciplines on laboratory planning. Due to heterogeneity of the laboratory world, a large number of authors get a word in edgewise. Sincere thanks are given to the authors listed. They gained the articles besides their daily work. The outline of the book follows a chronological order of a laboratory building. The beginning is made with planning issues followed by properties of technical equipment, laboratory furniture and individual equipment including safety aspects. The last chapter is addressed to operation and management. It comes down to meet safety regulations and standards as well as efficiency and quality.

In part two, a chapter in terms of sustainability aspect is prefaced as a joining element. It is embodied that sustainability can be achieved only in context with design, technology and users and mainly depends on the motivation of users. As a matter of fact, recent initiatives of sustainable laboratories, such as EGNATON e.V. European Association for Sustainable Laboratories, have come to a large number of conclusions, rudimentary presented in this book. Deserving of thanks furthermore the DGNB – Deutsche Gesellschaft für nachhaltiges Bauen – has developed a guideline for sustainable laboratory buildings. It has been extended and completed by the EGNATON certification system for laboratory plug-in units.

This book provides more attraction and productivity of laboratory locations and enables the operators to recruit high potentials within international competition with perfect design and optimized equipment.

A major part of sustainability is taken by human factors representing beside others sensibility and mood of human beings, correlating closely with their environment. Building design has a leading role here. Furthermore, architecture of research buildings is the bridge to the public area. As publisher in this respect, I would like to emphasize that architects are burdened with the responsibility to clearly signalize sustainability of the building as well as functionality and efficient technology.

Bensheim, Summer 2011 *Egbert Dittrich*

What has changed since?

Sustainability has taken the first hurdle and became a major topic within the past three years for laboratories as well as for other construction sites.

Leading entities in the research business take care of the issue more and more in a structured way. The pilot phase of the first laboratory building certification by DGNB has shown some minor changes that need to be carried out without crossing the red line.

EGNATON started to establish the certification of laboratory apparatus and plug-in units and succeeds to point out the importance of equipment and user behavior as a major role to achieve sustainability.

Even awareness in terms of sustainability has been raised.

The intuition clearly is: A single entity or organization has the ability neither to define nor to put into practice sustainable strategies for its own. Collaboration in the business, between research organizations, companies and not at least countries, is essential to establish a common understanding of sustainability. Although some steps had to be revised, the way must be imperturbably followed. Structures and definition of the issue must become manifest in order to create acceptance by a common trans-boundary sense.

Bensheim, Spring 2015 *Egbert Dittrich*

Part I
Laboratory Building and Laboratory Equipment – Subjects of Laboratory Design of Building and Equipment

Egbert Dittrich

In particular, the design and planning of research buildings, where the recruitment of researchers and their teams in the planning phase often lags behind, the planner faces with the daunting task of developing an adequate equipment of buildings without question rounds. The planner, therefore, has to rely on his experience or more neutral research in other similar institutions. The user-independent planning – and flexibility – is therefore of considerable importance. This is even more difficult than expected, especially when high-end user, in a later stage recruited, require equipment that relieves them of any technical problems in the use and leaves no wishes unfulfilled.

Modern research in the life sciences will be held today in large interdisciplinary teams, that is, only permanent exchange of knowledge and interaction can achieve results in a sufficiently short time. This requires a (not quite) new philosophy with respect to the size of the rooms, communication zones and zoning, and equipment. Laboratory planning in turn requires, due to the strong technical facilities, also planning teams and can be implemented by individual planners barely. Thus, also the successful overall planning is under the dictates of transparency, communication in real time, and the subordination of individual interests for the entire planning team. Only an integrated design can claim to build a building of the highest efficiency at the end.

So to speak, by means of the given severe impairment of users by laboratory work, safety aspects, and so on, the risk creates that negative effects, such as drafts, noise, or emissions by incorrect negatively planning effect the work of these users or even make work impossible. To achieve safety in the technical sense of planning, plans are to be verified by simulations, at least for large complex laboratories.

Furthermore, laboratory buildings provide a central point for many companies of their corporate identity, that is, the satisfaction of the users, and their work also correlates to the benefit of the whole society with the appearance of the building and the wellbeing inside. Sociologically, organizations continue to play a role that results in the highend area of the laboratory sciences requiring an optimally

The Sustainable Laboratory Handbook: Design, Equipment, Operation, First Edition.
Edited by Egbert Dittrich.

designed (liberal) environment. This circumstance also requires the planners utmost efforts and the high risks involved.

The approaches presented in the following chapter reflect different directions by means of their variety and diversity, partly again the difficult path finding.

The industry requests and the attempt was made to provide laboratory typologies, representing as it were a guide for directions on which they can draw repetitive again and again. The chapter of Heinenkamp expresses the difficulties, precisely because no laboratory is alike.

Not without consideration, the chapter of two users (Schiemens and Krause) has been adjusted at the beginning of the chapter with general planner considerations, in which the authors describe the process of decision making in a laboratory planning from their perspective. It should not be overlooked that the statements and wishes of users are essential to the successful planning. The separation of structurally not achievable needs from the state of the art in areas where possible developments can be realized and necessarily increase the productivity and the structuring of the user wishes, makes good planners. Therefore, the analysis of one's own job by the user is an invaluable early planning.

1
Introduction: Laboratory Typologies

Christoph Heinekamp

The term *lab* or *laboratory* is as multifaceted as its linguistic background: In Latin, "laborare" means to work, to suffer, to try. The term is used for buildings – laboratory building, rooms – chemistry laboratory, functional units – central laboratory, or virtual entities – web laboratory. Even in the restriction to rooms, there is a spectrum ranging from the chemistry laboratory to the sleep and language laboratory. Workshop and equipment rooms are not clearly separated from the laboratories – a situation apparently often misused in order not to have to meet safety requirements. In the chapter "area of application," the laboratory guidelines "Sicheres Arbeiten in Laboratorien *(Safe Working in Laboratories)*" BGI/GUV-I 850-0 (Issue: March 2014) include an extensive definition of the term for "laboratories," which also serves as the basis for the remarks to follow.

> Laboratories (labs) are workrooms used by specialists or people instructed to carry out experiments for the research and usage of natural scientific processes. ... Among these, we find, for instance, chemical, physical, medical, microbiological, and genetic engineering laboratories. [BGI 850].

The classification into lab groups and lab types can be performed and viewed from different angles and according to the following criteria:

- Purpose
- Type of use
- Science direction
- Field of activities
- Working methods
- Physical structure
- Requirements in Terms of Building and Safety Technology.

The Sustainable Laboratory Handbook: Design, Equipment, Operation, First Edition.
Edited by Egbert Dittrich.

1.1 Purpose

Labs and lab buildings are constructed for "public users" – universities, research facilities – and "private users" – industrial firms, and service providers. The buildings are constructed in order to be used by the owner or to be rented. With regard to the conception of the lab areas, distinction can be made between a "customized" area for specific usage corresponding to a specific space allocation plan and "potential sites," such as those in flexible general purpose buildings (Figure 1.1).

Laboratory areas can also be classified corresponding to the type of use:

- *Teaching:* practical training, teaching labs
- *Research:* basic research, applied research
- *Diagnostics/analytics:* contract laboratory, process analytics
- *Development:* pharmaceutical, lacquer, and process development.

All existing laboratory areas can be classified by means of these four groups. There are enormous differences regarding the surface areas required, the technical and mechanical equipment, and the occupancy rate, which makes qualitative and quantitative comparison of the areas impossible. Even a seemingly clear and distinct group, such as the group of practical training labs, differs hugely in terms of equipment and requirements, as can be seen in Figures 1.2–1.6. The separations within the groups are also not clear and distinct; this is because teaching can also take place in the research laboratory and research, in turn, can take place in the development laboratory. Independent areas of analytics are often implemented as a type of service in research and development departments.

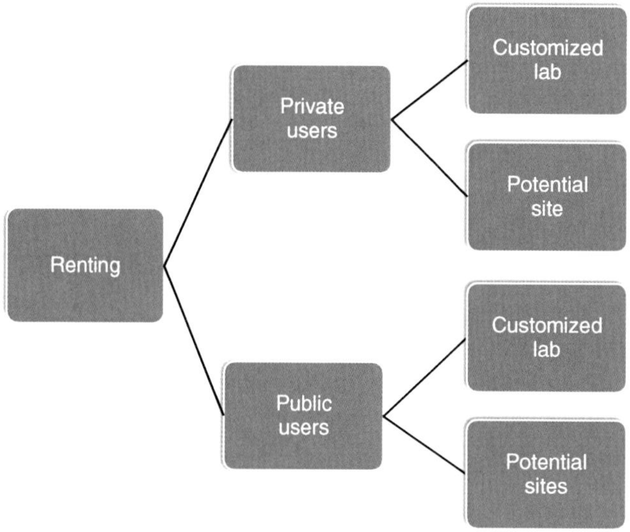

Figure 1.1 Purpose of the building.

Figure 1.2 Practical training chemistry.

Figure 1.3 Practical training physics.

1.2 Science Direction

It is obvious to group the lab areas or lab buildings corresponding to the basic science areas:

- Chemistry
- Biology
- Physics.

Figure 1.4 Practical training geological sciences.

Figure 1.5 Microscope room.

We find overlapping even when taking a two-dimensional look at the core disciplines, and they cannot, thus, be allocated clearly to one specific group (Figure 1.7). A broader grouping by science area and application, such as pharmaceutics, medicine, or engineering, will make the problem even bigger because some type of overlapping also exists for these fields.

1.3 Fields of Activities

Classification of the laboratories can be performed corresponding to the activities carried out in the laboratory.

Figure 1.6 Practical training anatomy.

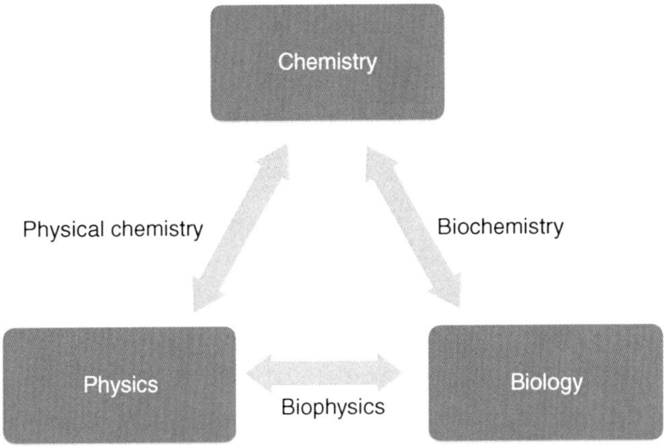

Figure 1.7 Biochemistry: biology or chemistry?

- Synthesis/preparation laboratory
- Analytics/measurement laboratory
- Process engineering.

Something is either

- produced, synthesized, prepared, or cultivated
- characterized, analyzed, or measured
- developed: Measurement methods or production processes.

Synthesis laboratories are engaged with the characterization and measurement of the substances, whereas most analytics laboratories are engaged with the preparation of samples and the development of measurement methods. Now, the question is whether these activities take place in the same room or in separate areas.

1.4 Working Methods

Grouping in the laboratory can be performed depending on the respective working methods:

- Working at the laboratory bench with small-sized devices
- Machine- or major device-oriented working
- Process-oriented working.

1.5 Physical Structure

Classification of labs into types can be carried out according to physical structures:

- Single laboratory
- Double laboratory
- Open-plan laboratory
- Combination laboratory/laboratory landscape
  - Files for physical structures.

The combining of the classifications results in an allocation structure with clear and distinct specifications, as can be seen in the example of a synthesis laboratory (Figure 1.8).

1.5.1 What is the Conclusion Resulting from the Evaluation of the Lab Allocation Tree?

A double laboratory in a public research institute typically consists of four to six workplaces. As a minimum, one chemical hood is required per staff member in a chemistry synthesis lab. Synthesis lab, organic chemistry or inorganic chemistry? Inorganic, organometallic, or solid-state chemistry? In some cases, solid state chemistry means synthesis isolators (Figure 1.9).

Decision criteria for the design concept are provided by each level. These criteria consciously or unconsciously flow into the planning process. However, the basic conditions for the building and the detailed information for the lab room have not yet been provided. The example shows that the air quantities required differ by a factor of 3.

Good laboratory buildings result from the inside, through an internal interactive process, and from the outside, through an interdisciplinary planning team. To a smaller extent, this requires types of laboratories and standard laboratories, but to a larger extent, this requires use-specific and building-related needs and requirements.

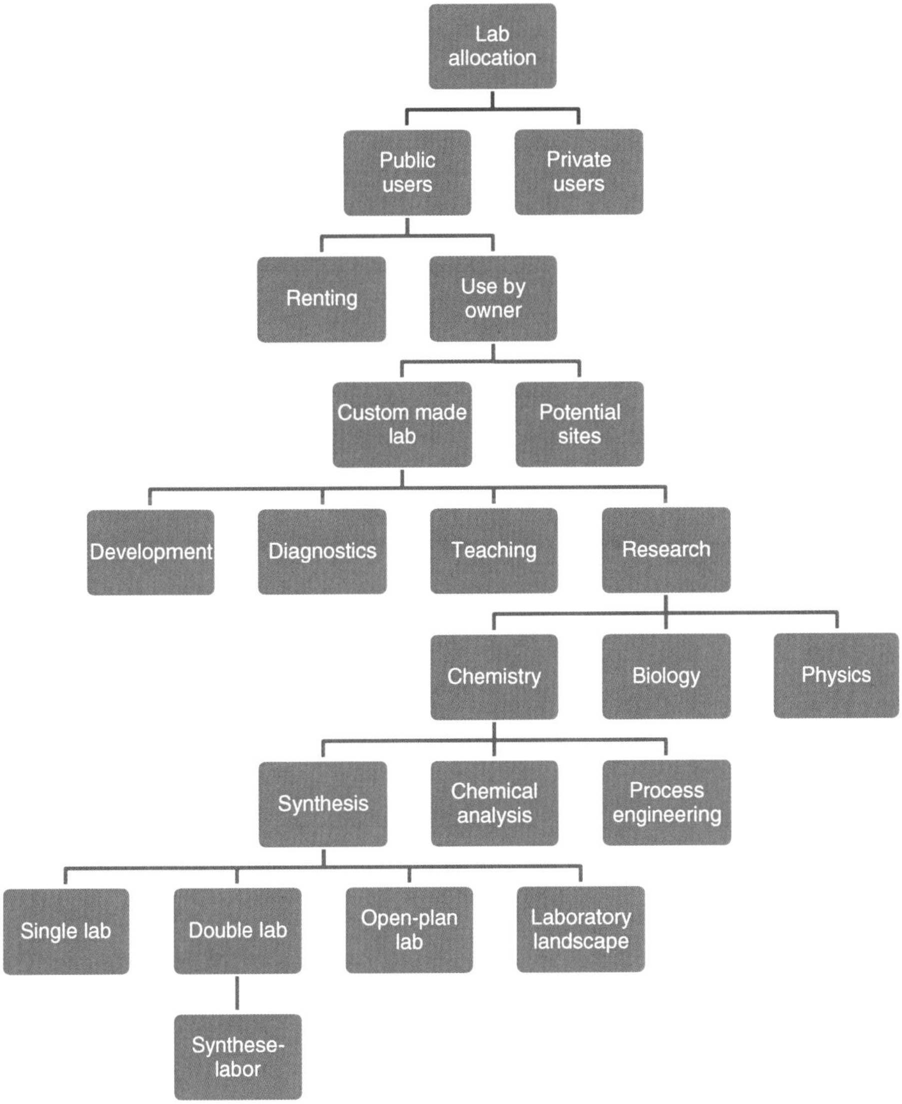

Figure 1.8 Laboratory allocation tree: example, synthesis laboratory.

1.5.2
Use-Specific and Building-Related Needs and Requirements

The determination of requirements for the use-specific and building-related needs and requirements does not primarily refer to rooms, but to workplaces and working areas. In an initial step, it is essential to define the special buildings or special components required which result from the intended use or the safety

(a) (b)

Figure 1.9 Inorganic synthesis lab (a) and synthesis isolators (b).

requirements. The second step concerns the characterizing of those special areas in the lab building which result from operator, object, or product protection. In order to find the technical system of the building concept, the basic conditions for the air quantities and the media equipment are determined in a final step.

1.5.3 Determination of the Areas for Independent Buildings or Special Components

Legal stipulations or building typologies of completely different types result in laboratory areas which should be implemented in the form of an independent building or part of a building.

- Safety laboratories with protection level BSL4 genetic engineering
- Handling of explosives
- High-pressure synthesis
- Ceiling height used >3.5 m
- Keeping of animals
- Clean rooms
- High-accelerating electrical fields, for example, NMR
- Sensitive to EMC and oscillation

1.5.4 Determination of the Areas as Restricted Areas

The laboratory areas are not accessible to any third parties at any time. This refers to the restricted access for lab staff members within a laboratory area, that is, only designated lab staff members have access to the locked area.

1.5.5 Areas with Locks and Access Area

- Safety areas BSL3 genetic engineering, biomaterial regulation
- Isotope areas at protection level 1

- Product protection or prevention of cross-contamination
- Handling of active or highly-active substances (Amount > mg)
- Safety area BSL S2 genetic engineering, biomaterial regulation
- Laser areas and exposure to radiation
- Handling of highly-toxic substances (Amount > xg).

1.5.6 Determination of Areas with Special Requirements Regarding Fire and Explosion Protection

Labs for endurance tests and kilo labs dealing with solvents of more than 2.5 l are typical rooms with special requirements regarding fire and explosion protection. These areas are either separated from the laboratory area or implemented within the laboratory area in a separated form.

1.5.7 Determination of Areas for the Laboratory Equipment

Lab equipment rooms with high thermal load and noise level, which are only operated for a short time, dark rooms, and storage rooms are allocated to the lab areas and implemented without daylight.

1.5.8 Determination of Areas for Special Laboratories

These are lab areas with daylight for joint use, for example, cell culture or for special equipment, such as a mass spectrometer. The areas for special laboratories are either integrated into the lab area or realized in the form of an autarkic zone.

1.5.9 Determination of Standard Laboratory Areas

The project-specific standard lab areas are formed by combining all lab workstations into one lab area. Standard lab areas may either consist of several standards or, in some areas, can be formed differently in a project. The development of standard lab areas, with validity for the entire project or only in some areas, creates the flexibility required for standard jobs. Evaluation and documentation working places have either been integrated into the standard laboratory areas or have been allocated directly. The special lab areas and the lab infrastructure areas result in laboratory areas in the form of usage units.

1.5.10 Conception Laboratory Building

The grouping of those lab areas with technological safety and building requirements is the basis for the conception of the laboratory building (Figure 1.10);

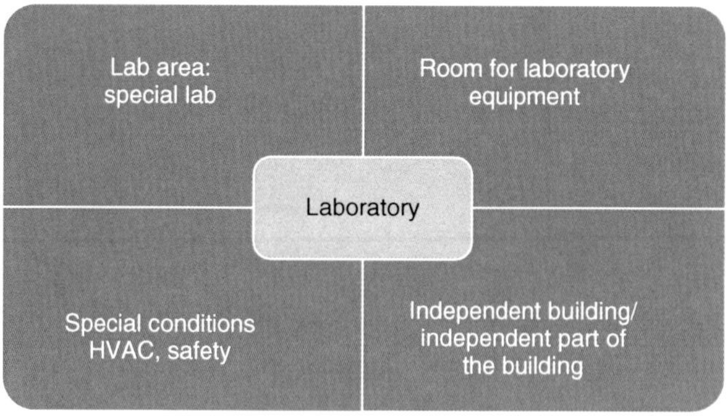

Figure 1.10 Lab building with type of lab area.

administrative areas and further functions, as well as communication and interaction in the future building are to be taken into account.

1.6 Conclusion

Laboratories for natural science mirror nature – they are apparently indefinitely complex and simply organized in detail. There are appreciable differences between the laboratories, and yet, they are just the sum of simply structured individual workplaces:

- Laboratory bench workplace
- Laboratory bench equipment workplace
- Chemistry workplace with chemical hood and hazardous substance storage
- Workplace with product and operator protection – Safety workbench
- Equipment workplace with flexible supply
- Supply workplace with storage cabinet and cooled storage rooms
- Evaluation and documentation workplace.

Project-specific individual laboratory areas are created by combining these workplaces; in the spirit of Aristotle (384–322 BC) "*The whole is more than the sum of its parts.*"

Often, the planning of rooms and laboratory types is nothing but looking back at existing areas. New laboratory buildings are not supposed to be a copy of the existing buildings. It is better to learn from the existing buildings and the experiences made in order to develop new and sustainable laboratory areas in the future.

2
Requirements and Determination of Requirements

Christoph Heinekamp

The initial step in the planning and new organization of laboratory units is always the determination of requirements. Qualities and quantities have to be defined for the determination of requirements. The determination of the room conditions and the requirements for the laboratories with regard to the classification into safety or hygiene classes is often not carried out at the beginning of a project. However, these determinations can be made soon after the first detailed adjustments have been made. The determinations on the space required take a long time because completely different results are calculated depending on the respective views. Seen from an economic point of view, the fewer square meters a laboratory has got, the better it is. However, more space means a higher degree of flexibility when talking about laboratory processes.

When taking a closer look at the areas, it is essential to differentiate clearly between the different definitions of area types. A gross floor area (GFA) is twice as large as a main usable floor area (UFA); this is because construction floor area, functional area, and circulation area take up a lot of space in the laboratory building. In particular, the size of the functional space, that is, the space used for building technology with ducts and building control systems, makes up quite a significant part of the laboratory building. Figure 2.1 displays the relations between the various types of areas.

The determinations in industrial laboratories are mostly made in the form of GFAs, whereas UFAs or main UFAs are defined in laboratories used for public projects (e.g., Universities, research facilities).

The demands of space must be appropriate; future developments have to be taken into account in a meaningful way. There are big differences in the ways the demands of space are determined. In many cases, this determination is calculated by asking the users, with the users measuring the space existing and then adding some subjective value to cover future developments. There are central area specifications for specific tasks, as can be seen in the following example of university construction in Switzerland (Table 2.1).

There are area specifications for research groups in some governmental research facilities: 350 m^2 for one research group and 800 m^2 for each independent department.

The Sustainable Laboratory Handbook: Design, Equipment, Operation, First Edition.
Edited by Egbert Dittrich.

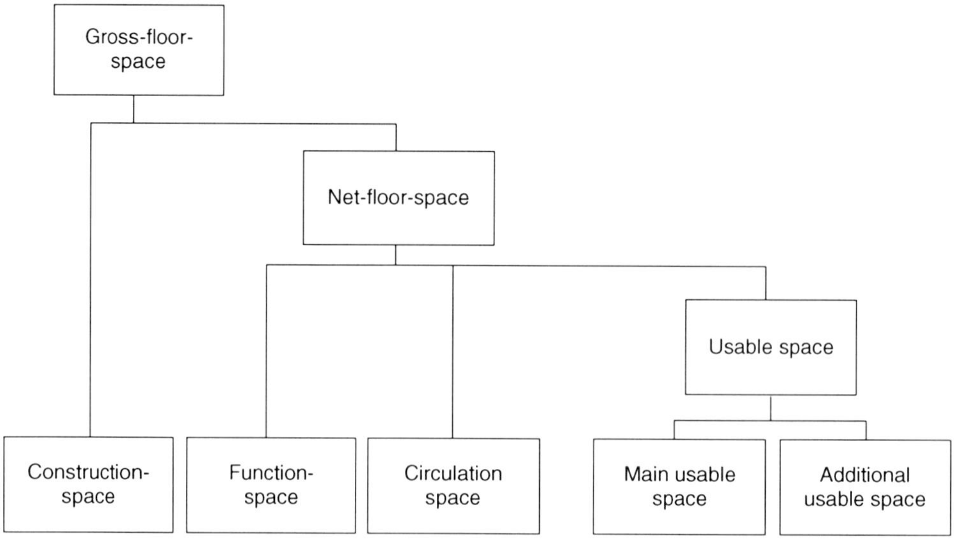

Figure 2.1 Types of areas.

Table 2.1 Central area specifications.

Work areas	m^2 Work space/occupant	m^2 Infrastructure/ occupant	m^2 Floor space/occupant
Office/administration/ conference, laws, literature, math	12	3	15
Theoretical IT, design, electronics, hygiene	12	8	20
Analytical and organic chemistry, electro-technics	15	10	25
Physics- and technical chemistry, microbiology, solid-state-physics	15	15	30

The existing areas are just divided again in the case of reconstructions; a determination of requirements will not be performed. The benchmarking determination is a quasi-arbitrary determination because the area per staff member ranges between 18 and 50 m^2. The actual requirements, therefore, cannot be determined by this.

The usability of the laboratory area is only rarely taken into account when the areas are being discussed. It is not the number of square meters which is decisive, but the number of workstations and how many pieces of laboratory equipment need to be operated. The building grid and the laboratory depth both have great importance with regard to the efficient usability of the laboratory area. Laboratory areas should follow a basic grid in a longitudinal direction in order to facilitate the methodical and technical development and the efficient utilization of the laboratory. Basic grids exist in values of 1.10, 1.15, 1.2, or 1.25 m, with the latter grid being an office grid, which is not economically suitable for laboratory areas. A basic grid of 1.10 m in the smallest room and a wall thickness of 15 cm results in clearance dimensions of 3.15 m. Together with the depths (wall-sided) of 90 cm for fittings, which are required for laboratory fume hoods, safety cabinets and, naturally, also for laboratory benches with media supply, a passage width of 1.35 m is achieved. A minimum distance of 1.45 m is required for opposing workstations.

Thus, the flexible use of laboratory areas with a grid of 1.10 m is not possible, because the maximum depth of one laboratory line is only 80 cm. During the 1980s, the basic grid of 1.20 m proved its worth in the University construction, with the corridor widths larger than required, especially in double laboratories. The basic grid of 1.15 m (Figure 2.2), resulting in a load-bearing grid of 6.90 m with a flat ceiling, has established itself as the standard grid in laboratory construction, because it ensures the minimum distances between laboratory lines with allowances and the effective use of the laboratory area.

We see in the four rooms in Table 2.2, that laboratory benches have been realized in rows of 26 linear meter each, and that the bench depths have been partially reduced by means of the 1.10 grid. By using the building grid of 1.25 m, we see an increase in the laboratory space of almost 9%, an increase which cannot be used.

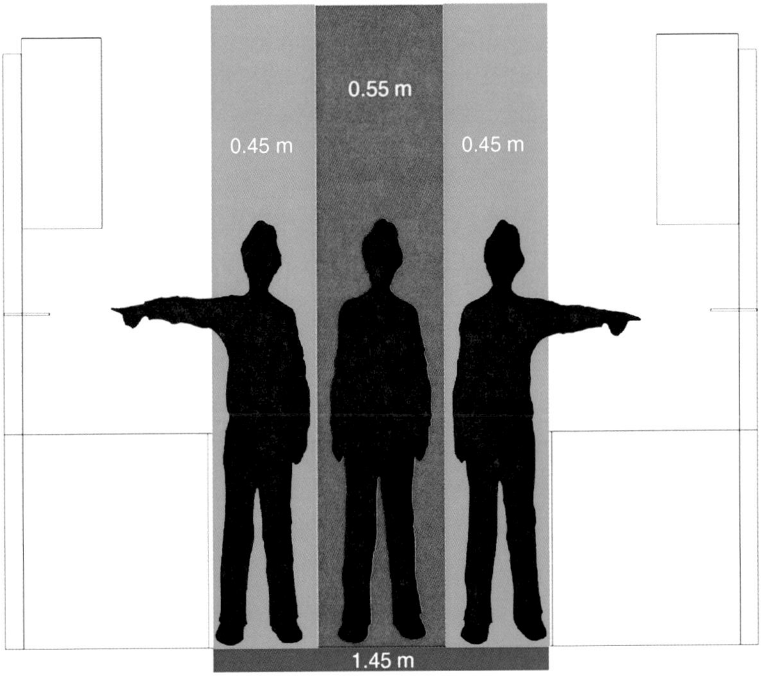

Figure 2.2 Distance areas for opposing workstations.

Table 2.2 Space misuse through wrong grids.

Floor size	Grid (m)	Space (m^2)	Change (%)
Depth 7 m; width 6.45 m	1.10	45.15	−4.5
Depth 7 m; width 6.75 m	1.15	47.25	—
Depth 7 m; width 7.05 m	1.20	49.35	+4.4
Depth 7 m; width 7.35 m	1.25	51.45	+8.8

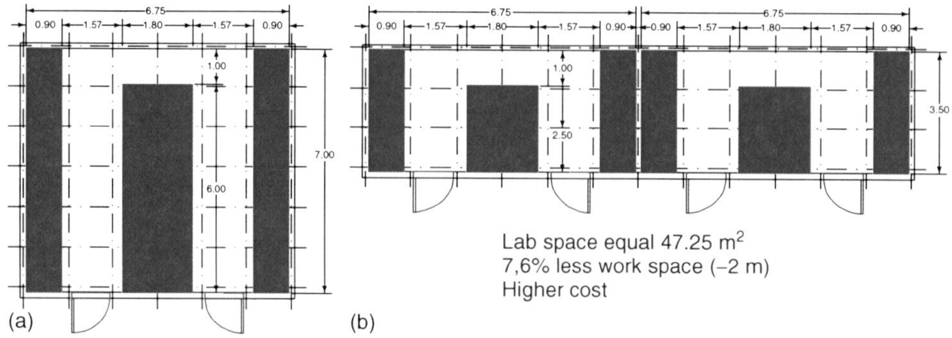

Figure 2.3 Room depths have an impact on laboratory workplace surface. (a,b) Same laboratory area 47.25 m^2 and (b) 7.6% less working space (−2 m), higher cost.

The realization of laboratory spaces with depths too small will reduce the work surface with the same basic laboratory surface; this is depicted by the simple example in Figure 2.3.

2.1 Area Misuse through Wrong Grids

If laboratory rooms are equipped with two or four grids, the area consumption for the same work surface will be even higher. Two laboratory rooms with two grids each and with a workbench of 7 m require 32.2 m^2 in total. A laboratory room with three grids and a laboratory bench of 14 m requires only 23.1 m^2 in total. There is a difference of almost 40% in the total area required, with the usage being the same.

2.1.1 Determination of Requirements of Workplaces and Storage Space for Extra Equipment

It is, therefore, important to consider the usable laboratory space for workbenches and extra equipment for a proper determination of requirements. By means of such consideration, it is possible to determine comprehensible requirements and demand-oriented laboratory areas. Seen from an economical point of view, both

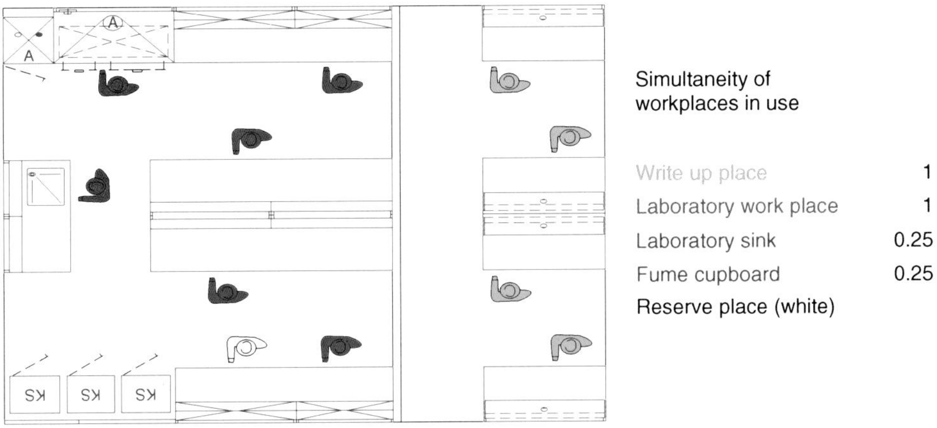

Figure 2.4 Laboratory workplaces with frequency of use.

kinds of areas – an area too small and an area which is over dimensioned – are disadvantageous.

As shown in Figure 2.4, the laboratory workplaces required are determined through the allocation of staff members, that is, allocated to people with a concomitance of 1 and jointly used by four staff members (concomitance of 0.25).

The total demand for the laboratory and the side rooms of the laboratory will be represented in this way, including the additional taking into account of the storage space for extra laboratory equipment. Table 2.3 displays the typical biomolecular laboratory of a research facility. Special areas, such as locks for safety or hygiene zones and special storage rooms for samples or materials, have to be determined separately. Administrative rooms – offices – and communication areas – recreation room, meeting point, conference room – have to be assessed

Table 2.3 Example for the area measurement of a work group.

Laboratory	Bench linear meter/occupant	Occupants	Total Bench length (m)	Bench space (m^2)
Space for plug-in-units	1.5	1.50	58.10	116.20
Write ups	2	2.00	19.50	64.35
Fume hoods	1.00	1.00	—	52.00
Storage space	—	—	26.00	0.00
Chilled rooms	—	—	0.00	20.00
Incubator rooms	—	—	12.00	24.00
Locks for humans	—	—	12.00	24.00
Locks for material	—	—	12.00	24.00
Locks for clean room	—	—	6.00	12.00
	—	—	6.00	12.00
Total			**161.6**	**348.55**

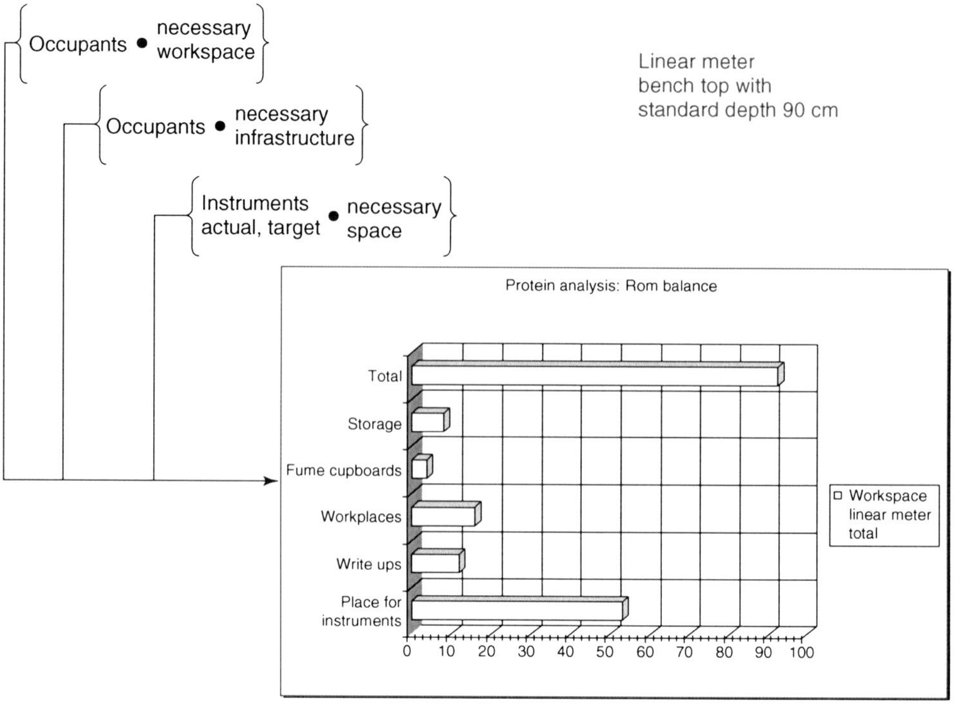

Figure 2.5 Model on the determination of space requirements.

in terms of the area depending on the staff numbers and the area specifications given for companies or organizations for such kinds of areas.

The space determined is compared with the existing area; the area demanded, compared to the existing area may result in a smaller or larger amount of area required. The analysis of the existing area shows parts which are used inefficiently, alternatively, lack of space is also shown. In some cases, the distance of the spaces may be under a specific limit, or functions are not depicted adequately. In addition to the determination of present space requirements, this model serves the determination of future laboratory space requirements by means of the planned increase in staff members (Figure 2.5 and Table 2.4).

The determination of requirements provides optimized useful floor area approaches which have to be merged to form a building in a reasonable way. The systematic technical development structure and the usage relations have to be defined for the building design. We find work flow and also working culture, including communication and interaction, among the essential requirements for a good laboratory design.

Storage space for equipment and working areas call for specific requirements for the room conditions and for the technical supply, including ventilation, refrigeration, media, and electricity, as well as data technology. In order to achieve sustainable laboratory design, the availability of these provisions has to

Table 2.4 Example for floor space calculation with forecast.

		Reality today		Demand today			Future demand		
Type	**Area**	**Occupants**	**Floor space (m^2)**	**Occupants**	**Bench (m)**	**Floor space (m^2)**	**Occupants**	**Bench (m)**	**Floor space (m^2)**
Laboratories	Chemical analysis	16	338	16	190	411	25	267	679
	Microbiology	30	665	30	243	587	33	246	750
	Research	29	849	29	346	749	40	468	1152
	Total lab	75	1852	75	779	1747	98	981	2581
Offices	Quality control	13	121	13		160	12		148
	Analysis	1	15	1		16	4		56
	Microbiology	2	43	1		32	2		80
	Research	11	154	11		176	15		212
	Total office	27	332	27		384	35		496
	Total	102	2183	102		2131	131		3077

be designed and total amounts for future alternative use have to be determined by eye. Here, the usable undeveloped areas in the ductwork and in the installation routes are of particular importance.

2.1.2 Flexible Laboratory Space

Users, builders, and architects mistakenly think that flexible laboratory space does not require any determination of requirements because, in their minds, these flexible laboratory spaces apparently meet all requirements. In reality, it is the budget which will then define the size of the project, and economic considerations will define the determinations on the heights of the storeys and the sizes of the ducts. Basic conditions will be determined unconsciously and, in the end, the only thing which can be moved flexibly will be the bench. No flexible laboratory space has been created; however, usage must flexibly adapt to a building situation which does not meet the requirements.

Given the fact that the costs for the new building of a laboratory amount to a meager 9% of total costs, and that personnel costs make up the essential part of total cost by amounting to more than 75%, the main focus – seen from the economic point of view – should be on an optimized flow of work.

The determination of requirements for flexible laboratory space is more elaborate than for fixed laboratory space. In the case of flexible laboratory space, the limits regarding the structural extension and the technical development must be defined going beyond the space required.

The size of a connected unit and the usable ceiling height are the only two requirements for the building. The amount of air from the ventilation system and the available amount of cooling may have a restrictive effect on flexibility. The investment budget available demands the requirements for the new laboratory space to be defined exactly. Nobody can afford or wants to construct a building with resources he/she is never going to use. The successful determination of the requirements for a flexible laboratory building is achieved by means of usage scenarios and an experienced planning team.

The determination of requirements and the correct requirements are fundamental for the planning of a new construction. Simply copying the requirements from old buildings without carrying out a future-oriented determination of requirements is just like building on sand. New laboratory space provides the opportunity for optimized work flow, an innovative and communicative working environment, and safe workstations. It will not be possible to build a sustainable laboratory building which operates economically and which does a successful job without resorting to a target-oriented determination of requirements, along with the definition of the requirements and the basic conditions.

3
Laboratory Concept and Workstations

Christoph Heinekamp

At the end of the last century, we saw the specialization of natural sciences into specific specialist fields. As a consequence, special buildings for the fields of biology, chemistry, and physics were constructed and economically optimized in the course of time. Laboratory buildings were systemized by means of grid structures and the individual functions were represented in zones. Zoning into installed and noninstalled sections led to the separation of office and laboratory work. Office buildings and laboratory buildings or building units with split levels were the result of the use-specific adjustment of floor heights; the floor height of the office area differing from the floor height of the laboratory area. Due to these approaches, the investment costs could be reduced; however, the experimental and theoretical tasks also became partially separated from each other and long distances between the departments were generated.

Systematization was adopted by many space allocation programs in the 1990s: Standard laboratories with 20 m^2 (Figure 3.1) or 40 m^2 and adjoining rooms with 10–20 m^2. During the implementation, flexible standard laboratory buildings with a building grid of 1.2 or 1.15 m^2 were created, which enabled the optimal user-specific partitioning and separation for special requirements.

Safety areas (S1 to S2) can be defined at any time by means of this structuring. The processes can be represented in different laboratory rooms in order to avoid cross-contamination or unnecessary risks.

Buildings for biological or medical research were usually built as a three-winged structure – laboratory room-corridor-adjoining room-corridor-office/laboratory room. The adjoining room zones in chemistry buildings require a lower amount of unlit areas and, therefore, these buildings were usually constructed as two-winged structures (e.g., Figure 3.2). Physics buildings are often composed in the form of two-winged structures and – for specific use – in the form of three-winged structures.

In today's laboratory buildings, you often find open laboratory doors and equipment such as a refrigerator standing in the corridor, even though this corridor is part of an escape route. On the one hand, this kind of misuse is due to a shortage of space, and, on the other hand, this is due to the fact that equipment rooms are incorporated into various operational procedures. The incorporation of the

The Sustainable Laboratory Handbook: Design, Equipment, Operation, First Edition.
Edited by Egbert Dittrich.

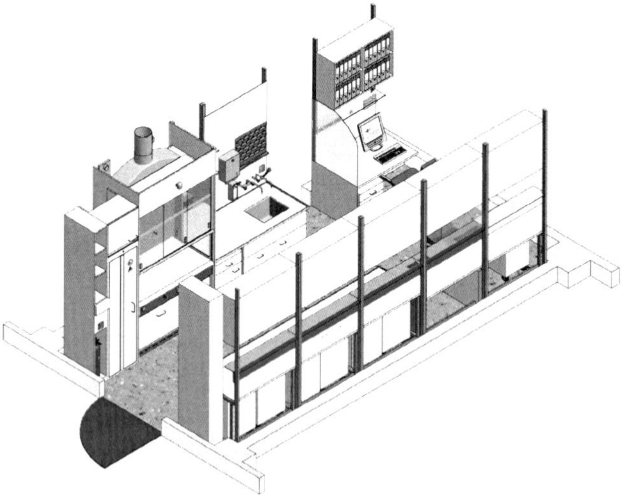

Figure 3.1 Single laboratory 20 m².

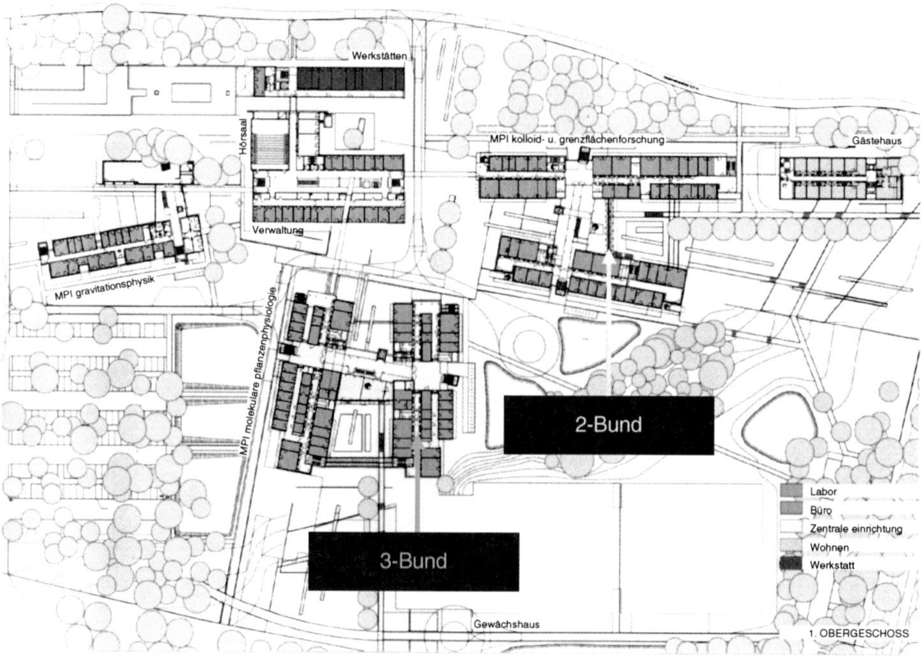

Figure 3.2 Chemistry building two-winged structure and biological building three-winged structure, architects ASPLAN.

functional development into the laboratory area and the direct connection of the equipment rooms to the laboratory rooms provide for short distances and enable the joint use of equipment and facilities. The connecting doors in the laboratory unit and to the equipment rooms are of exclusively functional importance. The door to the equipment room will be closed whenever this is required by the respective task: room darkening or noise or product protection. The planning concept displayed creates a connected laboratory zone with a length of 28 m and depths, with one-sided facade connection, ranging from 15 to 18 m – taking into account the appropriate technical development, the fire zones, and the lengths of the escape routes. Due to the rectangular placement of the laboratory equipment, the workstations will be provided with sufficient daylight even under these building depths. The compact structure guarantees short distances in terms of the operating processes, enhances the interaction between staff members and the number of safety installations required, and gross floor space will be reduced. All principles for economic laboratory buildings have been considered.

As a result, we achieve a zoning which covers – starting with the documentation zone – the laboratory workstations, the internal development, the laboratory infrastructure, and the equipment-room zone. The technical development will be performed exclusively according to the necessities of the usage zones. The documentation area is provided only with electricity and computing technology. The concentration of the laboratory sinks with water supply and wastewater is carried out in the infrastructure zone. The fact that the operational procedures and the staff interaction will be enhanced has been clearly confirmed by a case-study in the field of human factors and ergonomics. The scientists investigated the procedures of the same work group within existing areas and after the move to the laboratory landscape (Lab 2020).

Laboratory space which can be used in different ways is created in laboratory landscapes (Figures 3.3–3.7). This laboratory space guarantees a high measure

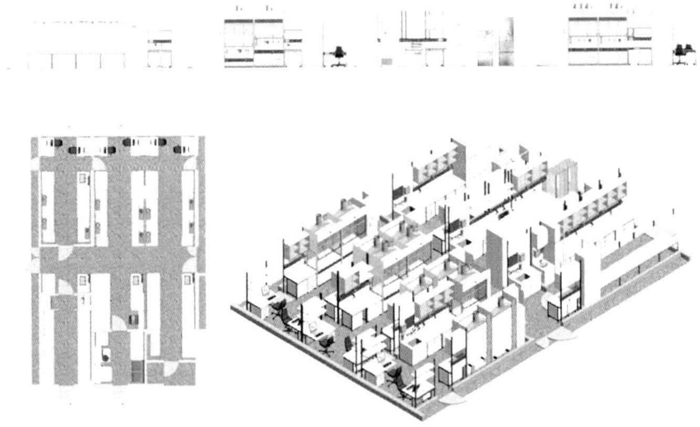

Figure 3.3 Laboratory landscape.

Figure 3.4 Laboratory depth 17 m.

of flexibility for future demands and it provides perfectly adjustable connecting areas, to be used, for instance, for robot installations and large-scale analysis equipment. The total depth of the laboratory modules reduces the facade surface and will, thus, have a positive impact on investment and operating costs.

Current natural sciences form interdisciplinary research teams composed of specialists who take a holistic look at the tasks occurring and are, thus, able to find completely new approaches to problem-solving.

These tasks require laboratory buildings which provide the technical preconditions for each specialist discipline. Laboratory buildings for chemistry, biology, and physics are no longer required; the demand is rather for laboratory areas providing workstations which meet the demands of the various specialist fields.

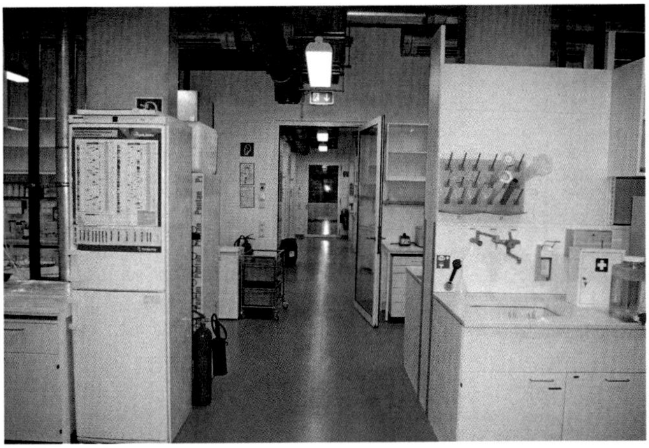

Figure 3.5 Internal laboratory corridor in laboratory landscape.

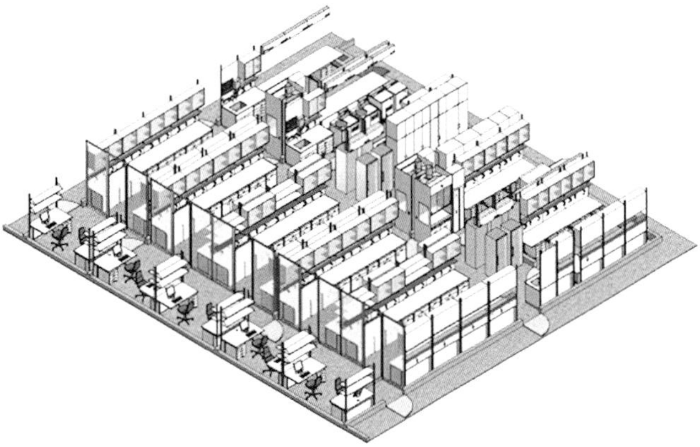

Figure 3.6 Isometry of laboratory landscape.

The technical development of the laboratory modules via central ducts offers the highest degree of flexibility with regard to the task available. The central duct will provide the air quantity for the chemistry workstations; biology and physics areas will be provided with air ducts of smaller dimensions. The installation of ceiling supply systems allows for adjustments only affecting the storey which is to be reconstructed. In order to achieve better accessibility for maintenance work and refitting, laboratory installations will be carried out without suspended ceilings. The fact that all installations are concentrated on the ceiling complicates the good cleaning ability, which is particularly required for laboratories of security level 2. The sewer lines in the laboratories require incline pipes which need a respective storey height or they restrict the installation options. When taking into consideration these two tasks, it makes sense to develop a hybrid supply structure for specific building types. The supply with incoming and outgoing air, and electro and

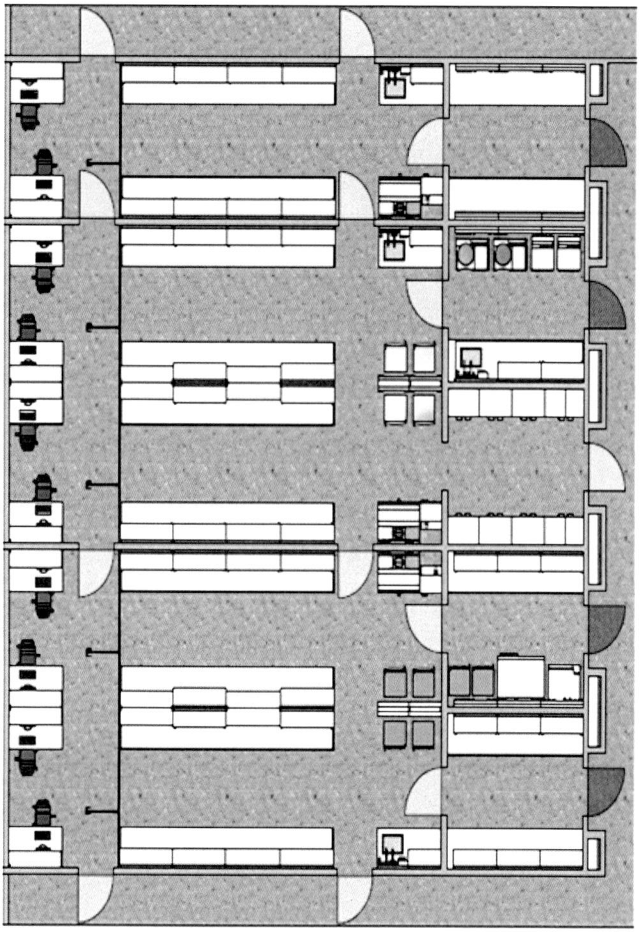

Figure 3.7 Floor plan lab module.

computer technology is carried out via central ducts or storey area supply rooms. Media supply and sewage engineering is carried out via decentralized single ducts, so that only a few isolated media lines will run openly through the room. If need be, development for the electricity supply will be executed via the documentation zone and supply with incoming air will be executed via the laboratory corridor. The installation concepts, such as are optimized regarding the building and the tasks, make the realization of an open installation for the laboratory modules possible (Figures 3.8 and 3.9): an open installation with just a few difficult-to-clean pipes left.

Work in a laboratory requires more and more elaborate documentation work and theoretical preparation of the experiments. Evaluation work places, in the form of window workbenches running parallel to the facade, have been created. The use of computers at the evaluation work places led to a change, because the

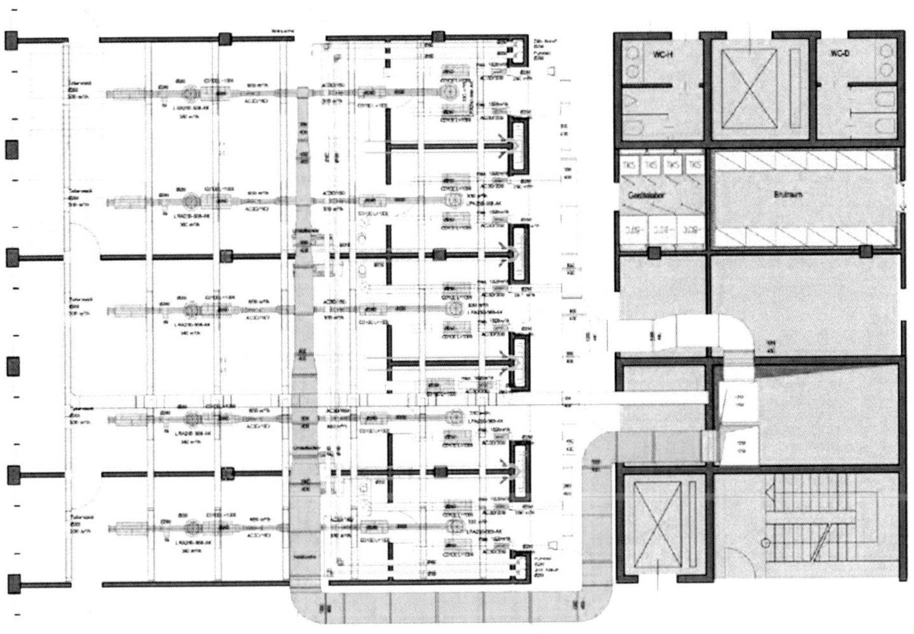

Figure 3.8 Technical development lab module.

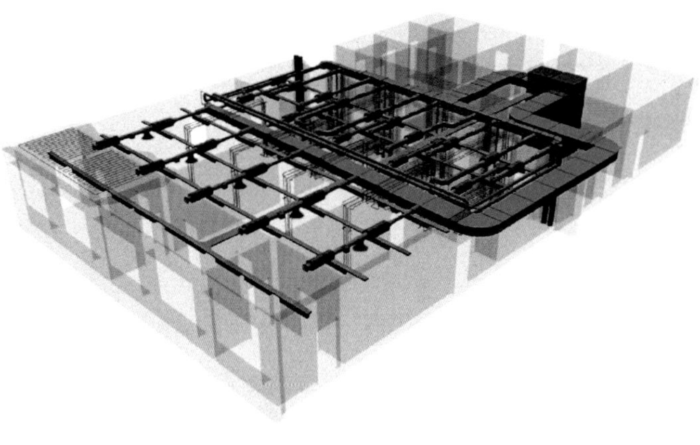

Figure 3.9 Isometry technical development lab module – engineering consultants, Feldmeier.

requirements of the screen workstation regulations were not met by these window workbenches. The amount of computer work performed for some applications in a laboratory has increased to take up more than 50% of the working time. However, supervision of the experiments requires these working places to be in the laboratory or in adjacent rooms.

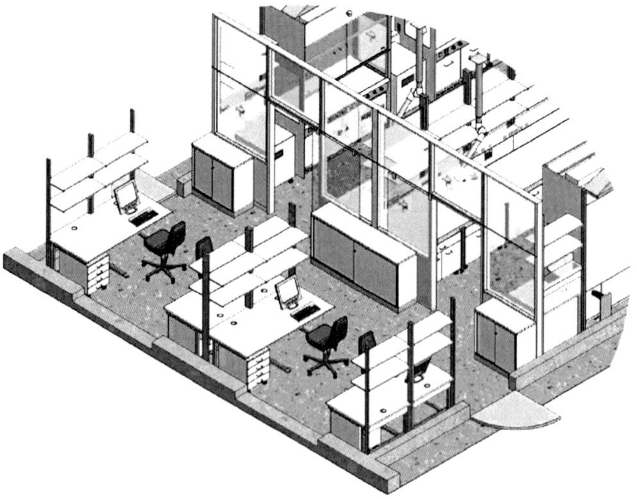

Figure 3.10 Documentation working place at the lab.

Protective clothing and goggles also have to be worn at the evaluation work places in a laboratory. Noise exposure constitutes another problem at these work places because the sound level in a laboratory may be too high to perform intellectually challenging activities. As a response to these new requirements for job safety and protection, work places for documentation separated by a glass wall (Figure 3.10) were created; these places provide an office standard with immediate proximity to the laboratory for instant intervention. The line of sight from the documentation zone to the laboratory reduces the degree of working alone, a type of working which is not allowed without additional organizational action. Access to the separated documentation workstations, however, is possible only via the laboratory. A laboratory coat and protective goggles have to be worn when crossing the laboratory; this is not often done in practice. The staff members in the documentation zones can become very isolated due to the fact that they can only be accessed by walking through the laboratory. Therefore, interaction by the members of the work group or the department is restricted. Furthermore, if visitors are led to the documentation work places, they will have to cross the laboratory unnecessarily.

In order to enhance communication and the escape route situation, the documentation areas should be separately developed via corridors or general purpose rooms – such as short-term archives and copy rooms – and the documentation areas should be given an internal corridor (Figures 3.11 and 3.12). The central communication points can be easily reached via this access, and the central communication points can connect the general purpose rooms and the laboratory, or respectively, the documentation areas. Mechanical ventilation of the documentation zones is required due to the occupancy rate and the direct connection to the laboratory. However, it is perfectly sufficient to provide the area with incoming air

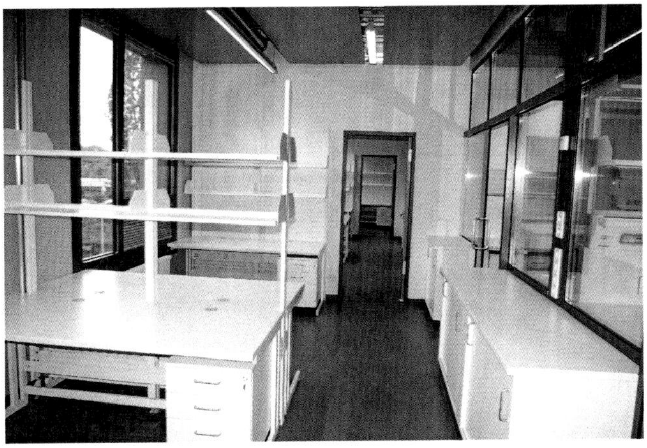

Figure 3.11 Documentation zone with internal corridor.

Figure 3.12 Documentation zone as an "open office."

only, and to realize the outgoing air by means of overflow direction to the laboratory. As a result of this technical option, the areas will not require any additional quantities of air and the mixing of the laboratory air with the air space of the documentation zone will not take place. Flow direction with overpressure from the evaluation area to the laboratory will always be ensured, even if the user does not close the connecting door.

4
Determination of User Needs – Goal-Oriented Communication between Planners and Users as a Basis for Sustainable Building

Berthold Schiemenz and Stefan Krause

In recent years, the sustainability concern has turned into a strong global movement that nations and companies cannot and do not want to elude any longer. Here, the term *"sustainability"* is understood as a permanently resource-saving guiding principle, which requires the avoidance of any use, yielding no direct profit. In this context, the question arises whether there are also sustainable laboratory buildings and if so, what they look like.

Often, the sustainability issue threatens to be reduced to the use of new, environmentally friendly technologies to realize proven traditional *constructional concepts*. This does not mean that new buildings already look old when they are erected – as a rule, the appearance suggests the opposite – but it means that, all too often, only environmentally friendly embodiments of beloved familiar concepts are considered for certain requirements.

This ignores the fact that maximum sustainability can be materialized by identifying alleged land requirements as unnecessary, thus rendering their realization superfluous. Space optimization is the key to sustainability. If the necessary critical discussion about the needs of the users is not consistently held, even the most sustainable construction elements turn into a "beauty patch" for outdated planning and architecture.

This traditional approach is tempting, since the future user receives a known and often proven constructional solution which will gain faster acceptance. On the other hand, this approach may also be more convenient for the planner, since coordination with the user may be quicker, and "sustainable" technologies can be applied anyway. However, this procedure circumvents the constructive and joint effort to reach the best overall solution. The user's needs seem to be optimally fulfilled, in reality, however, they are realized with a minimum of planning using new constructional concepts. This means "sustainable" realization of waste.

Thus, the primary task will not be the realization of an existing building concept in a more sustainable way – for example, by the use of solar energy or "green" building material – but to find a solution which helps reduce the *total building expenses* and which remains flexible in view of future changes in requirements.

Intensive communication between users, on the one hand, and planners, on the other hand, is required to ensure the successful realization of this commitment.

The Sustainable Laboratory Handbook: Design, Equipment, Operation, First Edition.
Edited by Egbert Dittrich.

Within the context of this communication, it is essential for planners and architects to have a good understanding of the kind and the scope of the intended use to be able to offer the best constructional solution. In the case of laboratory buildings, this is particularly demanding, since here highly specialized areas are concerned. In contrast to residential construction or office building, the majority of the public – and in laboratory building architects and planners without experience are no exception – only has a vague idea of the actual work flows and processes in laboratories. This is aggravated by the fact that the "standard" laboratory does not exist. Laboratories differ significantly according to the kind of work which is carried out there and thus, also in the detailed constructional requirements.

Therefore, without *communicating* with the user, even a technical planner will usually not have sufficient understanding of the specific work flows in the respective laboratory sector in order to identify optimum solutions. On the other hand, users often have limited knowledge regarding the planning framework, the technical possibilities, legal requirements, and the mindset of planners.

For this reason, it is necessary to find a way for successful communication between planners and users so that the respective needs are understood before constructional solutions are discussed. This effort to understand the needs is, a priori, not characteristic for a typical planning process. Users frequently express concrete ideas regarding the future structural realization already in the early planning sessions:

- *I need a laboratory with a floor space of at least 100 m*^{2}*!*
- *I need a floor with epoxy resin coating!*
- *I need a fume cupboard from company XYZ!*

This procedure is not ideal, as the immediate requirement is to obtain the ideal work environment for a certain activity. Thus, the laboratory of a certain size or a certain building material is never the original need of the user, but already the specification of a solution.

In an ideal planning process, however, the contracting entity or public agency provides a functional description of the requirements as a basis. This also includes details regarding current trends and possible future developments in its specific field which could have an effect on the required design of a laboratory space. Based on these data, the planner can elaborate a solution and present it to the user for comparison. The planner does not only provide the methodical and technical competence for the design and erection of the laboratory buildings, but in particular, the knowhow regarding new technologies and methods for the sustainable realization of the building projects (Figure 4.1).

Successful communication at the beginning of the planning phase should therefore be based on *functional requirements.* They are an essential part of the basic evaluation according to phase 1 of the fee structure imposed on architects and engineers (German HOAI).

Later, a method is outlined to explain how this communication process can be successfully designed

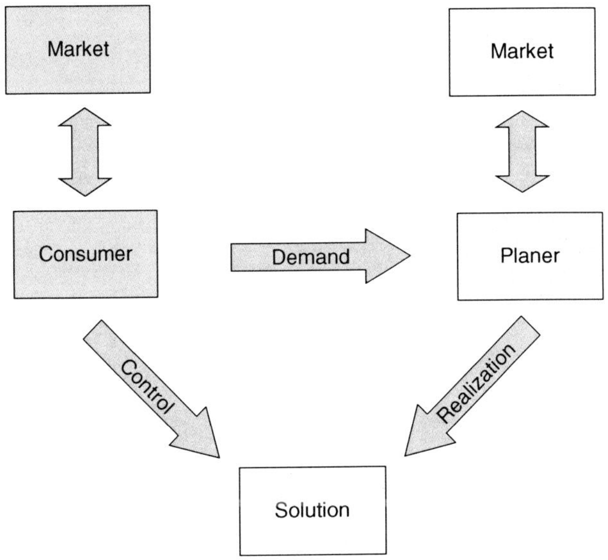

Figure 4.1 Planning process.

4.1 Work Areas

The initial aim of the planning is the preparation of a schematic showing the room groups according to the user data, taking work areas as the smallest possible units. In general, work areas shall not be distributable to two separate rooms; if required, however, they can be combined with further work areas in one room of suitable dimensions and design.

Examples of work areas

- Sample Receipt
- Analysis Laboratory 1
- Analysis Laboratory 2
- Office Workstations Laboratory Staff
- Office
- Administration.

The information will be supplemented by general requirements relative to the work area. As a rule, at least the following Information is required:

- Kind of work to be carried out (office, synthesis lab, analysis lab, sample storage, etc.)
- Number of staff and workplaces respectively.
- Kind and number of laboratory equipment required (fume cupboards, LF-benches, laboratory benches, etc.)
- Kind and number of large-scale equipment

- Particular approval requirements (GenTGes, BioStoffVO, etc.)
- Particular requirements (clean-room classes, Ex-requirements, etc.).

All necessary data, which will later determine the type and extent of the required surfaces, are identified in the result. Based on this rather abstract data and already at this stage of the project, the planner can provide a rough estimate of the areas required and thus, also of the construction cost to be expected.

A (preliminary) draft, however, should not be prepared on the basis of these data, since it requires precise knowledge of the work flows. Only then, the individual areas named can be optimally arranged to one another while synergies can be identified and finally realized.

4.2 Work Flows and Room Groups

The individual work areas will be arranged in a process scheme according to the functional work process. Based on the users' specification, the individual work areas will be graphically arranged in a chart making visible all essential interactions between the single areas.

For this purpose, a specific field ("box") will be created for each work area in the chart. If reasonable, a color code can be used here to roughly distinguish between the different types of use.

Subsequently, all work areas which must necessarily be in immediate proximity are combined in one room group. To facilitate further discussions, the room group can then be assigned a name (e.g., RG110: Scullery). It is particularly important to choose a room group as small as possible, since otherwise the creative leeway in the further planning process will be severely restricted.

In the next step, the workflows between the room groups and the individual remaining work areas are visualized. For this purpose, the elements are again linked according to the users' specifications. The increasing stroke width indicates increasingly intensive interaction; here again, the kind of interaction can be distinguished by different colors (Figures 4.2 and 4.3).

The result is a graphical representation showing all significant workflows in the laboratory area to be planned. The development and the joint discussion provide a basis for efficient mutual exchange. In this planning phase, it is important that the user briefly explain and justify the arrangement and the interaction presented. In this way, the process will once again be jointly scrutinized, but above all, also the understanding for the specific workflows on part of the planners will be enhanced.

In the next step, all essential workflows throughout the total work process should be analyzed with the help of these charts, whereby the flows of persons should be examined separately from material flows.

During such discussions, auxiliary areas are often identified that are not in the immediate focus and that are, thus, easily forgotten, but which will cause

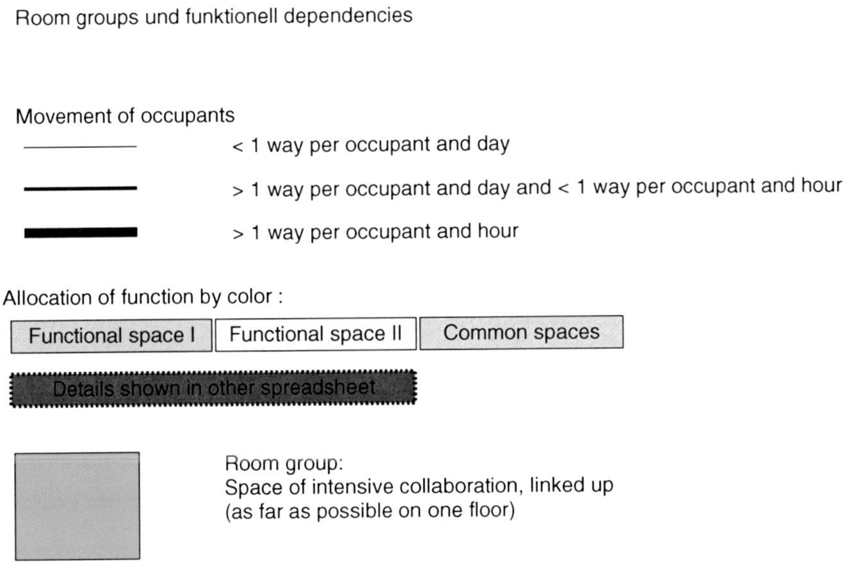

Figure 4.2 Room groups and functional dependencies.

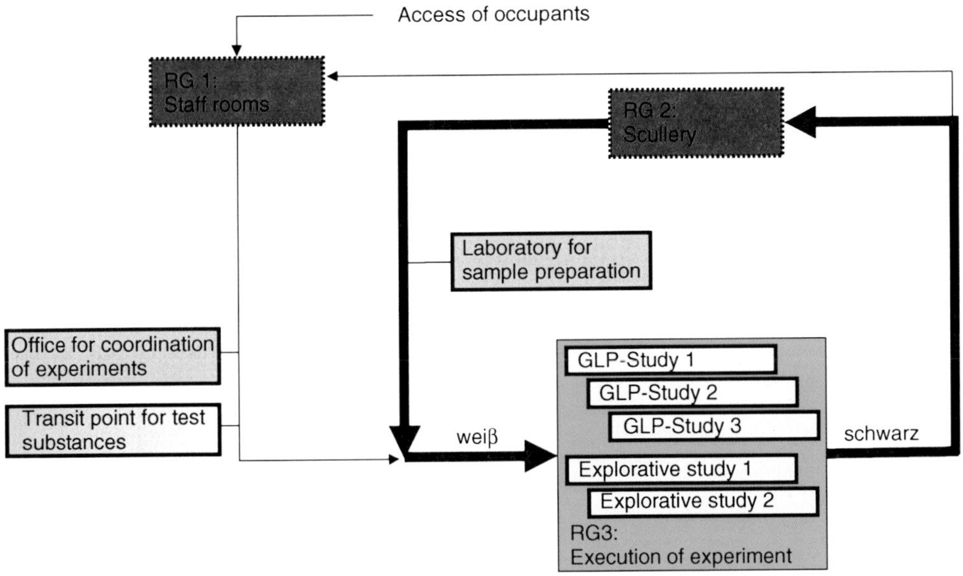

Figure 4.3 Room group schemes.

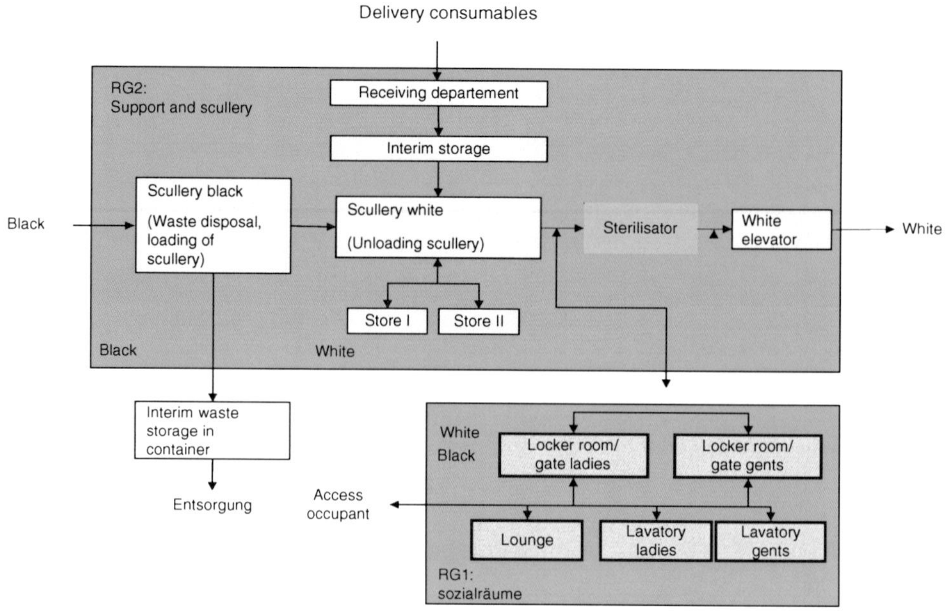

Figure 4.4 Workflows.

substantial costs – that is, due to major modifications in plans – if later on, their inclusion into the subsequent planning process is required.

Examples for this are provision areas for special supply goods or special kinds of waste, smaller storage areas, or even changing rooms in security gates (Figure 4.4).

From the data regarding the frequency of work processes, a planner can find criteria for dimensioning parameters (e.g., simultaneity factors) and for the appropriateness of automation (e.g., automatic front sash, centralized switch-off, etc.), which often provide a considerable energy savings potential in future operation.

Following the process described above, the planned building is laid out without a single architectural plan having been drawn up. Not only does this help make the planning process very efficient, it will also be avoided that, based on preliminary drafts, detailed discussions regarding the structural design arise and lead to unnecessarily limiting preliminary decisions.

The jointly developed room-group scheme as well as the person and material flow schemes can also be translated into a functional room book, in which work flows are described and technical requirements are consistently formulated.

These documents finally enable the fixation and documentation of the users' basic requirements for the planning process.

Based on this, it is incumbent upon the planners to design laboratory areas which do not only fulfill the current needs, but which are also flexible enough to meet future requirements and, with this, the requirement of long-term sustainable building.

5
Corporate Architecture – Architecture of Knowledge

Tobias Ell

In a market that is defined by innovation, research, and development, modern laboratory buildings become crucial and pivotal points in corporate architecture. The laboratory as a place of research and analysis fulfills three functions:

1) As the core of a scientific company, the architecture of a research space directly reflects its culture, value system, philosophy, and vision.
2) At the start of the value-added chain, the laboratory acts as the origin of knowledge and innovation.
3) Furthermore, in the future working world, the quality of the architecture also acts as a magnet for high potentials.

These three levels of meaning of corporate architecture directly interact with and mutually influence one another. At the center are the corporate culture, strategy, and vision which are directly reflected in the corporate architecture. Hereby, there is a dialog between the built corporate architecture and the corporate culture. A company defines its architecture but the architecture also defines the company in the long term.

> We shape the buildings, afterwards the buildings shape us.
>
> *Winston Churchill, 1943*

Here, all three levels of corporate architecture are examined with view to their significance for the modern laboratory. What measures arise from this for the research buildings of the future? (Figure 5.1).

The Sustainable Laboratory Handbook: Design, Equipment, Operation, First Edition.
Edited by Egbert Dittrich.

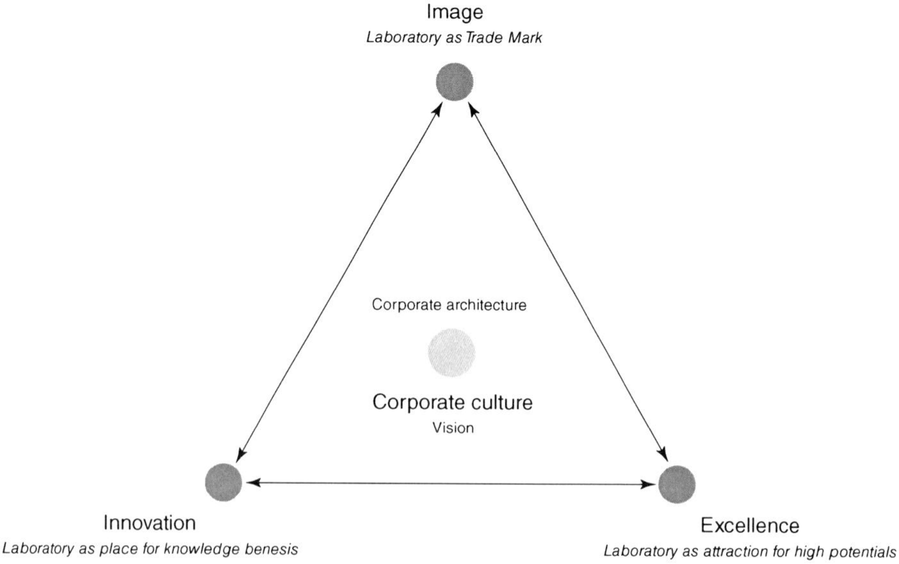

Figure 5.1 Corporate Architecture

5.1 Image–The Laboratory as a Brand

The architecture of a company is of essential public interest. It fulfills not only the company's own functional requirements but also has a direct influence on how the market, the customers, the public, and the employees perceive the company.

In this context, corporate architecture can be understood not only as branding and design object or as means to transfer content and information but also directly reflects a company's culture, value system, philosophy, and visions. What does the company stand for? What values does it convey? The architecture of a laboratory building conveys these messages to the outside world.

The architecture is, for example, a direct expression of a company's corporate social responsibility and says something about its sustainability concept. It is not only from the economic viewpoint that we, as planners, increasingly consider the interesting field of green building certification of laboratory buildings. The German Sustainable Building Council is currently doing all it can to establish also a German label on the market.

The sustainability of a laboratory building, therefore, offers a direct message on how a company deals with the subjects of ecology and economy, as well as sociocultural, socio-political, and functional aspects. Modern laboratory buildings are resource saving and energy optimized. They offer a long-term, suitable working environment and also express a company's appreciation of its employees. "To provide people with a human workplace," becomes the core task of planners.

Qiagen, the worldwide market leader in Sample and Assay Technology, has, together with the architect, developed a corporate architecture within the frame of an innovation initiative. The Qiagen research building of the future is global, dynamic, and technological. As the company's calling card, it conveys innovation, logic, excellence, and openness. The structure, materials, and form of the architecture are a direct reflection of these values. The research building becomes the "mouthpiece" and communication medium for the corporate culture.

5.2
Innovation – The Laboratory as the Origin of Knowledge

The future belongs to those who recognize where new knowledge originates. The assumption that "the most successful man in life is the man who has the best information" (Benjamin Disraeli, 1870) no longer applies on the path to the knowledge of society. Information is available at any time worldwide. We plan and build in a world which is increasingly being defined by the topics of *networking* and knowledge. The commodity that is knowledge becomes an essential resource of the future and man, as the knowledge carrier, becomes a guarantee for success and a driver of innovation of many companies.

> Innovation and creativity are central elements of Europe's modern, knowledge-based societies for effectively taking on the opportunities and challenges of globalization and growth.
> Federal Ministry of Education and Research, Creativity, and Innovation, European Year 2009

Laboratory buildings are rooted in innovation and research. The basis of innovation is the process of the "birth" or origin of knowledge. At the beginning of a value-added chain is always the origin of the knowledge, an idea that can be marketed and an optimized process.

In recent years, more than 90% of the innovative ideas which made it to market maturity were developed in the face-to-face communication, in direct social exchange. A study conducted by Massachusetts Institute of Technology (M.I.T.) initiated by Prof. Thomas Allen, confirms this hypothesis. The image of the brilliant researcher who comes up with the next "one billion dollar idea" has lost its appeal. Instead, we are seeing a paradigm shift in corporate culture and philosophy through which the shaping of knowledge and the innovation processes are increasingly becoming a central task of corporate architecture.

The architecture has the task of providing a solution for the contradiction between the highest technological requirements and a need for quiet, concentration, and contemplation, as well as the demands for a productive, communicative, and creative work environment.

If we assume that the corporate architecture has a direct influence on the corporate culture, it also defines the process of the origin of knowledge in the long term. Hereby, the influencing variables of knowledge and communication form a decisive factor in the design and planning of corporate spaces. The laboratory becomes the knowledge space of the future and the design of effective knowledge spaces becomes increasingly important for companies in international competition.

But, how can a company create spaces which initiate and support the origin of knowledge? *Innovation* and *inspiration* require *interaction* – a principle which also works in a similar way in the neural processes of our brain. Our brain can change the properties of synapses, nerve cells, and also entire areas of the brain in relation to their use – an ability which is described as neuroplasticity. Modern laboratories should be able to create a similar free space for change. In this way, further development and knowledge, therefore, only occur in the "neural" exchange of highly qualified top researchers.

Communication points such as an open documentation zone as well as thinker lounges and scientist cafes offer these points of neural association. The modern laboratory landscape resembles an urban city district which offers, besides the constraints of equipment set-up areas and process operations also communication density, diversity, and free space.

However, besides these places of communication, there must also be individual places in which to retreat or withdraw. In the modern laboratory, these areas create space for concentration and contemplation. Besides scientist cafés and innovation lounges, thinker cells and think tanks are also becoming increasingly important in everyday laboratory life. The laboratory of the future is not a large, open plan room but must instead offer a wide range of opportunities – a space of opportunity which becomes a space of reality through the researcher. Communication and concentration, quiet and exchange define modern laboratory buildings.

> With time, the soul becomes dyed with the color of its thoughts.
>
> *Mark Aurel, 152*

Our work environment has a direct influence on our motivation and effectiveness while an improvement of the work processes only influences efficiency. With this, the laboratory space of the future directly influences a company's innovative strength through these diverse opportunities. The significance, therefore, of these innovation and knowledge spaces increases the more the companies focus on self-organization in networks in their processes.

5.3
Excellence – The Laboratory as a Magnet for High Potentials

Worldwide "talent gap" can be noticed not only in top research and analysis. Operators of public as well as free-enterprise laboratories vie for the most talented researchers. There is tough competition for high potentials on the market and they are well looked after.

The pharmaceutical company Novartis, for example, regularly checks by means of a checklist whether its top talents in China are satisfied and if they plan to leave the company. Other companies do similar things. A new generation of high potentials has, meanwhile, arrived in the business world and new communication styles define everyday life.

In 2009, most patents worldwide were registered or initiated by German researchers although more than 75% of these were abroad. The ideal work conditions and an equally appealing as well as motivating work environment are evidently playing an increasingly important role in the choice of the workplace.

At the same time, employees are also the *core* of knowledge and innovation of a company and, with this, its greatest asset. The laboratory of the future must, therefore, act as an *attractor* for high potentials on the market. In addition, it should also ensure the loyalty of the experienced employees.

However, the choice of workplace is also being increasingly determined through a company's soft skills. The corporate culture and whether a company is appreciative of employees play a major role in this decision. Just as a product represents a company, the architecture is also a direct expression of the corporate culture and, with this, a direct expression of the company's appreciation of its employees.

The added values in the direct work environment determine the choice of workplace in the long term. The laboratory of the future must, as a basic requirement, form an adequate working space with regard to its functionality and depicted processes. However, we as planners are also realizing a growing number of projects in which a more attractive workplace should be created for researchers: a running track around the technology center or the laboratory offers leisure and relaxation potential, fitness rooms in research buildings ensure a balance between physical and mental activity, and scientist cafés in R&D laboratories are in keeping with the times and offer modern urbanity. Google offers a good example of this idea – the company headquarters in California is defined by a relaxed and informal university atmosphere. The highly talented graduate from nearby Stanford University, for example, will feel at home in a familiar environment after his graduation and, in doing so, become part of the Google corporate culture.

As a company, you must, therefore, always ask yourself why a talented scientist should change to your company site.

Perhaps, the site is on the outskirts of a city. How, for example, can I ensure that a highly talented M.I.T. professor is loyal to my company? Create added values! Offer the laboratory building of the future as a modern place where knowledge is developed. It is partly through this idea that projects such as the Novartis Campus and Qiagen Campus receive increased media presence and a lead in the search for high potentials in today's employment market.

6
Scheduler Tasks in the Planning Process

Markus Hammes

Process quality, in the context of sustainable construction, refers to the quality of the entire planning and construction process, as well as the subsequent use.

In the case of the quality of our built environment, the planning and construction phase is the crucial moment that has an impact on the economic and sustainable operation of the building. It is to this end, deemed as plausible to place the optimization at the beginning of the planning process of a sustainable building.

The quality of the planning is influenced by the project preparation, an integrated planning, the proof of optimization, and the complexity of the approach. In the execution phase, the securing of the sustainability aspects in the tender and in the award, as well as the establishment of the prerequisites for the optimum use and management, must be ensured. Within the framework of the construction work, the management of the construction site, and the construction process, the quality of the contracting companies and their *execution quality*, as well as the systematic commissioning, is crucial. Through the planning preliminary work and its realization, the basis is created that allows for controlling and management, systematic inspections, maintenance, and servicing to be carried out by trained operating personnel for the purpose of the further sustainable life cycle of the laboratory building.

The influence on the building projects within the planning phases and in particular, with regard to the conceptual decisions that have to be made early is extremely high. Conceptual errors can hardly be corrected in the execution phase or in operation, and if at all only coupled with a considerable amount of time and money.

The regulation on the professional fees for architectural and engineering services (HOAI, as amended 2009), specifies alongside the professional fees, also the services and the quality assurance standards that have to be provisioned by architects or engineers.

Within this manual for sustainable laboratory buildings, a wide variety of planning aspects will be explained in detail. The following sections are mainly focused on the priorities that have to be set with regard to the basic procedures and attitudes in the task area of the scheduler, which affect the quality of planning

The Sustainable Laboratory Handbook: Design, Equipment, Operation, First Edition.
Edited by Egbert Dittrich.

services, for the purpose of realizing the goal of the design of sustainable laboratory buildings.

6.1 Project Preparation

The project preparation is the responsibility of the building owner or the operator. In special cases, he can also hereby procure help or expertise from external planners.

The quality of the project preparation by the building owner, the user, or the operator of the laboratory building is of fundamental importance to the work of the planning team. The requirements planning and goal agreements serve as the underlying basis for the formulation of the requirements of the building owner with regard to the start of a construction project. Goals, resources, and requirements have to hereby be documented without prejudging the planning, whereby, the former then constitute the basis for the activities of the planners and can serve as a reference in the course of the planning process.

The task of the planner in essence entails among others, the critical evaluation of the requirements from the project preparation. The analysis of the task has, moreover, to be carried out with great empathy for the requirements of the building owner and the users. An unprejudiced view from the outside in this case constitutes an opportunity for effective input at the start of the planning process. This can also lead to improvements within the project preparation itself, pursuant to which the workflows, routes, and connections and room sizes can, for example, be redefined or adapted within the overall framework.

6.2 Integral Planning Teams

Within the integral planning, which includes the entire life cycle of a building from project preparation to demolition, the integral planning team is in essence of special importance. The complex dependencies between architecture, structural engineering, technical building equipment, and laboratory facilities are particularly evident in the case of laboratory buildings. The availability of highly qualified expertise alone is, however, not sufficient. It also in the future has to be possible to provide for a further development of the quality of cooperation between the various technical disciplines.

Alongside the teamwork and communication skills, the willingness to participate within the framework of a conceptual cooperation has to this end be regarded as a basic requirement.

The task of the planner is to contribute conceptually in the early quality assurance phases from the perspective of all technical disciplines. The decision-making process moreover also has to be evaluated in a sustainability-oriented manner and optimized, under the consideration of alternative solutions. In order to allow for these findings to be adequately taken into account, the acceptance of all technical disciplines within the planning team, and on the part of the building owners and the operators, is in essence imperative. It is incumbent upon the coordinator of the integral planning team to ensure the *timely conceptual participation* of all technical disciplines, as well as to ensure that the guidelines and their implementation are ensured through the participation of the users/operators in the planning process.

6.3 User Participation

The participation of the users/operators in the planning process is also a constituent part of the integral planning. In the case of the laboratory buildings, the timely and conceptual integration of future users/operators or their representatives is imperative for the successful integral planning. The requirements and needs of users should be considered by a panel of representatives that is independent of the planning team.

It is for the purpose of this approach, also not imperative, that any given subsequent user is already specified at the time of planning. Within the framework of the project development as well as the requirement planning and goal setting, the interests of potential users can in essence be timely and conceptually integrated at an early stage through a variety of representatives.

The task of the planner is to, in essence, participate by way of an open, transparent, and targeted communication with the user/operator, as a planning partner in the development, execution, and implementation of a sustainable building concept.

The decision-making processes have to be clearly documented.

6.4 Planning Process

The optimization and complexity of the planning process is an extensive subject area. Alongside the numerous legal requirements, numerous concepts are also required to be drawn up, which do not need to be listed in detail at this point. However, the basis of all further considerations and concepts in the laboratory building is a *safety concept* that takes into account special protective measures for the activities in the laboratories as well as the professional and project-specific risks that thereby arise.

A central planning tool within this framework is the development of alternative concepts or variants within a concept. The technical, economic, and environmental parameters have to be comprehensibly taken into account in a transparent manner in all planning steps, and moreover, also evaluated and analyzed with regard to all aspects of sustainability.

In order to test the effectiveness of the concepts as well as to optimize them, further developed hand tools in the area of building simulation will in the future be made available to the planners. Already as of today, fire protection concepts and safety concepts can, for example, be verified through evacuation time calculations as well as smoke and fire virtually simulated in close approximation to the real conditions. Energy, ventilation, and lighting concepts can be used to map all aspects of energy efficiency including all those that concern the comfort of the user. The planner is also required to participate with his knowledge in the further development of the programs. In particular, with regard to the technical aspects, such that the assessments of variants can in essence no longer be deemed as dependent on assumptions, estimates, and other uncertainties.

6.5 Execution Phase

The integration of sustainability aspects in the selection of the company and the tender can only be successfully configured within the framework of an interaction between the building owner and the planner. This is where the basis for a qualitative execution at a high-quality level is established. The criteria established in the planning, for the selection of the building materials, have to be integrated into the tender technical specifications and should be capable of being comprehensibly verified in the offers. The deployment of the products has to be tested and documented on the construction sites.

In this phase, the prerequisites for optimum operation with regard to use and management are also specified. Due to the required flexibility in the case of future-oriented laboratory buildings, the recording of all essential data and product information in object documentation is of particular relevance. This includes a user guide, service, inspection, operation, and care instructions. For the purposes of integral planning and use specification in the laboratory building, a forward projection through all phases up to the completion of the construction project and in the case of subsequent alterations, is of particular importance.

6.6 Commissioning

In the case of a laboratory building, the systematic commissioning of the building and laboratory equipment serves as a vital contribution to the functional optimization.

From the viewpoint of the integral planning and the subsequent user/operator as a planning partner, the timely involvement of the latter in the commissioning, in essence serves as the basis for a profound building handover. The commissioning has to be structured in the previous planning phases and prepared with the help of performance records and documentation prior to the acceptance.

The systematic commissioning requires a concept for the adjustment and readjustment. The buildings and laboratory equipment have to be coordinated and adjusted. In connection with the complexity of a laboratory building, a readjustment after about a year, for the purpose of an optimized operation, can be deemed as reasonable. All setting values have to be documented with their place of the settings, the setting data, and measuring points in a structure according to the facility and the plant.

6.7 Conclusion

Sustainability needs to be further established as a desirable goal for all those involved in the construction process. The architects have in particular on account of their professional ethics, an obligation with regard to the configuration and quality of the built environment. The planners have to bear the responsibility for their actions.

On account of in essence not being bound to ideologies and dogmas, the planners can contribute to raising the awareness with regard to this issue, in order to allow for a trusting relationship to develop between the building owner, the users, the operator, and the planners.

In order to raise the awareness among all participants and to introduce transparency in the planning process, a conscientious study of all aspects of sustainable building is in essence required. The knowledge of the individual aspects and their *interrelationships* is necessary in order to be able to credibly elucidate as well as actively shape the planning process. Education and further training, active action and the integration of all aspects of sustainability in the planning process are the tasks of the planner.

6.8 Best Practice

6.8.1 Project: Center for Free-Electron Laser Science CFEL, Hamburg-Bahrenfeld

- *Principal*: Free Hanseatic City of Hamburg, Agency for Science and Research (BWF)
- *Architects*: hammeskrause architekten bda, Stuttgart (Figure 6.1).

Given the dynamic structural requirements measuring rooms have to meet, the approach to building laboratory and office complexes differs both, formally and topically. A brick-covered mezzanine level provides the basis for the round,

Figure 6.1 CFEL.

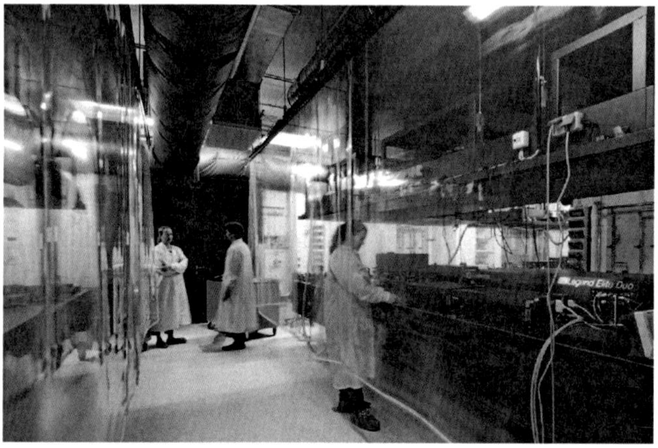

Figure 6.2 CFEL, laser measuring.

three-storey office complex with its lightweight façade consisting of linear rows of windows (Figure 6.2).

The laser measuring rooms on the ground floor are sitting on top of the foundation slabs that are about 1.5 m thick to ensure that vibration-free impulses can be sent in a time span of a trillionth of a second. Completely shielded from the exterior, the laboratory space is highly flexible on the inside. Each laser unit consists of one laser measuring room, one physical measuring room, and a room housing the technical equipment. It is possible to combine these units for complex experiment sequences (Figure 6.3).

To be able to master the challenges of today's scientific laboratory work, multi- and interdisciplinary arrangements that comprise various departments are now

Figure 6.3 CFEL, communication.

frequently necessary. This approach requires room for intellectual professional discussions, scientific disputes, the sharing of information – to sum it up: Places to communicate. Hence, the changing work environments and work patterns do have an impact on the conventional laboratory designs.

Three seminar rooms, 24 laser measuring rooms, and a cafeteria surround the bright entrance hall, and 2 central garden courtyards. The foyer in the heart of the building, which stands as tall as a house, is unique because of its special geometry: Horizontal office levels rotate around a spherical triangle, which are vertically linked with each other by bridge-like stairwells and allow various views from one level to the next. Toward the top, the atrium opens up, thanks to a skylight featuring a filigreed steel structure filled with foil cushions made of ETFE (Figure 6.4).

Figure 6.4 CFEL, atrium.

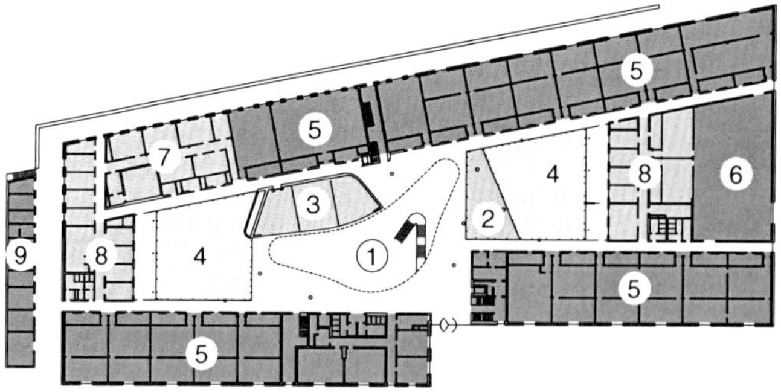

Figure 6.5 Floor Plan Top Floor. 1 - Laboratory landscape; 2 - Offices + special laboratories; 3 - Meeting rooms; 4 - Infrastructure; 5 - Guest apartments.

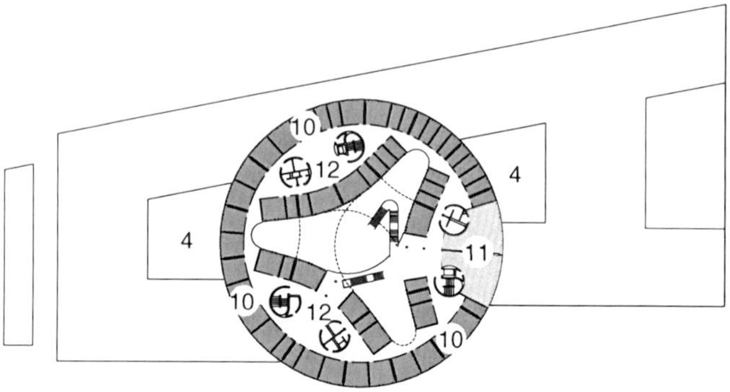

Figure 6.6 Floor plan top floor. 4 - Garden courtyards; 10 - Cell offices; 11 - Meetingrooms; 12 - Open space work space with communication zones and tea kitchens.

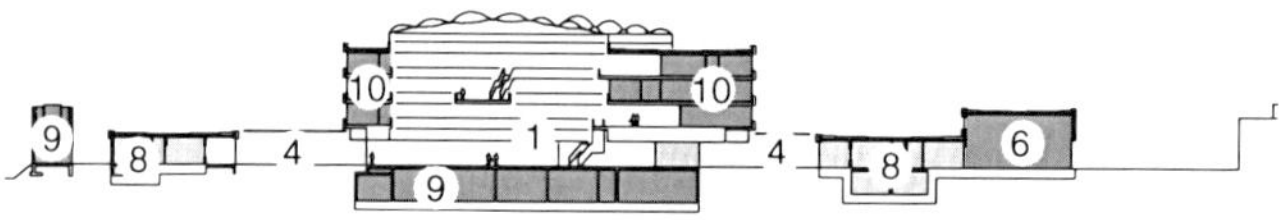

Figure 6.7 Cross Sectional View of the Building. 1 - Foyer; 4 - Garden courtyards; 6 - Experimentation hall; 8 - Chemistry/physics labs; 9 - Infrastructure; 10 - Cell offices.

6.8.2
Project: Max Planck Institute for Aging Biology, Cologne, Germany

- *Principal*: Max-Planck-Gesellschaft zur Förderung der Wissenschaften e.V., Munich
- *Architects*: hammeskrause architekten bda, Stuttgart (Figure 6.8).

The offices that will accommodate the scientists are located alongside the exterior façade on the first and second floor. Traditional, rectangular floor plans promise their future occupants spaces to retreat in and focus. Natural light and ventilation further support the ambience. A hallway separates the all-around

Figure 6.8 MPI Cologne.

office tract from the departments' laboratory clusters located on the interior of the building.

Beginning in the geometrically more stringently dimensioned office tract alongside the exterior facades, the architecture of the building's focus turns toward the interior of the freely and lively designed communicative center via the laboratory clusters. Hence, the entire design concept of the structure is based on the conviction that communication is a prominent driver of scientific progress (Figure 6.9).

Other areas are within easy, quick reach, spatial density, and vastness alternate in an interesting way, while a variety of visual contacts encourage informal meetings and stimulate spontaneous exchanges. The mixing of the functions and

Figure 6.9 MPI Cologne, labs.

Figure 6.10 MPI Cologne communication.

Figure 6.11 MPI Cologne, atrium.

the elimination of out-of-scale functional monocultures are of great importance (Figure 6.10).

The main entrance leads directly into the center of the newly constructed building – the atrium. It provides the interface for internal and external communications within this compact, dense building. The entire room structure and the access routes are identifiable from here. The public cafeteria and the auditorium of the institute, both of which are located on the mezzanine, can also be accessed from here (Figure 6.11).

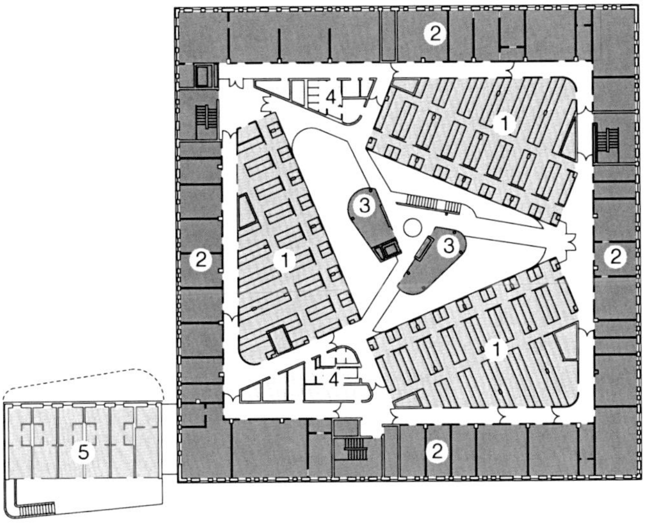

Figure 6.12 Floor Plan Top Floor. 1 - Laboratory landscape; 2 - Offices + special laboratories; 3 - Meeting rooms; 4 - Infrastructure; 5 - Guest apartments.

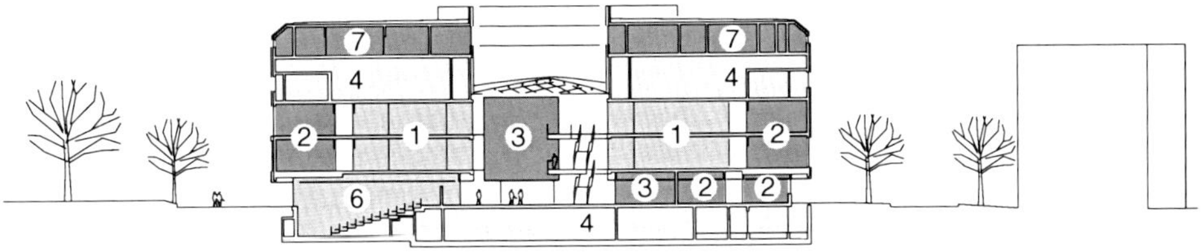

Figure 6.13 Cross Sectional View of the Building. 1 - Laboratory landscape; 2 - Offices + special laboratories; 3 - Meeting rooms; 4 - Infrastructure; 6 - Auditorium ; 7 - Vivarium.

7
Space for Communication in the Laboratory Building

Markus Hammes

A paradigm shift with regard to communication is currently taking place in laboratory buildings. Space for communication is no longer considered an unnecessary luxury but rather a *necessary constituent part* of a sustainable laboratory building concept. The following article explores the definition, development, and importance of communication in research – and laboratory buildings and their spatial conditions. Based on this approach, it will be demonstrated that the laboratory operation and communication have a long shared history that needs to be rediscovered and integrated into modern laboratory buildings.

7.1
Definition of Terms

A standardized and generally accepted definition of the term *communication* does not exist. Various academic fields deploy different approaches in their exploration of the subject of communication. A variety of communication models are explored in the information sciences, media studies, the speech act theory, linguistics, as well as semiotics. The access can be based on own experiences, or else on action-theoretical, problem-theoretical, scientific signal-theoretical, as well as on biological, psychological, and behavioral – theoretical, and systems-theoretical basic assumptions. Thus, communication can, among others, be understood as a technical, mathematical, cybernetic, biological, evolutionary, psychological, or social process.

This, among others, is accountable for the fact that there is hardly a term, which is used as inflationary and in most varied contexts as the term *communication*. In everyday life, everyone is aware that the term is used and understood in different contexts, for example, with regard to technological aspects in communications technology or with regard to target groups in the marketing sector.

In the scientific exploration of communication, the question is asked, as to how communication can be explained by way of the conditions under which it takes place, what are the criteria for successful communication, and how can reliable

The Sustainable Laboratory Handbook: Design, Equipment, Operation, First Edition.
Edited by Egbert Dittrich.

models be compiled, from which predictions and procedural instructions can be derived.

Two well-known quotations clearly serve to demonstrate the breadth of scientific, philosophical, and theoretical consideration.

"One cannot communicate" (Paul Watzlawick, Menschliche Kommunikation. Formen, Störungen, Paradoxien (*Human communication. Forms, disorders, paradoxes*) – Bern 1969, ISBN3-456-82825-X. – p.50–53).

"The human being cannot communicate, only communication can communicate" (Niklas Luhmann, Die Wissenschaft der Gesellschaft (*The science of society*), Frankfurt am Main 1992, p. 31, Suhrkamp Taschenbuch Wissenschaft, 1001).

In the case of the consideration in the context of research and the laboratory building, it is purposeful to agree on a definition and define it as the basis for further consideration.

"Communication," which is a term derived from the Latin word "communicare," means the activity of conveying information through "sharing, informing, enabling participation, making together, merging." Communication, in this original sense has to be in essence understood as a social act between living things: (In this case: The human being). This social action is described through the exchange or transmission of information. This most original approach to the term *communication*, today, among others serves as the basis in the theory of action.

7.2 Historical Development

Although the historical development of the laboratory construction is described elsewhere, it is important to again briefly establish the casual connection to the subject of communication.

The historical origins of the "laboratory construction" can be traced back to the thirteenth century, in which the basic principles of a medical order envisaged the separation of the profession of the medical physician from that of the pharmacist. This led to the setup of pharmacy offices, alchemist kitchens, and tasting courtyards of the smelters. Whereby, the apparatus and equipment such as retort, mortars, and chemical stove with hood, which serve as the forerunners of today's fume cupboards, can be traced back to this period.

In the seventeenth century, a further separation of craftsmanship from science took place. Especially in the case of the smelters and for the purposes of the mining operations in the mining sciences, chemical laboratories were set up for the study of ores, minerals, and chemical products. An increasing level of adverse side effects in the melting processes, however, made the separation of these work spaces necessary (Figure 7.1).

The illustration of an alchemist kitchen from the thirteenth century serves to illustrate the two parallel main features of this early development, which today still play an essential part in the consideration of laboratory buildings.

Figure 7.1 Alchemist laboratory, thirteenth century.

1) The main reason for the set-up of a laboratory is for the safety of the people working in it and for the purpose of ensuring a smooth operation.
2) There are places of intellectual technical talks, of scientific dispute, for the exchange of information, in short, of communication (Figure 7.2).

Around the 1750s, G.B. Piranesi drew up the ideal plan of a university. The plan shows the typical formal, aesthetic order and constructive regularity of architectural fantasies of classicism.

It is, moreover, interesting to note that the spatial integration of all functions and the mixing of all areas of life in an overall ensemble were considered a spatial ideal state for the generation of knowledge and innovation: living, working, researching, and teaching beneath one roof. From the lab to the wine cellar, all functions are in essence housed in one and the same building complex (Figure 7.3).

The illustration from about 1840 of Justus von Liebig's laboratory at the Chemistry Department of the University of Giessen serves to illustrate the laboratory as a place of communication.

7.3 Development in the Modern Age–Why and When Were These Ideal Conceptions Lost?

The mixture of functions in a complex of buildings or even in a room has changed fundamentally in the modernism of the twentieth century, whereby, this does not apply only to the area of the laboratory buildings and communication.

Undeniably, whether consciously or unconsciously, is the influence of the Bauhaus and CIAM (Congres Internationaux d'Architecture) on generations of planners.

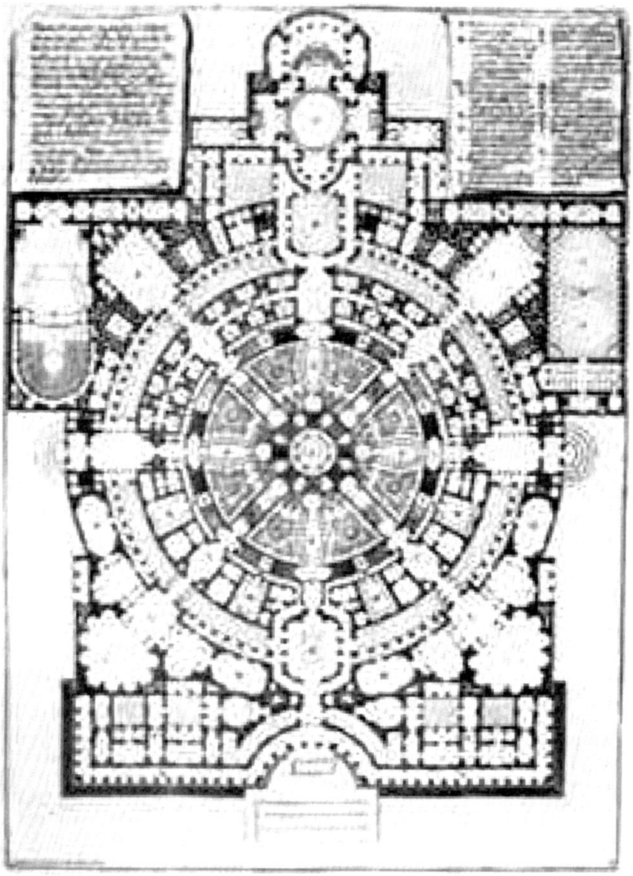

Figure 7.2 Ideal lay-out of a university, Piranesi, Around the 1750s.

In 1933, the IV Congress of the CIAM was held in Athens. In 1941, the Athens Charter was published, which was translated into the German language in 1962.

Core focus was the unbundling of urban functional areas.

The separation of the functions was understood and formulated as an ideological dogma. Pursuant to this urbanistic approach, the separation of functions and zoning has to be introduced as a system for the spatial order in all areas of life. In the post-war period, these ideas were updated and taken up for planning purposes. Examples to this end are well known. The dormitory towns (satellite towns) of the 1950–1970 years for "housing," the commercial areas for "work" or the newly established campus universities on the "*grünen Wiese*" for the "teaching and research" are based on this theory.

Figure 7.3 Laboratory of Justus versus Liebig, approx. 1840.

These general social development and planning theories, of course, had an influence on the planning of laboratory buildings. The separation of laboratory and office space in separate buildings or parts of buildings serves as an example of a planning approach that has been pursued for a long time.

Economic aspects also spoke in favor of the separation, such as the lower floor height and the smaller amount of functional areas of the office building parts which would lead to a reduction of the investment costs in comparison to the requirements of laboratories.

This was likewise intended to provide for a clear physical and organizational separation of the functional areas of the safety-relevant sections and to be optimized exclusively under this aspect. Each function was spatial zoned and separated and could thus be seemingly "perfectly" solved separately.

On account of these developments, the hitherto space for the informal exchange, in essence for communication was obviously lost. This was not intended under the circumstances, but rather a consequence of other considerations. However, the consequences are still noticeable today with regard to our considerations and evaluations.

In German-speaking countries, the term *communication* has only been in use since the twentieth century. As late as 1969, Paul Watzlawick described the term *communication* in the preface to the aforementioned work as "unusual in German."

This circumstance serves to substantiate the hypothesis that the loss of communication in general was neither properly perceived nor recognized as a quality loss.

Only gradually were the awareness of the qualities and the necessity of communication again clearly perceived, and the lack of communication and communication skills assessed as a loss. The modified work environments, the development of linear processes in networked processes, accelerated processes, and new scientific findings have in essence led to the need for a new assessment.

7.3.1 Why Is Communication Important in the Laboratory Building?

As demonstrated within the context of the historical development, space for communication in connection with laboratories is not an invention of the modern era. It is rather the opposite that holds true, space for communication was also an integral (and unconscious) constituent part of a laboratory.

If laboratory work is to be considered part of a *complex process* that involves the generation of knowledge and innovation as an objective, the ability to communicate has in this case been regarded to be of outstanding importance.

7.3.1.1 Communication Promotes Knowledge and Innovation

In the theory of action, the underlying definition of communication as a social act is described in more detail. A further feature of communication is creativity. In communicative social action, new thoughts, ideas, and solutions to problems evolve that would not exist without communication.

"Knowledge is created at all organizational levels, but each with a different contribution to innovation capacity. In particular, the interactive and thus the communicative aspects play a significantly high role."

"In terms of rationality in the innovation context, this in essence means more and better ideas evolve through collaboration in the team, on account of communication, ... "

(Thesis research on the topic inventor research of the HANS SAUER, chair for Metropolis and Innovation Research. Dissertation paper by Katarina Bobkova: Generate knowledge and manage innovation: development and testing of a questionnaire for the measurement and analysis of evolutionary knowledge creation in innovation processes, 2008).

At the Massachusetts Institute of Technology (MIT), human communication is increasingly being more accurately decrypted with technical assistance under the direction of Alex Pentland.

Based on his findings, Pentland has been able to demonstrate that human communication, to a considerable extent, takes place on an archaic, nonverbal level. This has enabled communication within a group even before the evolution of language. Even today, more importance is attached to how something is said over what is actually said. In a study conducted among the employees of a large German company, he was able to prove that teamwork that is implemented through "face to face" contact is up to 30% more productive than that of a lone fighter in the home office. Employees find new solutions considerably faster when they look at each other and communicate interactively with each other.

In order to cope with the challenges of today's scientific work in the laboratory, multi- and interdisciplinary work that goes beyond individual disciplines is increasingly becoming necessary. As explained in the previous chapters, these changing work environments and working practices also affect the conventional typifications of laboratories.

7.3.1.2 Communication and Safety in the Laboratory is Not a Contradiction

The knowledge of the users about the processes in the laboratory, both in terms of work safety as well as in terms of optimized and sustainable operation are considerable. Organized training and informal experiences are a constituent part of a comprehensive communication process.

To avoid misunderstandings, it still goes without saying that space for communication must not be confused with interference of highly concentrated work, whether in the laboratory or elsewhere.

7.3.2 How Does Space for Communication Evolve?

We can consider two different levels, namely those that affect the interaction between people and those that affect the interaction between people and space.

At the level of spatial organization. Following the negative findings of the separation of functions, function mixing is today applied as a re-discovered principle also to modern urban developments. The mixing is in essence regarded as an opportunity.

With regard to laboratory buildings, this means that space for communication has to once again be regarded as an integral part of interior design concepts for laboratory buildings. This can lead to connected laboratory landscapes with different zonings or to the close proximity of the most varied functions that are spatially completely separated from each other. What is important is the mixing of the functions and the removal of un-scaled functional *monocultures.*

Short distances, a fascinating alternation between spatial density and distance as well as a variety of *visual contacts* allow informal meetings and stimulate spontaneous exchange.

The other level of consideration concerns the design of the space itself. The University of Applied Sciences Mainz offers as of 2010/2011 the degree program "Communication in Space." Account has hereby been taken of the fact that "the design of architectural space as a space of communication in recent years, increasingly gaining economic importance".

Alongside the interaction between man and space, the integration of analog and digital media for the conveying of emotional content and messages about room atmosphere, materials and surfaces is a key objective of the study.

7.4 Conclusion for Future Concepts

Communication is perceived as a *necessary social act* between people across and beyond all the technological aspects and innovations of information and knowledge transfer and generation. Only the awareness of the quality and necessity of communication leads to its consideration and in turn allows new laboratory building concepts, room layouts, and laboratory landscapes to be established.

This means that alongside the physical space, ideational space has also got to be available for communication. On account of the fact that space for communication was functionally separated and disappeared from consciousness, it has to laboriously be created again. Space for communication must once again become an integral part of space approaches.

Space for communication does not automatically mean additional space for the other functional areas. Rather, it is the first step to raise awareness and acceptance of suitable areas for communication, which is then followed by the integration of such areas in the traffic areas or use areas within an overall concept. The different requirements, arising from the use and operation in the context of the safety aspects of the laboratories, can hereby also lead to different spatial concepts with regard to communication-promoting spatial structures.

Short distances, spatial density, and the function of mixing for the promotion of communications are not aspects that speak against compact, economical, and sustainable laboratory buildings.

Space for communication in the laboratory building, in essence, only serves as a synonym for how our work world will also change in the laboratory, in the future. The laboratory building of the future will be imbued with the most varied activities and functions.

8
Fire Precautions

Markus Bauch

8.1
Preventive Fire Protection

8.1.1
Scope

In terms of jurisdictional building requirements, preventive fire protection must ensure the safety objective given in Section 14 of the German Model Building Regulation.

8.1.1.1 Fire Protection

Buildings and structures shall be arranged to be built, modified, and maintained such that the occurrence of fire and the spread of fire and smoke (fire propagation) are avoided and, during a fire, the rescue of people and animals as well as effective fire-fighting is possible.

How these objectives are achieved is described in the building code and possibly in special building codes. Following these codes is thus the building law perspective, and the fulfillment of these protection goals is the statutory minimum requirement for sure.

Deviations from these rules are possible in individual cases. In this case, the deviation must be justified and it should be demonstrated through which replacement the protection goals are achieved in an equivalent manner.

Especially with special structures, but also partly for normal buildings, a fire protection plan may be required as part of the application documents, in which the fire prevention and, in particular, compliance with the building regulations required for the concrete construction projects are presented.

It is important to note here that up to this point only the required jurisdictionally relevant building regulations as objectives have been considered. For insurance purposes and in particular also for operational reasons (protection against business interruption, protection of research results, etc.), the protection objectives may be significant, in particular the protection of property and minimization of the risk of *business interruption.* These protection objectives may require

The Sustainable Laboratory Handbook: Design, Equipment, Operation, First Edition.
Edited by Egbert Dittrich.

or make economic sense, and therefore are well above the building regulation requirements.

Furthermore, fire protection requirements in specific regulations involving laboratories are observed.

Below are the details of the measures of *preventive fire protection.* Organizational measures for preventive fire protection are as follows:

- Selection and installation of fire extinguishers
- Training and designation of fire protection assistants
- Training and appointment of a fire prevention officer.

Everything is required in professional association regulations or partially in building lease, but not explicitly discussed here.

8.1.2
Legal Framework – Construction Law

In Germany, building regulations are governed by state law (competing legislation). In order to ensure substantially uniform building regulations in each state, model ordinance and model building code regulations are developed and updated by a transnational body (the ARGEBAU). Even though these model rules have no legal force, they serve as a basis for the respective local construction law. Therefore, in the following we exclusively discuss these model rules. These rules are publicly available on the Internet (*www.is-argebau.de*).

8.1.3
Model Building Code

The model building code describes, among other things, the fire protection requirements for the so-called regular buildings; it is meant for those buildings that are primarily of residential and office/administrative use and similar use, and not, for example, due to their size or as special structures to be assessed.

These "regular buildings" are classified, depending on the size of the functional units and the height of the building, into different building classes. Laboratory buildings, in rare cases, belong to the building class 4 (building with a height up to 13 m and use units with no more than 400 m^2). But, typically when the building's body exceeds the limits of the building class 4, class 5 buildings (other buildings including underground building) are used because of the exceeded height and/or area. Laboratory buildings can also be special structures, particularly where one or more of the following definitions for special constructions are met:

1) High-rise buildings (buildings with a height of the floor of the highest habitable space of more than 22 m);
2) Buildings with more than 1600 m^2 floor space of the floor with the largest expansion, excluding residential buildings;
3) Schools, colleges, and similar institutions;

4) Buildings and structures where the use is associated with handling or storage of materials with increased risk of fire or explosion;
5) Facilities and premises that are not listed in items 1 and 4, but with a usage associated with comparable risks.

For these "special buildings," special building codes may apply, in particular, in which the special fire protection needs of these buildings are taken into account.

In the model building code, we will specify in particular

1) *the land and its development*
   a. accessibility (for the fire brigade)
   b. distances to other buildings;
2) *buildings and structures*
   a. stability
   b. fire protection
   c. heat, sound, and vibration protection
   d. traffic safety;
3) *construction, building types*
   a. approval, certificates, proofs in individual cases
   b. types, demonstration of compliance.

The protection goal is that the spread of fire and smoke (fire propagation) is prevented and effective firefighting is possible, which is in the fire protection concept of the model building code, and in the first place achieved by the separation of spaces/usage units/floors and fire sections through walls and ceilings with satisfactory fire resistance. Therefore, a section of the code is devoted to these components:

8.1.3.1 Walls, Ceilings, and Roofs

- General requirements for fire behavior of building materials and components
- Bearing walls, columns, exterior walls, and partitions. In laboratory buildings, is important to note that, among others, partitions are required between functional units and completion of spaces with increased risk of fire or explosion. These partitions must have the fire resistance of the supporting and stiffening components of the floors, or at least be fire resistant.
- Fire walls separating fire compartments, that is, entire buildings or parts of buildings, from each other. They are not relevant here explicitly.

8.1.3.2 Ceilings, Roofs

Ceilings must be stable and long enough as supporting and separating elements between floors, and, in case of fire, resistant to the spread of fire. Therefore, they must be fire resistant (class 5 building) or highly fire retardant (class 4 building).

The other essential protection objective of the building code is the rescue of people during a fire, which is achieved by the rules relating to the number, type, and length of escape routes. The section on escape routes, ports, and balustrades are of particular importance from the perspective of laboratory design.

8.1.3.3 Section 33 (MBC)

First and Second Escape Route For use of units with at least one work room, at least two independent escape routes must be available to the outside on each floor; both escape routes may, however, lead within the floor on the same corridor if necessary. For use units that are not located on the ground floor, the first escape route must be through the main stairway. The second escape route may be another necessary stairway, which should be accessible with rescue equipment by the fire point-of-use unit. The second escape route is not required if the rescue through a securely accessible stairwell is possible, where fire and smoke cannot penetrate (safety stairwell).

Building whose second escape route leads through the rescue equipment of the fire department, and where the top of the parapet is more than 8 m above the ground if a ladder is put up, might only be built if the fire department has the necessary rescue equipment such as rescue vehicles with lifting platforms. For special structures, the second escape through the fire rescue equipment is permitted only if there are no concerns for the rescue of persons.

For the route length, the following applies: From any point in an occupied room and a basement, at least one exit to a required stairway or to the outside must be accessible in a maximum of 35 m.

In the model building code, further requirements are mentioned:

- Stairs
- Necessary stairways, exits
- Necessary corridors, open corridors
- Windows, doors, other openings, and balustrades.

This applies the principle that the escape route should become safer ad safer. Within the utilization unit (i.e., laboratory room within the laboratory), the building code makes no requirement for the emergency exit except the length limit of 35 m. If the "necessary corridor" is reached, the escape route becomes safer, that is, the "*necessary corridor*" requires a particular design (walls, ceilings, doors, fire loads, length of the corridor, design of smoke compartmentalization) to match the established guidelines. The route length of maximum 35 m is measured including the path on the "necessary corridor." The second escape route must run as the first escape route over the same "necessary corridor."

"Necessary corridors" are not required if the functional units are not more than 200 m^2 within operational units, or which serve as an office or for administrative use with no more than 400 m^2. This wording opens up the possibility of multiple spaces to a certain extent – laboratories as well – to summarize a unit of usage without "necessary corridor" and build "open area laboratories." The question of the boundary surface of "open area laboratories" will be discussed again later.

When you reach the "necessary stairway," which is always in a "necessary stairwell," you reach, from the perspective of construction law, already a very safe area. The design of the escape route terminates here. To make this staircase sufficiently safe, high demands are placed on its openings, walls, exit to the outside, possibility of smoke funnel, and the absence of fire loads. Since, however, it cannot be ruled out that this stairwell may become unusable in the event of a fire, a second escape route independent of this staircase is required. This can be made possible by the fire rescue equipment (ladders) or must be provided structurally. The second escape becomes obsolete only if the first escape route through a "safety staircase" with particularly high requirements to prevent smoke ingress is provided. In laboratory buildings, it is common to plan a second escape route in general.

The question of the permissible area of laboratories is controversial in practice. Some authorities argue against the limitation of "utility units serving the office and administrative use" and limit labs to 400 m^2 (or even less, on the grounds of higher risk). Other authorities interpret the building code in such a way that this limitation formally refers to the provision of "necessary corridors" in "utility units serving as an office or for administrative use," and, therefore, it has nothing to do with the interpretation of laboratories of larger areas. Thus, only the building code limit of the fire compartments (1600 m^2) would be applicable.

Conclusion from the perspective of laboratory buildings is as follows: the building regulations law calls for space-enclosing ceilings, walls, and doors with fire resistance for laboratory space, and requires two (usually structural) escape routes with necessary corridors and stairways in appropriate stairwells. The laboratory area can be up to 400 or 1600 m^2, depending on local laws and regulations.

8.1.4 Special Building Codes

Depending on the construction, special building codes, for example, the high-rise building directive, apply. Other than pointing out the resulting features that are essentially determined from the fact that the second rescue is necessary to ensure the mandatory structural and firefighting requirements due to the building height, we will not discussed them here.

8.1.5 Other Rules and Regulations Including Structural Fire Protection Requirements for Laboratories

The following list and commentary does not claim to be exhaustive. The rules can be roughly divided into building regulations, professional trade association regulations, and standards.

8.1.5.1 TRGS 526/BGR 120/BGI 850-0

In these specific regulations for laboratories, a number of organizational and structural/technical fire protection measures are proposed:

- Emergency procedures, escape, and rescue plan
- Fire protection

- Fire-fighting equipment equipped with fire extinguishers according ASR 12/1.2[1)]
- Conducting fire drills
- Behavior in case of fire of fire-fighting equipment and gas cylinders
- Storage and holding of hazardous substances
- Workplace design
- Operating and communication areas.

8.1.5.2 Escape and Rescue Routes

This section refers to structural fire protection, which is important because here specifications about the building code and beyond can be found.

Taking into account local conditions, the materials used, and the procedures in laboratories, sufficient number of escape routes must be available. Escape routes shall only lead through an adjacent room if this room, in case of danger during the operation, allows a safe exit without assistance. It is better to have a second chance to escape in any laboratory space (see also building regulations of the countries). The maximum escape length must not exceed 25 m.

See also workplace directives ASR A2.3 “escape routes, emergency exits, escape, and rescue plan.”

8.1.5.3 Doors

Here again, requirements in addition to the building code are applicable. Doors of laboratories must open outward and be fitted with a viewing window. Sliding doors are not permitted for laboratories. Laboratory doors shall be kept closed.

8.1.5.4 Shut-Off Valves

It is also very important for firefighters to be able to turn off, from a safe place, certain services such as fuel gases and other services (e.g., outside the laboratory at the access door).

For specific laboratory areas such as laboratories that deal with genetically modified or pathogenic organisms or radioactive substances, special rules are applicable, but they cannot be discussed here due to space limitations.

8.1.5.5 Fire Alarm Systems

Laboratory equipment with automatic fire alarm systems is now common and can be considered “state of art,” even though there is no explicit mandatory requirement to do so. In spite of the increased risk of fire due to the handled substances in laboratories and the activities being performed therein, with the help of a fire alarm system in conjunction with a building alarm system the problem of “trapped spaces” can be compensated.

Fire alarm systems shall be installed in accordance with DIN 0833 and DIN 14675 or other national standards. This is associated with a sectoral planning by a qualified planner. During the installation of automatic fire detectors, special

1) Workplace ordinance.

aerodynamic conditions (e.g., near fume cupboards) are to be taken into account in order to ensure that potential fire smoke can also reach the detector and not extracted earlier.

8.1.5.6 Air Ventilation Units

Laboratories are typically equipped with sophisticated air supply and air extraction technology. The fire protection requirements of ventilation ducts and ventilation systems and their working principles are described in national regulations such as the German directive M-LüAR, *www.is-argebau.de*. The objective of this directive is the prevention of the spread of fire and smoke through the ventilation system(s). The directive distinguishes between compartment solutions in the fire protection of unprotected ventilation ducts at each passage through building elements (walls/ceilings) provided with fire-resistance requirements with fire dampers and shaft solutions in the ventilation ducting through the floors in their own fire protection separate shafts through the ceiling, requiring fire dampers at the entry of lines in the duct only. In principle, also for each room a separate shaft with its own extract could be designed, in this case without fire dampers.[2)]

Explosion protection requires regulations beyond the standard directives. Potentially explosive extracted air must be kept in separate ducts. In case of extracted air from safety cabinets with flammable gases, the extract duct must be kept in separate fire-resistant ducts and shafts.

8.2 Fire Protection Solution for Laboratory Buildings

For better orientation, each of the following texts is provided with a ground plan. These sketches are not to scale and serve as guidance only. The different wall thicknesses have the following meanings:

- *Thick line*: Wall with requirements for fire resistance
- *Thin line*: Wall without requirements for fire resistance
- *Dotted line*: Wall without requirements for fire resistance, with large glass surfaces
- *T30GA*: Fire-resistant doors with glass inset
- *RS*: Smoke-protection doors
- *T30RS*: Fire-resistant doors with smoke control function.

2) Comment by Egbert Dittrich: This is not a very sustainable solution, needs a lot of material, and does not allow heat recovery. The solution was favored for a while due to the fact that no special fire dampers for laboratories are in the market for the duct behind fume cupboards. Today, the use of fire dampers behind the fume cupboards is accepted because the dilution of corrosive substances is considerable and, in particular cases, scrubbers must be installed and/or periodical maintenance is required.

8.3
Fire Protection Solutions for Laboratory Buildings – Examples

8.3.1
Classic Laboratory

The "classic design" for laboratory buildings. The building has two planned escape routes through two independent, necessary stairwells. The access corridor is a necessary corridor, which is divided from a length of 30 m by smoke doors in smoke compartments. Each laboratory is separated from other rooms – the neighboring laboratories – by fire-resistant walls and fire doors. The second exit mandated by the laboratory directive is provided from each laboratory via the write-up area in the lab next door. The neighboring laboratory is to be evaluated as a safe area for the purposes of laboratory directive, as it is separated by fire-resistant materials. From the view of structural fire protection, it is still a better concept to guide the second exit to a surrounding building and connected to the stairwell escape balcony. This option is particularly important when the labs are to be leased to different users because, in this case, building regulations and practical usability of the second exit in the neighboring laboratory can be found in question (Figure 8.1).

In the sketch, a write-up for laboratory managers is located. It must be designed in a way to allow good visibility in the laboratory, in order to avoid the problem of "trapped space," that is, a room that is not connected to a necessary corridor but can be exited only via different user areas. Though supportive, in practice these

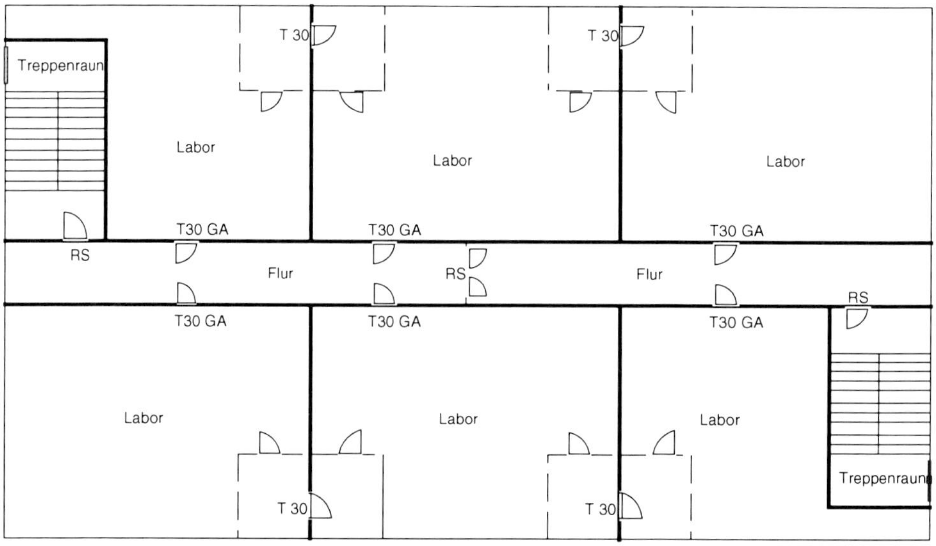

Figure 8.1 Schematic design of a "classical laboratory floor plan."

glass elements are often bothersome and will therefore be imposed or glued, and they lose their purpose; it makes sense to install a comprehensive automatic fire detection system with a coupled audible alert of the user.

Since the corridor is a necessary corridor, this should be kept fire load free (which also means that no closets made of combustible material and/or with combustible content (e.g., laboratory equipment or files, etc.) are allowed. Installations, for example, in the ceiling area, shall be separated with regard to fire protection against the corridor.

The disadvantage of this concept is its rigidity and the problems that the many walls require, with requirements for fire resistance, consuming compartments or fire dampers at penetrations with wires or pipes.

By means of legal approvals, this construction is safe and therefore usually easily approved.

8.3.2
Laboratory Units

By means of legal approvals, this construction design takes advantage of the fact that the building law allows the combination of similar rooms for use as units. But this is also a disadvantage. If renting of single laboratories to different users is planned, building regulations require the use of two units, which are not given here. Depending on the local licensing authority, maybe also the 400 m^2 limit of building regulations for usage units has a limiting effect for office use. Otherwise, the two escape routes of the structural building should have two independent *necessary stairwells*. The *access corridor* is also a necessary corridor, which is divided into a length of 30 m by smoke doors in smoke compartments. Each laboratory is placed against other rooms – with the exception of neighboring laboratories within the use unit – separated by fire-resistant walls and fire doors. According to the recommended laboratory directive, the second exit is given from each laboratory via the write-up area in the lab next door. Here, however, approval from the relevant authority for occupational safety may be required as to whether the lab next door is yet to be assessed as safe area in the sense of the laboratory directive, as it is no longer separated by a fire-resistant wall. Again, a better escape route concept is that the second exit leads to a building surrounding and connected to the stairwell escape balcony (Figure 8.2).

In the sketch, again write-up areas for the occupants are located. Its purpose has already been described earlier. In this construction, it is particularly useful to install a comprehensive automatic fire detection system with coupled audible alert of the user, since the alarm in case of danger is raised not only in the write-up area but also in the lab next door.

The corridor is necessary with all its described restrictions.

The advantage of this concept is its greater flexibility and the elimination of a number of walls with requirements for fire resistance, so that their penetration is not a problem.

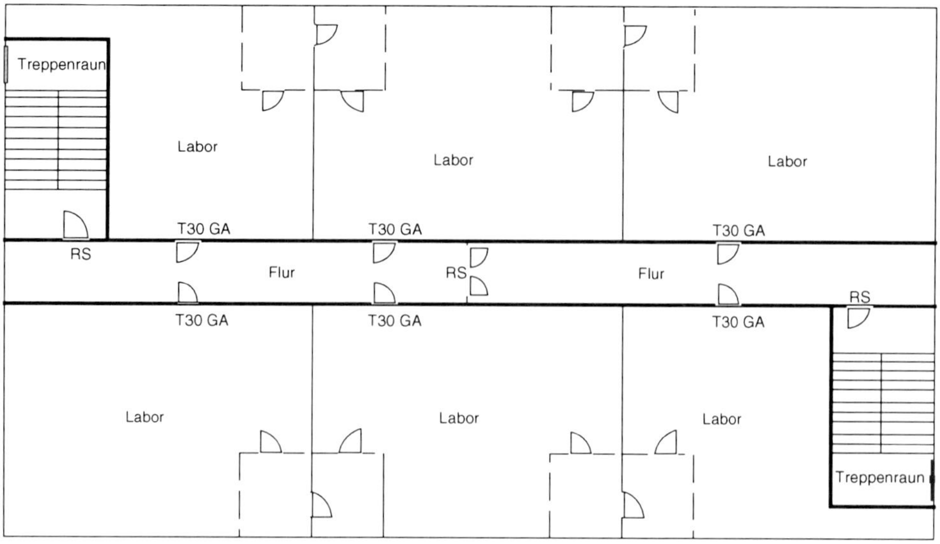

Figure 8.2 Layout.

By means of legal approvals, this construction is fraught with potential limitations and should, therefore, be coordinated as part of the planning approval with the authorities.

8.3.3
Open Architecture Laboratories

This design allows very high flexibility. Restrictive effect here, however, is that the escape route is no longer limited to catch the required corridor of the laboratory directive of 25 m, but this length is now to be considered to catch the required stairwell. This may have the stairwells "closer together," and many more stairwells are required. The escape route and the escape route lengths must be considered, as well as the furniture. Since the stairwells are not connected via necessary corridors but will lead directly to the laboratory, the access doors have fire-retardant and smoke-shielding (T30RS) design (Figure 8.3).

Especially in such large, open area laboratories, there is often a need to create shielded write-up workstations or separate rooms for devices. It has already been described in the topic of "trapped space". Even with this design, it is very useful to install a comprehensive automatic fire detection system coupled with acoustic alarms because the user has to be warned early in case of danger when the large space is not entirely clear. Especially, the potentially "trapped" areas can be evacuated in this construction solely via the open laboratory area; an exit through other (safer) areas is not possible.

By means of legal approvals – if the escape route guidance is in accordance with regulations – it is possible to clarify up to what size the authorities will not allow

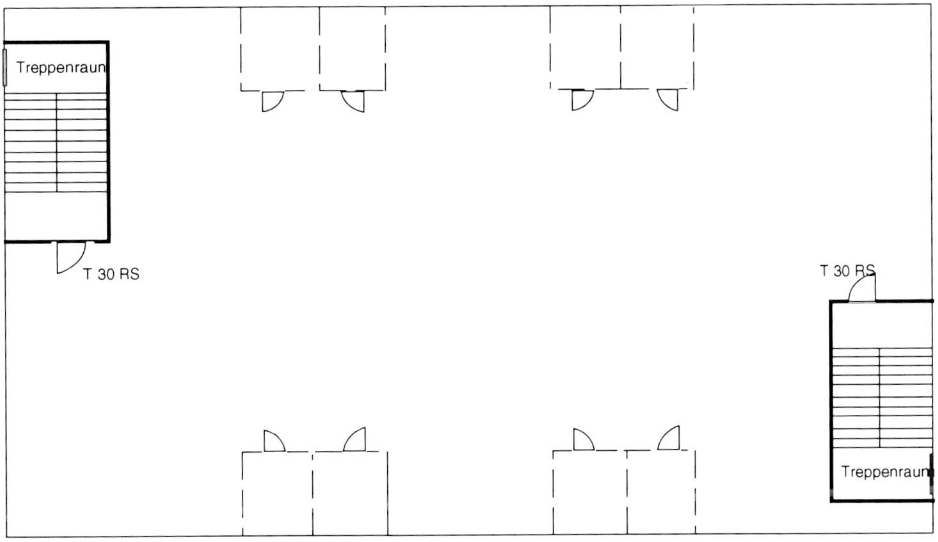

Figure 8.3 Schematic design of an "open-area floor plan."

the open area laboratory. As described earlier, there is a maximum of 1600 m^2 in principle eligible for approval. But, if necessary, the "400 m^2 usage control units regulation" needs to be clarified.

8.3.4
Particular Cases

The problem of separated side rooms was discussed using the example of "write-up areas." Here, ensuring self-rescue is important, and the user of such separated "trapped rooms" must be able to recognize a danger of his or her escape route and be warned sufficiently early.

If in laboratories experiments are run unattended overnight, it must be proved in each case whether additional safety measures are required. In addition to an existing automatic fire detection (or, even more so, if no automatic fire detection is available), monitoring can be carried out with mobile, radio, phone, or networked devices. These devices can detect and report faults or more characteristics of fire.

Laboratories with increased risk (e.g., night laboratories) can also be equipped with automatic fire extinguishing systems. Suitable here are – especially when water as the extinguishing agent is not suitable – gas extinguishing systems. The extinguishing gas today minimizes the risk to persons while triggering the extinguishing system.

For the operation of a laboratory building, also special rooms such as chemical warehouses and warehouses for flammable liquids are required. For these rooms, additional rules are applicable, which cannot be considered further here.

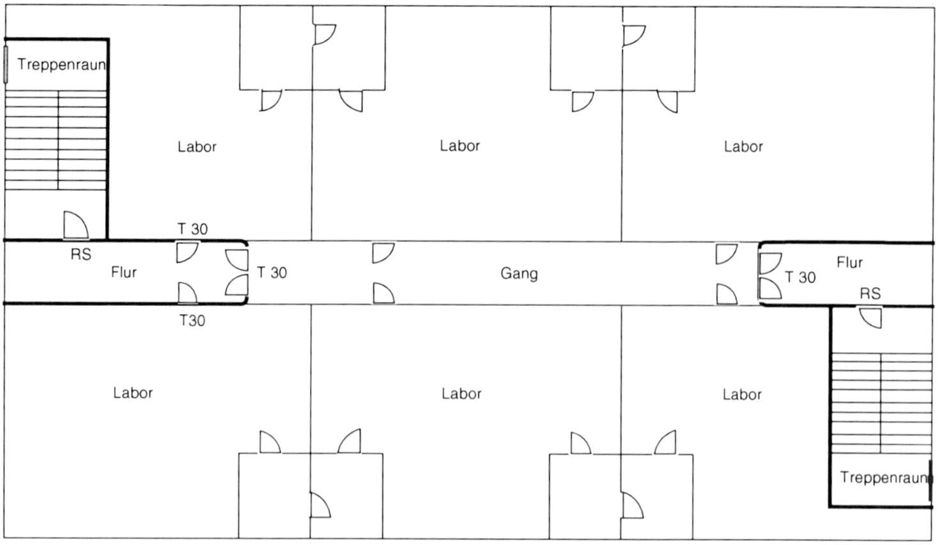

Figure 8.4 Schematic design of improved fire protection of an existing building with a necessary corridor incapable for improvement.

8.3.5 Problem of Existing Buildings

In practice, one is often confronted with the task of strengthening the existing laboratory building's fire protection system. This need may arise for various reasons:

- A new construction law may require change, by which the "grandfathering" of an approved existing building expires because the building does not comply with current building regulations.
- As part of an official inspection, deficiencies have been identified.
- In the course of "creeping use change," the building has become so far removed from the original approval status that a new or modification permit is required.

The finish of the partition walls and especially the corridors is, in practice, often critical. The partition walls – and sometimes the corridor walls – do not meet the requirements for the required fire resistance. Often, the corridor is simultaneously used as "supply channel": that is, all pipes and services are installed in the ceiling area of the corridor.[3)]

If you have the opportunity to achieve the authorization capability, by implementing a building concept described earlier, this is obviously the desirable way.

Sometimes due to local conditions (e.g., lower corridor and the inability to install a qualified suspended ceiling), it is impossible to renovate the building comprehensively.

For these cases, the following sketch will give suggestions (Figure 8.4).

3) Note of Egbert Dittrich: If the corridor is used as a fire tunnel, this solution is not useful either.

In this design, a "laboratory use unit" is accessed via short dead-end corridors. The dead-end corridors are designed as necessary corridors, with all the associated requirements and restrictions. You are therefore required primarily to meet the escape route length requirement of less than 25 m in the laboratory. The intermediate "corridor" is part of the laboratory unit and therefore not a "safe area" in the sense of a necessary corridor. Only via this dead-end corridor are the laboratories accessible, and this leads to the repeatedly discussed problem of "trapped spaces." In practice, it is essential to discuss this concept with the authorities involved. It is not suitable for new buildings but is a compromise in the inventory. Comprehensive fire detection with alarm for the user again is mandatory. For security reasons, it is always important to design the corridor as far as possible in the sense of "necessary corridor."

Part II
Layout of Technical Building Trades

Egbert Dittrich

Laboratory buildings are characterized by a high technical equipment, which is connected to the building and thus a part of the building.

A laboratory building not only is a conditioned containment but also provides a structure and depository for many technical installations, making the desired laboratory operation possible. Loads must be reflected, high pressure lines attached to the building, and various technical trades are fastened securely considering vibration.

The vast number of technical building trades is poorly or not used by the building as such, but more by the laboratory and work processes, thus are the result of prevention and safety systems for the benefit of users.

Working with hazardous and noxious substances, thermal loads, noisy devices, or other working conditions like negatively influencing factors, require systematic technically more or less sophisticated countermeasures, without which the work would not be possible in the laboratory buildings.

Certain technical devices serve only secondarily typical for their purpose. For example, the laboratory ventilation is of course welcome, but primarily is a measure to dilute dangerous gases inhaled in the range of users and no comfort improvement only. Thus, in the dimensioning of the relevant equipment and selection of materials, especially safety aspects and/or resistance, play a role.

Due to the technical complexity considering the demanding laboratory processes, the sum of the professional planners who work on a laboratory building is likely to be outbid by other types of buildings in individual cases only. The successful completion of the planning is, therefore, only guaranteed if the planners interact in an integrated design process. The interface problem and the typical continual changes for laboratory buildings in the planning process necessitate a very close cooperation between all parties involved. Here, the involvement of users and the operator comes to very high importance.

The technical building equipment has a decisive part to contribute to the realization of any kind of flexibility in laboratory buildings. Historically, "inflexibility" in

The Sustainable Laboratory Handbook: Design, Equipment, Operation, First Edition.
Edited by Egbert Dittrich.

the laboratory began with the use of pipe-networked gaseous and liquid services. But just-use changes are accompanied by changes of the services.

In order, therefore, to ensure a high degree of flexibility, the engineering technology planners have the responsibility to implement technical building systems that realize this same requirement for flexibility. Starting with the position of the service fixtures in ergonomic gripping space of the laboratory users, there is in this context a variety of detailed tasks, as design and quality of pipeline networks, partial shut-offs of areas, upgradeability, design of the electrical energy, spread of the incoming air, and their influence on the wellbeing and Occupational Health and Safety.

Last but not the least, the space requirement of the engineering devices, mainly has an influence in terms of the building volume, conditioning, energy consumption, and thus, operating costs over the entire life cycle are crucial for the interpretation of the building.

If less complicated and easier and therefore faster, the technical trades can be rebuilt, dismantled, and upgraded, and the sooner they will be adapted to changing requirements and thus, contributes significantly to the efficiency in laboratories. Long-term experiments but also annoying interruptions of work often inhibit in terms of ergonomic and productive working, and important adaptation of the building technology to new situations.

The experienced laboratory expert knows the importance of a properly functioning technical structure, which requires the user more or less because it is a part of his instruments and in general, he gives little attention. Only when deficiencies arise here, one is aware of its importance.

In particular, the ventilation is one of the most important factors influencing comfort and safety. On one hand, high air change rates are necessary – at least that's the unanimous opinion – to prevent harm to the users by diluting and extracting dangerous substances, but on the other hand, it also plays a crucial role for the comfort and well-being. Laboratories with strongly-felt drafts are not accepted and boost discontent or even refusal to work.

Opinions differ on technical building trades. There is no trade in the laboratory construction which is stronger in both the planning process and in the operation in focus, that decide on functioning and non-functioning. Not least, the discrepancy of life (technical trades 15–30 years, building about 100 years) helps the technical building trades gain more attention in terms of planning and execution quality in the context of sustainability.

The total energy consumption of a laboratory building and the amount of time spent in the building's life cycle material depend largely on the technology concepts.

Laboratory buildings are characterized by an exorbitant energy consumption, which can only be throttled if intelligent ventilation and cooling systems are planned, which can be adapted to any laboratory situation. Undeniably this correlates the energy-consumption of a laboratory building with the airflow. The high thermal loads require, in any case, even in temperate zones, cooling;

otherwise, the room's climate most of the year is outside the comfort zone. Therefore, heat recovery is important.

Just the density of the technical building trades makes the difference from other commercial buildings, and is the leveling rule. It places the highest requirements on the planner. The high consumption costs but also the obligation to safe energy and resources and compliance with sustainability criteria require a sensitive consideration of user requirements and benchmarks. Thus, such systems are to be integrated not only in the planning but bring in the evaluation of the results as an important building block in the operational management. Here, a standardization of the detection – sub-metering – and evaluation methodology proves to be particularly helpful, because due to the heterogeneity of the buildings, comparisons and evaluations of positive and negative peak values are valid only in conjunction with largest possible basic population, that is, we need many lab buildings to compare.

The following contributions to the engineering building trades not only come from the pen of an experienced lab expert but reflect the current state of the art, that is, it is expressed in the considerations, the consensus either from the changing content of the laboratory's scientific applications and the required constant adaptation and further development of the technical trades.

9
Development in Terms of Building Technology and Requirements of Technical Building Equipment

Hermann Zeltner

The technical maintenance structure represents the spine of the laboratory buildings, with all the essential lifeblood occurring in these buildings. All rooms are supplied with and disposed of central media, ventilation, and electric installations by means of the technical development structure.

The way a building is to be developed must already be determined at an early stage of the project, because this will have an impact on the essential parameters of the building:

- *Use:* Requirements regarding hygiene, media diversity, flexibility in case of change of use
- *Technical*: Maximum of media available in the rooms; central or decentralized supply of media
- *Constructional:* Floor heights, corridor situations (single ducts), and facade views (external and internal central ducts).

As a rule, we distinguish between two variants of technical development, mainly relating to the position and number of the supply and exhaust ducts.

Under the "technical development of single ducts" (Figure 9.1), supply ducts are arranged at regular distances, directly at the laboratory areas for each laboratory grid. From these ducts, it is possible to vertically distribute and support the supply of HVAC systems (ventilation and sanitation). A distribution like this does not work in electrical engineering, because the fields of "high-" and "low-voltage current" always require horizontal development.

Central duct supply concerns central supply ducts, arranged per lab area or storey, ducts from which the systems are horizontally distributed to the lab areas. This is arranged by means of large distribution channels or pipes running in a diagonal direction to the lab structure and – in an ideal case for the media and per lab grid – supplying the lab furniture once through vertical, downwards directed outlets. With regard to ventilation, the laboratories are approached from the distribution channels in a comb-shaped manner under lab grid separation (Figure 9.2).

In order to make the right choice among the technical development concepts described previously, when determining the basic conception of a lab building, some decision criteria with a critical measure of impact as to this are presented

The Sustainable Laboratory Handbook: Design, Equipment, Operation, First Edition.
Edited by Egbert Dittrich.

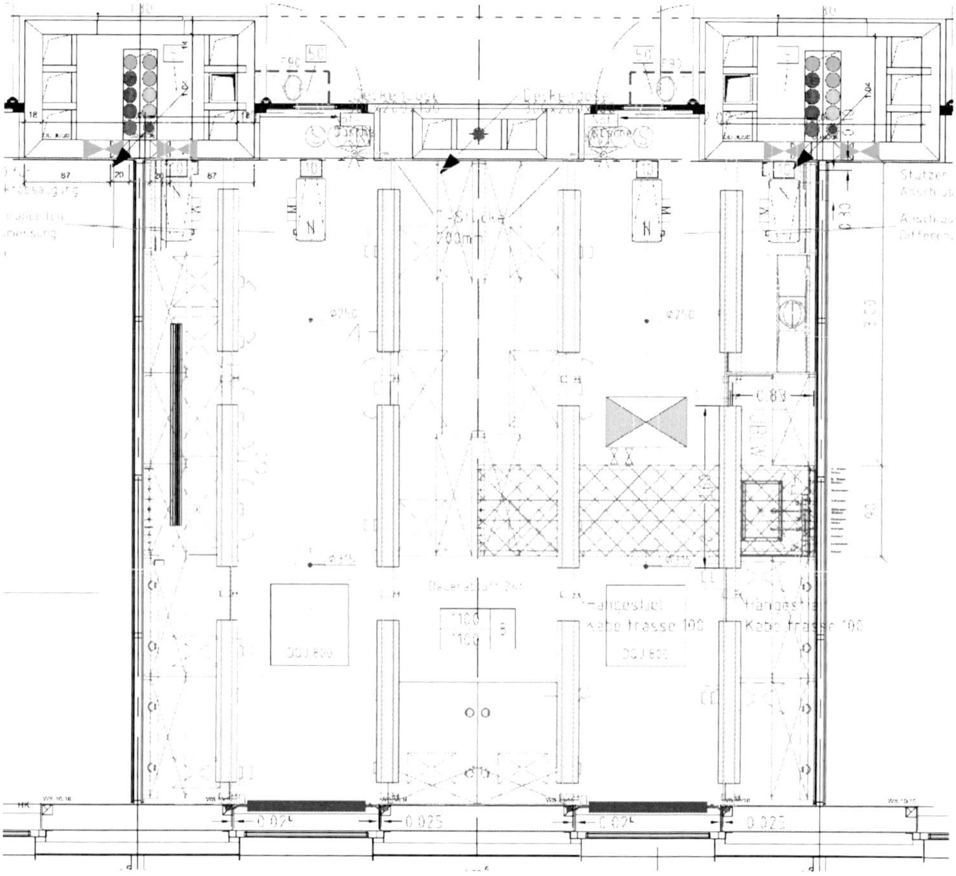

Figure 9.1 Single duct development media and ventilation.

in the following. Taking a closer look at the design of the ventilation system is of particular importance, because of the discipline's significance for the structure of maintenance in lab buildings resulting from the large diameters of its pipes.

9.1 Field of Research

The first step is the determination of the fields of research, which are to be represented in the building. This determination is necessary because flexibility – that is, suitability for all kinds of lab types – cannot be built in a cost-effective manner. Specific requirements, such as large air quantities and multi-faceted need of media in the field of chemistry (Figure 9.3), resulting in dense ceiling installations, are opposed to the high hygiene standards in the field of biomedicine (Figure 9.4), which must be met by easy-to-clean installations.

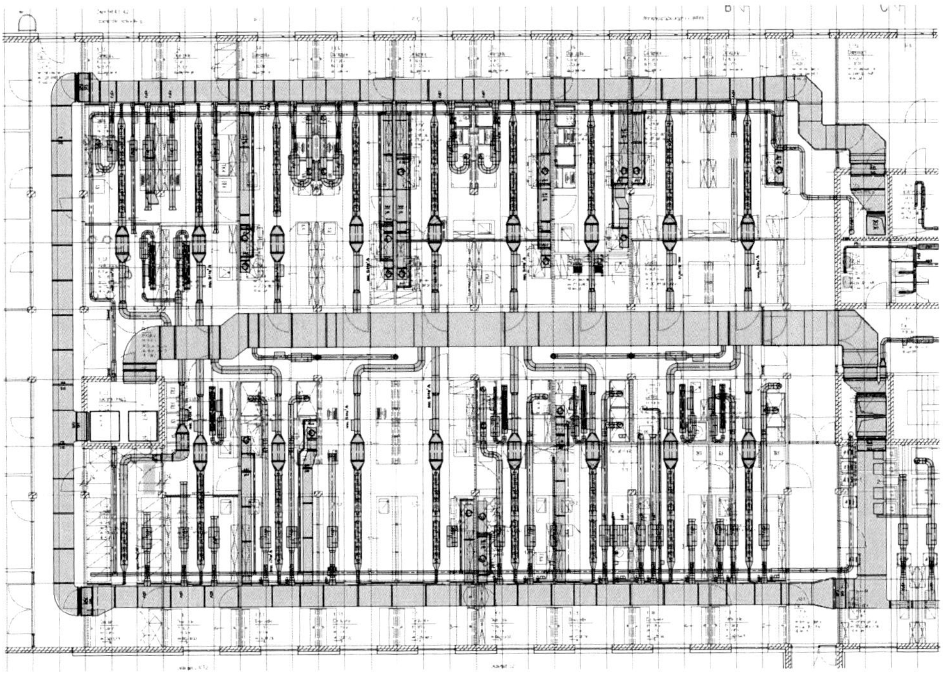

Figure 9.2 Central duct development ventilation in a laboratory landscape.

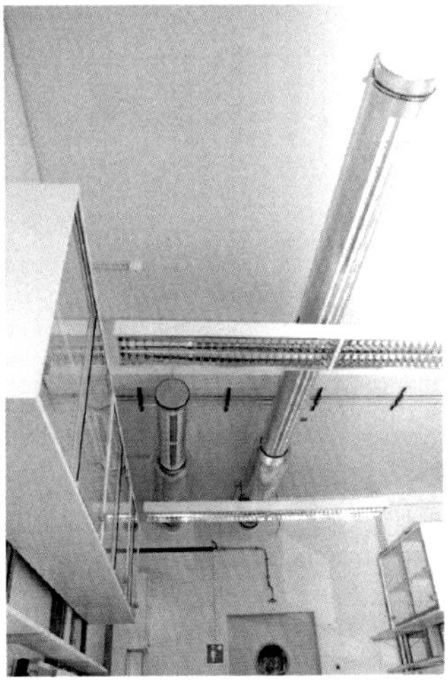

Figure 9.3 "Flexibility" of lab landscapes.

9.2 Required Flexibility of Laboratory Areas

The flexibility of the future building regarding conversion is already determined in the concept design of the building by means of the development structure. Taking

Figure 9.4 Single duct development medical research.

the physical structure of the *lab landscape* in a building as the basis, unproblematic placing of one room zone after another in a lab grid by means of the selected type of development is possible, with each room zone being completely different from the other: Figure 9.5 shows an example of such possible divisions of a grid.

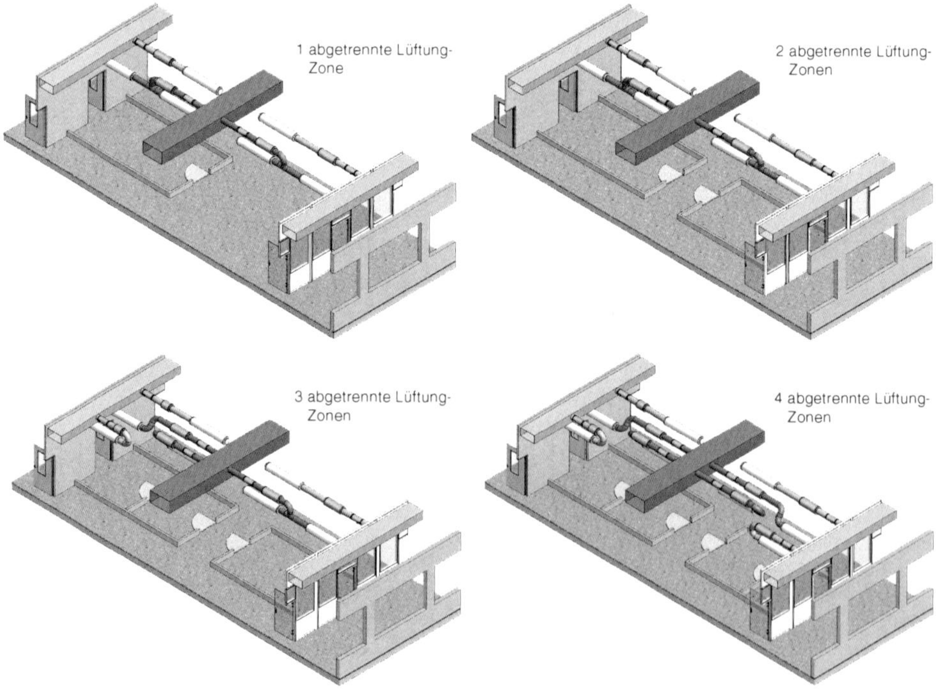

Figure 9.5 Visualization "flexibility" of laboratory landscapes.

- *one separated ventilation zone:* Side room (equipment), lab working zone, documentation zone
- *three separated ventilation zones:* Cold room, side room (equipment), internal lab corridor, special laboratory (e.g., BSL2 laboratory), documentation zone.

As can be seen in Figure 9.5, all meaningful variants are covered by the identical route concept of the ventilation; conversion works have to be carried out only within the lab grid concerned.

9.3 Number of Floors, Height of the Floor, and Development Extent of the Laboratory Area (Laboratory Landscape)

By looking at the essential geometrical parameters of a lab building, such as floor heights and room depths, we can see the strong impact exerted on them by the technology to be incorporated into the building.

Looking at the parameter "floor height," we learn from Figures 9.6 and 9.7 that a nonintersecting development is hardly feasible in practice with central duct development and, thus, a higher measure of clearance regarding interior height (of approximately 3.70 m) is required.

It is easy to save on some degree of clearance room height (up to approximately 3.3 m) in single duct structures, as shown in Figure 9.7. Constructional limits will soon be reached under this type of development; the ventilation ducts in the riser ducts will soon have reached sizes too large, so that installation in single ducts will no longer be economically meaningful. As a rule, the limit in laboratories with a depth of 8 m is reached after three lab floors; the limit in laboratory landscapes with a depth of 17 m is reached after two lab floors. Hence, this development variant is only suitable for use in a limited way.

As a matter of course, the field-specific parameters have to be checked more closely in order to enable the complete examination of the ideal technical development structure in a building. Due to the complex nature of the matter, it is necessary to develop a solution, which is individually adapted to the needs present and which exactly covers the specific parameters of the respective building and its use.

Such field-specific aspects will be listed and explained in the following.

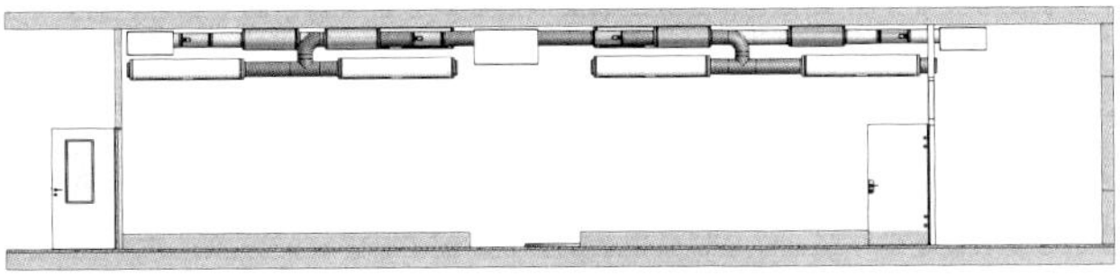

Figure 9.6 Sectional view standard laboratory with single duct development.

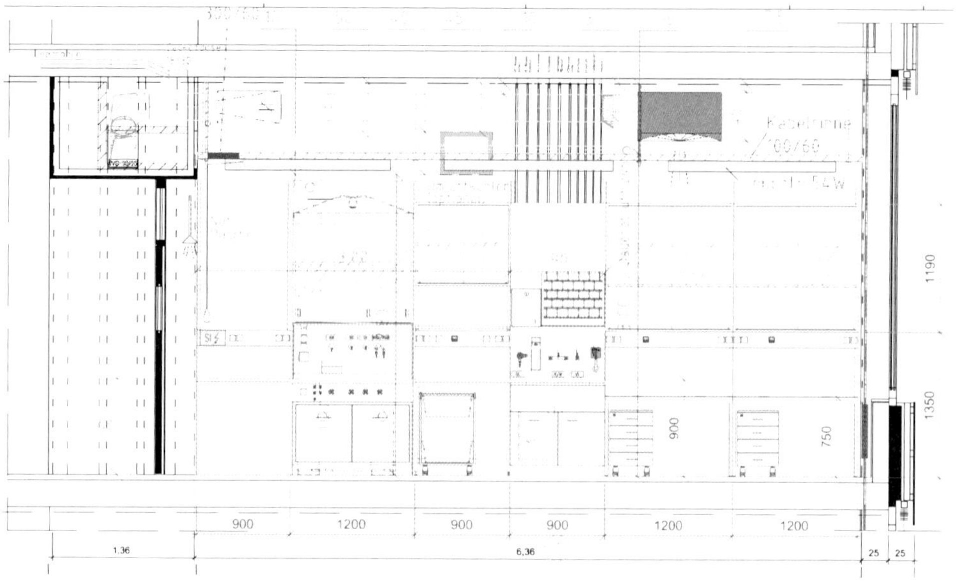

Figure 9.7 Sectional view laboratory landscape with central duct development.

9.4 Plumbing Services

There are different decision criteria of importance in the field of service supply concerning finding the most economic form of development.

The question whether services can also be supplied area by area and in a decentralized manner has to be answered. This may apply to high-purity gases, vacuum or deionized water, media, which, as a rule, are horizontally distributed in the areas (Figure 9.8).

Plumbing services, which are supplied in a central manner (e.g., cooling water, cold, and hot water, etc.) can be installed exclusively according to use-specific specifications, such as hygiene requirements or countability. However, it has to

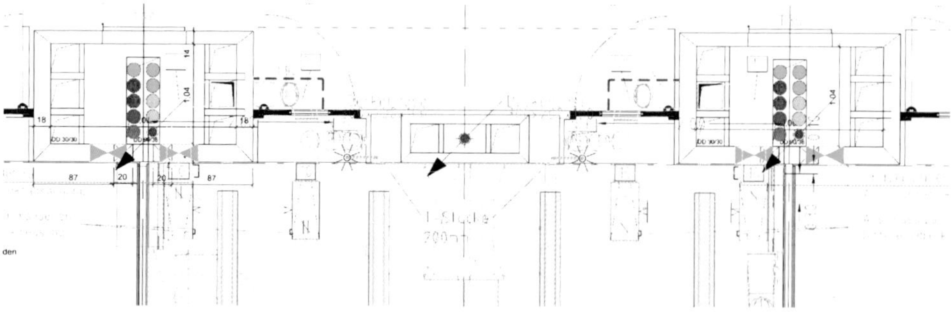

Figure 9.8 Single duct development media and ventilation.

be taken into account that areas in those laboratory landscapes with a higher amount of depth are pervaded by internal routes, and opportunity has to be provided to enable the separate media development of those laboratory lines resulting from these internal routes or to create an internal distribution of the lab furniture bypassing these routes (see Figure 9.9).

If the main focus in a lab area is on hygiene requirements, it might make sense to bring all centrally-supplied services into the laboratories via single ducts, with the incorporation into the lab furniture taking place in a covered form and directly in the installation room of the lab equipment (see Figure 9.10). It is also possible to direct the lab's sewage water by means of this form of connection.

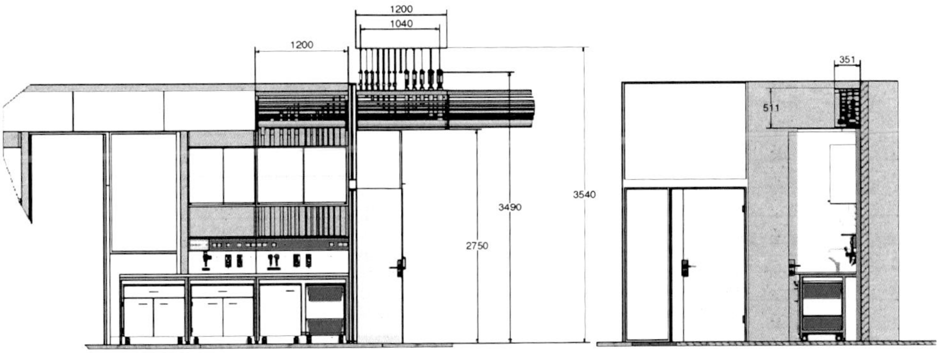

Figure 9.9 Plumbing services distribution within laboratory furniture.

Figure 9.10 Threading of services into a single duct.

In the case of no hygiene requirements to the installation of plumbing services existing, the horizontal distribution within an area turns out to be, by far the most efficient option. Only a few constructional requirements (ducts, fire protection regulations) have to be met and the flexible, easy-to-upgrade connection of the lab equipment exists.

9.5 Electrical Installation

Regarding the basic considerations of the fields "high-voltage power" and "electronic data processing (EDP)," it does not matter whether the building is to be developed vertically or horizontally – both fields are typically horizontal distributions carried out "area by area" and out of a storey distributor or an area distributor.

The following questions mainly arise in terms of the development of high-voltage current:

- Where and how is the fuse of the laboratory furniture guaranteed? Lab users must or can operate the units.
- How can additional electrical connections in the lab equipment be retrofitted in an efficient way?
- How can electrical wiring in zones, critical regarding hygiene, be implemented in an easy-to-clean way?

As a rule, there is only one approach to solving these questions in constructional practice:

The logical installation, which is the most flexible and easiest to update regarding high-voltage current, is that each lab line is supplied with current by a cable – grid by grid per type of network and area by area from sub-distributor rooms. Under this type of installation, the feeder cable is fused in the sub-distribution for cable protection and all laboratory outlets of the lab furniture are fused via small, so-called "electro technical fuse protection units" by means of FI and line safety switches. Alternatively, in the case of higher electrical loads or accumulation of major electrical consumers in the lab, it is possible to enable the supply of the lab lines by means of a busbar running on the ceiling and an outlet box.

In areas noncritical in terms of hygiene, the distribution of the wiring within the labs is implemented on open cable racks (see Figure 9.11a–c) with cable duct outlets into the lab furniture. There are several technical solutions available if an easy-to-clean execution, for example, in genetic engineering laboratories, has to be produced:

- Encased and padded cable ducts on the ceiling (see Figure 9.11a)
- Distribution via closed suspended ducts, for example, with lamp bands (see Figure 9.11b)
- Closed cable racks suspended from ceiling (see Figure 9.11c).

(a) (b)

(c)

Figure 9.11 (a–c) Examples power supply laboratory.

9.6 Ventilation

A situation such as that depicted in Figure 9.12 is the typical result in many existing buildings when the duct systems are examined more closely. There are no straight or modular lines, and adaptation in case of redesign or conversion requires a great deal of technical effort. Nonenergy efficient implementation, for example, high duct pressures, results in high costs of consumption and maintenance.

It is required to determine certain parameters for the field lab ventilation in the early stage of the planning in order to create the basis for the construction of a future-proof and – most importantly – economical laboratory building.

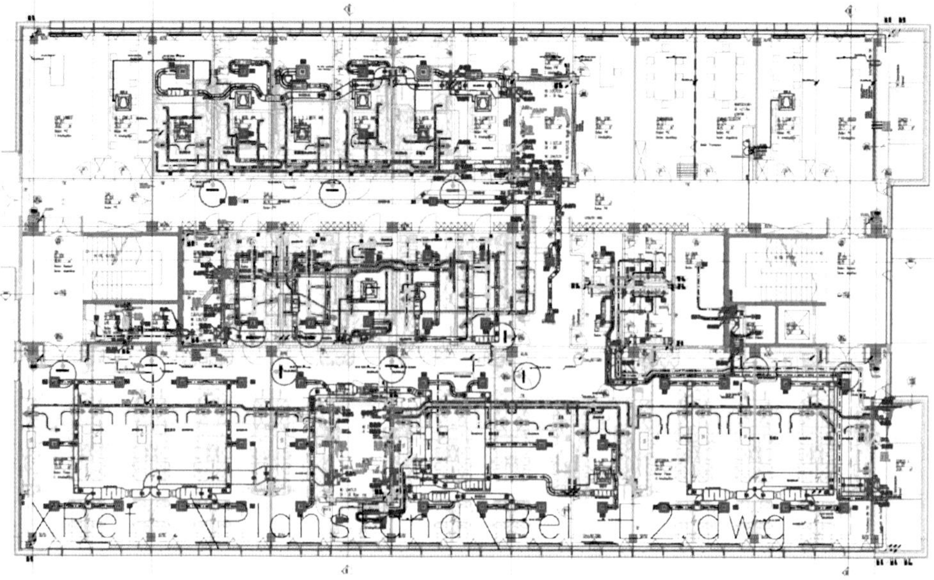

Figure 9.12 Duct course in need of improvement with central duct development.

The key points to be observed are as follows:

- Realization of physical structure enabling workplace-specific lab ventilation, thus, guaranteeing the basic air quantities to be reduced.
- Definition of ventilation systems really required and of the systems' parameter.
- Creation of a modular, straight-lined, and power-optimized duct system.
- Creation of circular duct systems to generate flexibility.

9.7 Determination and Optimization of the Air Changes Quantities and Definition of Air Systems Required

In order to enable the design of a highly efficient lab ventilation adapted to the necessities existing, it is required, for example, in a laboratory landscape, to zone the individual sections within the area and, thus, to structure the area (Figure 9.13).

If this is done in a consequent manner, working areas reserved for specific activities will be created. It is then possible to individually define the technical necessities in terms of media and ventilation, and to spatially structure them in a clear way. Due to the zoning, it is possible to determine the workplace-specific quantities of air needed and to reduce these quantities to the actual necessities. As a rule, this kind of observation goes along with the reduction of the total quantities of exhaust air, amounting to up to 50% in such areas, maintaining the same or increasing safety at work.

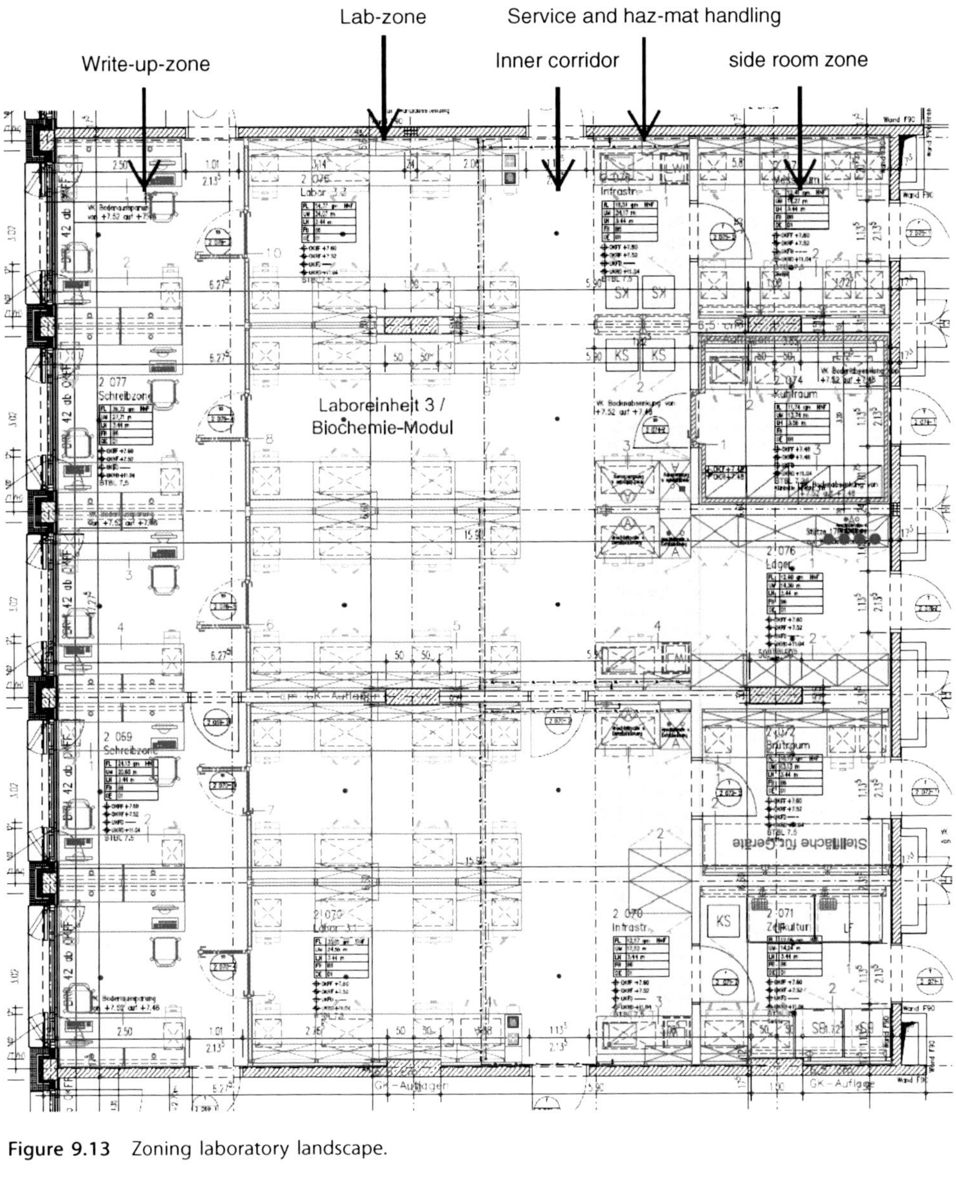

Figure 9.13 Zoning laboratory landscape.

It is, for example, possible to completely deduct the areas from the total change of air ventilation by the consequent planning of documentation zones. However, the areas are provided with ventilation with a clear separation from the lab area with regard to technical safety at work (Figure 9.14).

Needless to say, it is important to define the systems required as additional support for the optimization of the air quantities. Similar extract air emission sources

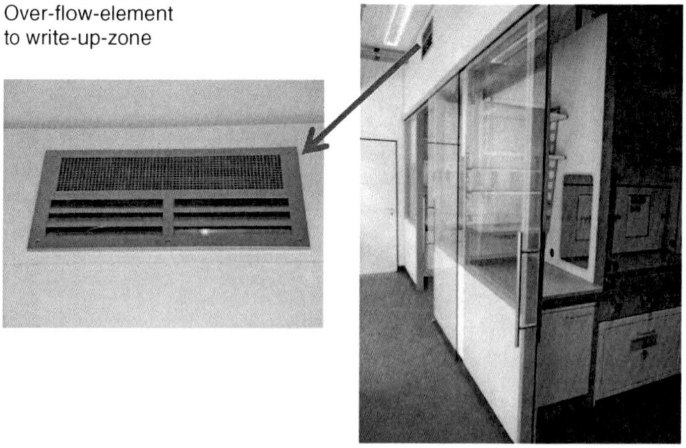

Figure 9.14 Connection to ventilation system documentation zone.

are merged to form a system, with – in addition to the type of exhaust air – the respective pressure losses of the individual consumers being taken into account. One possible example is listed as follows:

- Fume cupboard extract air, room, and spot extraction air
- High pressure extract air, for example, for filter cabinet housings and acid scrubber
- 24 h exhaust air for safety storage cabinets (hazardous material), acid and base and gas cylinder safety storage cabinet.

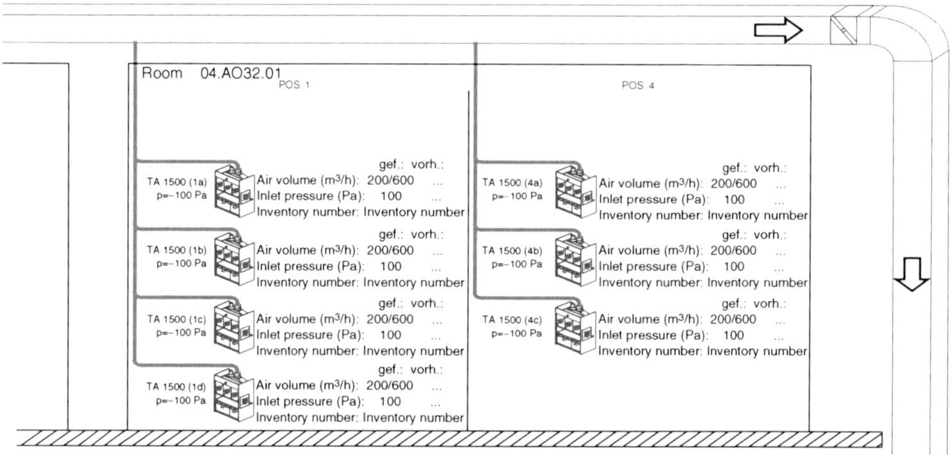

Figure 9.15 Extract: Ventilation scheme on interpretation calculation.

9.8 Creation of an Energy-Optimized Duct System

When comparing ventilation concepts in Figures 9.2 and 9.12, we see, that those straight-lined grid ventilation systems, which are given a considerable degree of flexibility in the design of the lab furniture due to the harmonious pressure profile within the circular duct systems make sense.

Hydraulic balance and optimized flow rates are the most important aspects to be created in order to make a well-structured ventilation system also function in an energy-optimized sense. The duct network has to be dimensioned in a way, which guarantees that the ducts' flow rates will never exceed a speed of $4\,m\,s^{-1}$ – regardless of the operating conditions. A pressure calculation of the duct network provides the interpretation of the systems with as few worst and best points as possible (Figure 9.15).

If all points mentioned are taken into account in the design process, the central ventilation appliances can be operated under optimized duct pressures, so that both the operating costs and the noise level occurring in the laboratories can be reduced drastically.

10
Ventilation and Air Conditioning Technology[1)]

Roland Rydzewski

10.1
Introduction

The general task of ventilation and air conditioning technology is to ensure indoor climates which are beneficial to the physical well-being of humans. This implies requirements for temperature, speed, and intensity of turbulences, as well as for the composition of the breathing air. In particular, in laboratories, there is a second task, that is, the protection of people against the influence of harmful substances that may be released during lab work. A third task of ventilation and air conditioning technology may be to ensure that special indoor climate conditions possibly required for the execution of lab work are reliably and consistently fulfilled; this may refer to requirements regarding temperature, humidity, or flow speeds, for instance. When implementing these three tasks, certain boundary conditions will have to be respected:

- The air volumes to be introduced are relatively high, for example, in comparison to the office rooms. Therefore, draft freeness requires particular attention, since the staff working there does not wear special protective clothing, as for example, in clean rooms, where for such reasons even air speeds of $0.45\ \mathrm{m\ s^{-1}}$ in laminar flow areas are accepted.
- Lab ventilation is energy-intensive due to the high airflow throughput; thus measures to reduce energy consumption have considerable effects on the operating costs of the building.
- Supply air volumes must be variably adjusted to the rapidly changing exhaust air volumes. Not only does this place special demands on control engineering it also implies that the air diffuser used must be able to exhaust without draft and without impermissibly high noise levels.
- High internal cooling loads may have to be discharged (e.g., in biotechnical laboratories).

1) Alfred Mayer provided major contributions to the Sections 10.5 and 10.6, for which I want to thank him.

The Sustainable Laboratory Handbook: Design, Equipment, Operation, First Edition.
Edited by Egbert Dittrich.

- Due to the large variety of working processes in laboratories, the suitable ventilation solutions vary from case to case. In general, the optimum is not achieved by standard concepts. Usage requirements can change; adequate flexibility reserves should be available.

Thus, for the reasons mentioned earlier, lab ventilation represents a rather demanding task within the scope of ventilation and air-conditioning.

10.1.1
General Note

This article does not seek to comprehensively and exhaustively analyze the subject, nor should be given a general presentation of ventilation engineering. The article will rather address some topics which, from the point of view of sustainability, are of particular interest for ventilation engineering in laboratories. In particular, the provisions of the numerous regulations can only be reflected to a limited extent; the reader will be referred to the relevant original documents.

10.2
Air Supply of Laboratory Rooms

Examinations of existing laboratory buildings show that air supply by fans accounts for 30–50% of the primary energy demand ("Leitfaden f. d. energetische Sanierung von Laboratorien," Forschungszentrum Jülich, 2007). Since the supply air is completely fed from external air (no circulating air proportion), relatively high energy expenditures for temperature control and probably for humidity treatment will arise, too. This situation, along with the necessity induced by energy policy to use energy more efficiently, results in the requirement to get by with the lowest possible supply and extract air flow while fulfilling all safety related requirements.

Due to the clear classification of the room areas and the explicit determination of hazardous material workplaces, it is possible to minimize areas with high air exchange rates. Evaluation areas, generally arranged on the façade and separated from the laboratory areas by glass walls, are influenced by the external climate and require cooling in summer and heating in winter. Internal cooling loads are of rather minor importance. The air exchange rate, typically 3/h, is determined according to the requirements of the hygienically necessary air exchange. If this is not sufficient to cover the cooling load, cooling sails can be taken into consideration.

As a planning basis for the layout, the ventilation and air-conditioning system must be dimensioned according to DIN 1946-7 and result in an extract air volumetric flow of 25 $m^3\,h^{-1}\,m^{-2}$ at the ratio of the laboratory floor space. Depending on the specific use, a lower value may be chosen in planning or operation, in case risk assessment indicates that the lower extract air volumetric flow is sufficient and effective for the intended laboratory activities.

These reductions in the extract air volumetric flow or the usage restriction resulting from it must be documented and marked at the entrance of the laboratory (e.g., Note: "Attention: Reduced air exchange!"). In individual cases, the risk assessment may even require higher air exchange rates. While the leeway for the operation of laboratories thus provided enables higher flexibility, it also increases the personal responsibility.

To date, the fact that the exhaust air volumetric flow stated in DIN 1946-7 is misunderstood as mandatory and applied without further consideration is still reality in the planning and operation of laboratories. The possibility to determine and implement the suitable air flow within the context of risk assessment is hardly ever used. As a result, laboratory rooms are often operated at unnecessarily high, sometimes even at excessively low air flow, with the first being at the expense of an energy-efficient operating mode, the latter at the expense of personal security.

Basically, it should be excluded that hazardous substances, when used according to the intended purpose, will be emitted beyond the sphere of fume cupboards or other collecting devices. Therefore, the general room exhaust air (e.g., via exhaust air ceiling diffusers) should not be understood as a protective measure against hazardous substances. The minimum extract air flow according to DIN 1946-7 is only meant to cover the residual risk probably arising from the use of hazardous substances against the intended usage. For this reason, operating modes in which extract air is exclusively exhausted via special extract systems are absolutely permissible.

10.2.1 Extract Systems

The central task of laboratory ventilation is the *protection of occupants* from contact with hazardous gases, vapors, and dusts. Various extract systems which determine the extract air requirements of a room are used for this purpose.

Fume cupboards: The extract air demand of fume cupboards according to DIN EN 14175 is given in the type examination report of the manufacturer. Typical values at open front sashes are as follows:

- For bench-mounted fume cupboards 400 $m^3\,h^{-1}\,m^{-1}$ of fume cupboard length,
- For low-level fume cupboards 600 $m^3\,h^{-1}\,m^{-1}$,
- For walk-in fume cupboards 700 $m^3\,h^{-1}\,m^{-1}$.

Fume cupboards of older designs are operated at a constant air-flow rate resulting in very high energy consumption. Throttling of the air flow occurs by means of hand valve or mechanical volumetric flow controller. Fume cupboards with variable air volume control equipped with sash position sensors are state-of-the-art. In case of changing fume cupboards operation, the room supply and extract air must immediately be balanced by a quickly responding control.

- Extracted workplaces or housings
- Bench-mounted fume cupboards
- Under-bench extraction
- Floor extraction systems for the collection of gases, vapors, and dusts which are heavier than air.
- Hoods, for example, above atomic absorption spectrometers, must be precisely dimensioned and adjusted to fully collect the thermally directed exhaust gas flow. Cross flows in the room must be avoided or taken into consideration when designing the extract air volumetric flow.
- Source extraction for the collection of emissions arising during work that cannot be carried out in laboratory fume cupboards or in housings. They are only effective when arranged in immediate proximity and must therefore be positioned as closely as possible at the emission source.

Cabinets for solvents and chemicals require constant aeration and ventilation. Special regulations apply to the following components:

- *Solvents cabinets, chemicals cabinets*: According to DIN 1946-7, ventilation systems in solvents cabinets must produce an air exchange of at least 5/h to avoid hazardous explosive atmosphere. Permanent negative pressure toward the surrounding rooms must be ensured. The filling or open handling of flammable liquids requires additional measures (e.g., increased air exchange, source extraction, extracted workplaces).
- *Compressed gas cylinder rooms*: Ventilation systems in rooms below ground level, in which compressed gas cylinders are used, must show an air exchange of at least 2/h, either in continuous operation or automatically controlled by a gas warning system.
- Thermal rooms
- Storage spaces for cryogenic gases.

While general room extract air is not mandatory, it is advantageous, in particular in rooms with major heat sources. Due to thermal buoyancy, the warm air concentrates preferably under the ceiling and can be effectively collected and discharged by ceiling air intakes. Otherwise, it remains in the room and contributes to the heating of the occupied zone.

10.2.2
Removal of Room Cooling Load

If the air volumetric flow, determined according to the extract air consumer, is too low to extract the room cooling load, the exceeding thermal output should be covered by cold-water based systems (circulating air cooler, cooling bar, cooling ceiling, cooling sail). However, if the exceeding thermal output is low, even an increase in the air volumetric flow should be taken into consideration to avoid additional investment costs.

As the following line of thoughts is showing, it is the winter not the summer that is critical for the removal of the thermal load:

- Laboratories are often inside areas (with upstream analysis area) so the cooling load is determined by internal heat sources and thus independent of seasons. There is no thermal-driven outflow via façades in winter.
- In summer, room temperatures of 26 °C are accepted; in winter, however, values around 22 °C are expected (among others due to the clothing). With a supply air temperature of 16 °C, this results in a temperature difference of 10 K in summer and 6 K in winter.
- Since the heat volume absorbed by the ventilation system is proportional to the temperature difference, it follows that at a given air flow in winter, the removable heat is 40% lower than in summer. Thus in winter, either a higher volume flow than in summer or additional cooling by means of water-based systems is required.

10.2.3 Supply Air

The inflowing air usually requires a supply air system (except for very small installations). The supply air volume flow rate of a room is measured on the basis of the total exhaust air volume flow and, as the case may be, on the requirements for the differential pressure with respect to the surrounding rooms. The adjustment of the supply air volume flow to the variable extract air volume flow places special demands on control engineering.

10.3 Air-Flow Routing in the Room

This is the central issue of chapter 10 in contrast to this, engineering systems, which are required for the treatment and the distribution of air in the building, is only a means to an end. The real target object is the indoor climate, thus the distribution of temperature, humidity and, if necessary, even the distribution of hazardous substances as well as the flow in the room. There are requirements made on this target object, the fulfillment of which decides whether ventilation technology can provide suitable conditions for the work processes in the respective laboratory room.

Nevertheless, ventilation planners traditionally deal more with systems engineering than with indoor climate. Regarding temperature, humidity, and hazardous substances, usually only mean values (strictly speaking, the values in the extract channel on the assumption of ideal mixed ventilation) are calculated, whereas the distribution in the room is not taken into consideration.

As far as the flow is concerned, operators rely on the information of the air diffuser manufacturers who provide speeds in certain intervals for their products. These values refer to an empty test room, which means, interactions of air diffusers with room geometry, furniture, and heat sources are not taken into account. A spatially differentiated picture of the indoor climate conditions is not available.

Frequent consequences occur when the building is used, for example, employees' complaints about draft, impaired functioning of fume cupboards, emission of hazardous substances. Here, a field report may be cited as an example: "On-site examinations always show that the positioning of the air diffusers, constructional design, supply air volumes, and supply air temperature have an impact on the retention capacity of fume cupboards." (Training days DECHEMA House 22/08/2007, Presentation Bernd Schoeler). With the problems being identified, it is tried to mitigate the situation afterwards – at a pinch, even with cardboard box and adhesive tape.

In the end, this defect is due to the fact that suitable tools to cope with the issue of indoor air flow were lacking in the past. The set-up and the climate-related surveying of real models at full scale or in reduced form have only been carried out in special cases, because of the high expenditure. However, the situation has changed. Today, the indoor climate can be calculated in detail, reviewed, and optimized using the instrument of Computational Fluid Dynamics (CFD). This provides the planner with comprehensive forecasts for subject areas such as draft, temperature stratification, flow dead zones with increased concentrations of hazardous substances, comfort parameters according to ISO 7730 (operative room temperature, DR, PMV, and PPD), air residence time in the room, ventilation efficiency, exposure to hazardous substances in the breathing area of people, and so on. Need for change will be noticeable already during the planning process (and not only after commissioning), thus room-climate conditions can be actively planned. The following chapter deals with this issue in more detail.

Requirements for the routing of the air flow in rooms are set out in the relevant regulations:

- "The ventilation system should be designed in a way that draft is avoided while all the safety-related ventilation requirements are fulfilled" (DIN 1946-7).
- "Supply and extract air must be routed in a way that the laboratory room will be completely flushed. Wrong layout or installation of the plant can lead to flow-related short-circuits, leaving areas of the laboratory room unflushed" (BGI 850-0-Sicheres Arbeiten in Laboratorien (Safe working in laboratories)).
- "Supply air diffusers must be designed and arranged in a way that does not affect extract-air installations. This requirement also applies in case of changes in the air volume flow due to controlling or switching of the ventilation systems" (DIN 1946-7).

When planning and designing the ventilation system, "probable interactions between supply air diffusers in the lab room and the fume cupboards need to be taken into consideration, for example, speed, direction, and volume flow of the supply air diffusers" (DIN 1946-7).

- "Supply air diffusers shall be designed and arranged in a way that no hazardous substances are flushed out of the fume cupboards" (Fortschreibung der Nutzungsrandbedingungen für die Berechnung von Nichtwohngebäuden, issued by the Federal Ministry of Transport, Building and Urban Development).

The implementation of these requirements proves to be a demanding task to which more importance should be attached during planning. In view of the variety of lab rooms and different requirements, it is not possible to give generally applicable recommendations for the airflow in rooms. Two ideal airflow concepts will be discussed later which, however, are hardly ever realized in their pure form.

10.3.1 Mixed Ventilation

Mixed ventilation is the prevailing concept in laboratories. Air is introduced into the ceiling area at high speed, often by using swirl diffusers. The intensive mixing of supply air with room air results in a more or less uniform distribution of the temperature, probably also of hazardous substances. This provides advantages: Locally released hazardous substances are immediately distributed around the room and thus become diluted; and temperature differences in the room are small. The ventilation effectiveness reaches medium values. The disadvantage is the fact that high air velocities may arise in the occupied zone and that people might perceive it as draft. Another possible drawback has already been mentioned earlier: Hazardous substances may be flushed out of the fume cupboards; the capture of hazardous substances by means of flues and source extraction may be affected. When selecting air diffusers, it is to clarify whether they are able to realize the intended extract behavior across the entire volume flow range to be covered. For example, the cool supply air "falls down" in a compact jet from insufficiently pressurized swirl diffusers instead of dispersing in horizontal radial direction. It should also be considered that a flush-mounted air diffuser will probably blow off completely differently from those freely suspended. Examples for it will be given in the following section.

10.3.2 Displacement Ventilation

Displacement ventilation is basically a vertical bottom-up displacement flow with a cool-air layer near the bottom and a layer of thermal and often even contaminated air in the ceiling area. This system is sometimes also called *layered ventilation*. Natural buoyant thermal plumes around the heat sources transport air from the ground-level layer to the ceiling area. The supply air is introduced into the room via extensive air diffusers. Displacement ventilation in its ideal form is realized by means of air supply diffusers arranged on ground level; but also supply-air diffusers arranged on a higher level, on the wall or ceiling, are compatible with the displacement ventilation principle. The extract air is extracted near the ceiling.

Displacement ventilation has proven its worth in many types of rooms since it provides various advantages:

- Due to the low air velocities, draft problems can be excluded to a large extent.
- The buoyant thermals in the area of persons provide high-quality breathing air.

- The air exchange efficiency (a measure for the effectiveness of air renewal in the room) is higher than for mixed ventilation.
- In case the sources of hazardous substances coincide with the heat sources, as is often the case, the ventilation effectiveness (a measure for the effectiveness of the removal of hazardous substances from the room) will also be higher than with mixed ventilation.

The result of high values of both air exchange coefficient and ventilation effectiveness is the fact that the required removal of heat and hazardous substances can be achieved with relatively low air rates. Displacement ventilation is thus characterized by high energy efficiency.

The RELAB study (RELAB, Energy Saving in Laboratories through Reduction of Air Flows, 1998) came to the conclusion that displacement ventilation is not suitable for lab rooms. This statement is doubtlessly valid, subject to two conditions: First, the exhaust air must be more or less completely extracted via extract-air extraction devices on the level of the occupied zone, thus via digesters, flues and source ventilation, and so on and second, the supply air is introduced at ground level. In this case, the air layer charged with heat and hazardous substances really extends down to the level of the suction opening of the exhaust-air extraction devices. The situation is different if at least one of the conditions mentioned does not apply.

- The higher the percentage of extracted ceiling air, the higher the lower limit of the contaminated air layer, probably even above head height. This often occurs, in particular, in biotechnical labs, where the extract air consumption by digesters and appliances is rather low, while the cooling load is relatively high.
- In case the supply air is introduced at low impulse above the occupied zone, the supply-air flow thermally driven in upward direction induces a circulation of the room air in the occupied zone, where the supply-air jet mixes with the room air. Depending on the room dimensions, a mixed flow with relatively low air velocities and an almost constant concentration level of temperature and hazardous materials can develop, while in the air layer above higher and increasing temperatures, concentrations of hazardous substances exist.

In the following section, this will be illustrated by an example.

10.4
Numerical Flow Simulation (Computational Fluid Dynamics (CFD))

Numerical flow simulations are based on the Navier–Stokes-equation, which has already been set up in the first half of the nineteenth century. Physically, it represents the principle of linear momentum for Newton's fluids (e.g., air, water, and oil). Mathematically, the Navier–Stokes-equation is a system of coupled, nonlinear, partial differential equations. However, all attempts to achieve an analytical solution have not yet led to success. Therefore, numerical calculation methods were developed at the beginning of the 1950s. The best known are the Finite Element Method (FEM) and the Finite Volume Method (FVM). With the development of commercially distributed computer programs from around the 1980s, user-friendly and practically feasible tools for the calculation of flows, even in complex geometrical configurations, are available. Today, these programs are used on a broad front in aeronautics, in the automotive industry, in process engineering, and other industrial fields as an indispensable tool for product and process development. The application possibilities in the field of ventilation and air-conditioning have been known and frequently demonstrated for a long time; however, they are only used in special cases. This is due to the fact that the modeling expenses are still high and that, in contrast to the development of industrial series products, costs can only be passed on to few units. And, it has to do with certain inertia of traditional processes and standards in the building industry.

Flow simulation in buildings is applied, in particular,

- when increased demands are made on the indoor climate;
- when one and the same room is built in large numbers;
- when innovative indoor climate concepts shall be realized, with little practical knowledge being available;
- when planning security is particularly important (e.g., because subsequent renovation measures are excluded);
- where high demands are made on energy efficiency.

Flow simulation is a demanding task which belongs in the hands of experienced specialists. Otherwise, forecasting errors are possible, in which people may trust blindly ("it has all been scientifically calculated") and on the basis of which serious wrong decisions may be made.

Flow simulation provides the greatest benefit when it accompanies the planning process. Usually, it begins with the presentation of the initial concept elaborated by the ventilation planner. It is only in rare cases that this already results in a satisfactory solution. As a rule, the first simulation shows where improvement is required. The optimum solution will be gradually identified considering several variants in coordination with other project participants. Sometimes, however, the decision to carry out flow simulation is made late in the course of the project, if for example, doubts about the planned ventilation concept arise shortly before the tender. Then, the possibilities to implement the results of the flow simulation are limited accordingly. Sometimes, flow simulation is only commissioned when problems have arisen after the commissioning of the building, for example, when employees' complaints about draft become more frequent and clarification of causes as well as remedies is required.

A particular topic within the context of flow simulation is the treatment of air diffusers. Since indoor air flow strongly depends on their discharge behavior, the simulation must concentrate on their aerodynamic properties.

It is only in exceptional cases that manufacturers have measured the velocity profile at the air passage aperture and provide the result for simulation purposes.

Usually, the only remaining way is the expensive development of a 3D geometry model of the diffusers including supply air tube, terminal box with fixtures, air deflection blades, and so on, and the calculation of the discharge behavior in the course of the flow simulation. Often, a key issue of simulation is the influence of the selected air diffusers on the indoor climate and the distribution of hazardous substances.

Later, the application possibilities and the knowledge possibly gained from the numerical flow simulation will be illustrated by some case examples. Even suboptimal situations, which highlight the respective need for improvement, are intentionally shown.

A hint for the understanding of the figures: In most figures, the values of velocity, temperature, or tracer gas concentration in selected sectional planes through the room are shown in color. The values can be read using the boundary scale. Areas which either do not belong to the airspace (e.g., the inside of devices) or in which the value of the relevant state variable is outside the selected scale are consistently white.

10.4.1 Case Example 1: Comparison of Supply-Air Systems: Swirl Diffuser + Ceiling Sail/Textile Diffuser

The initial situation is the conversion of an existing building with relatively low clearance height into a lab building. Due to the existing features on site, suspended ceilings were not provided. Different variants of air diffusers were under consideration, among them swirl diffusers in various installation situations (suspended, integrated into ceiling sails) and textile diffusers.

The suspended swirl diffuser provided a generally acceptable flow pattern only in connection with a widened edge offered by the manufacturer. In some areas, the swirl diffuser in connection with ceiling sails above the laboratory aisles resulted in excessively high flow velocities along the furniture (Figure 10.1). The textile diffuser proved to be the solution with the least risk of draft in the occupied zone (Figure 10.2).

10.4.2 Case Example 2: Comparison of Supply-Air Systems: Swirl Diffuser, Flush with the Ceiling/Displacement Diffuser on the Ceiling

In this lab room (see geometry model, Figure 10.3), the air is supplied via swirl diffusers installed flush with the suspended ceiling. In certain places, excessively high air velocities occur in the occupied zone (Figure 10.4). Due to the suboptimal arrangement of supply-air and exhaust-air passages, ventilation short circuits occur occasionally (Figure 10.5).

Under undisturbed conditions, the air diffuser Trox-Hesco Procondif produces a bell-shaped flow profile with relatively low air velocities; the contraction and the acceleration of the supply-air jet known from perforated plate diffusers does not

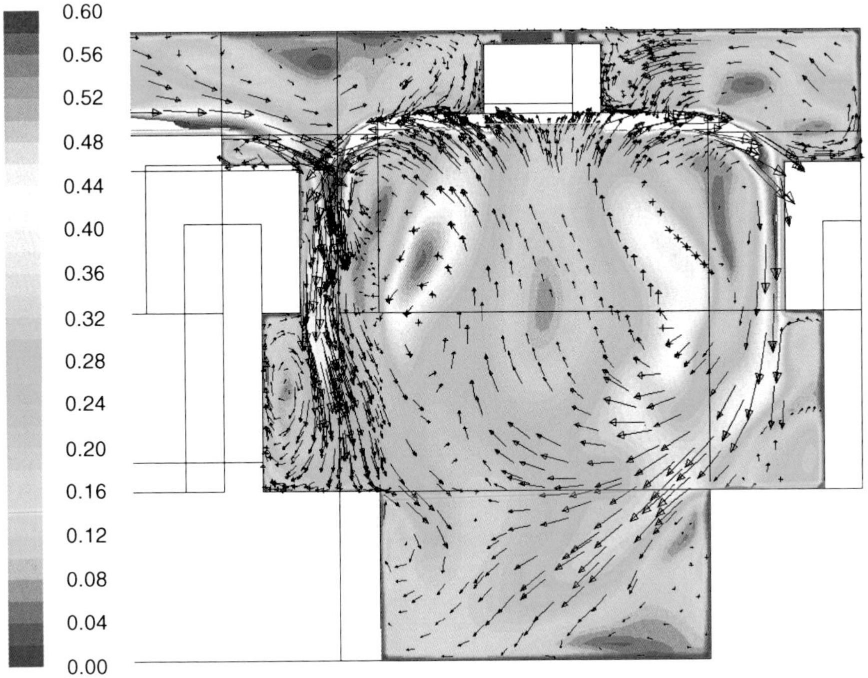

Figure 10.1 Ceiling sail with integrated swirl diffuser, centrically arranged above a laboratory aisle. The vertical sectional plane runs transversely to the laboratory aisles and intersects the swirl diffuser (visible by the terminal box, white) in the center. Flow pattern: Below the ceiling sail, the supply air flows radially to the outside, meets the radial jet of the air outlet of the neighboring laboratory aisle, and is thus diverted downwards along the furnishing. Relatively high air velocities over the worktable with increased draft risk. If digesters were arranged there instead of cabinets and worktables, the collection capacity at opened sash would be affected.

occur. Due to the disturbing influences in real lab rooms (heat sources, furnishing), the flow is diverted in many ways (Figure 10.6). In total, the air velocity in the occupied zone is lower than in the previous variant and lies in an acceptable range.

The room air flow principle can neither be classified as mixed ventilation nor as displacement ventilation, but lies in-between.

10.4.3
Case Example 3: Ventilation Optimization of a Model Lab Room

Subject matter of the simulation is the ventilation optimization of a model lab room (Figure 10.7). For situations with low to higher air exchange rates (4–12/h), an air routing system shall be developed, which is superior to mixed ventilation in terms of ventilation effectiveness while fulfilling the requirements for thermal comfort. The extensive air diffuser over the lab aisle discharges the

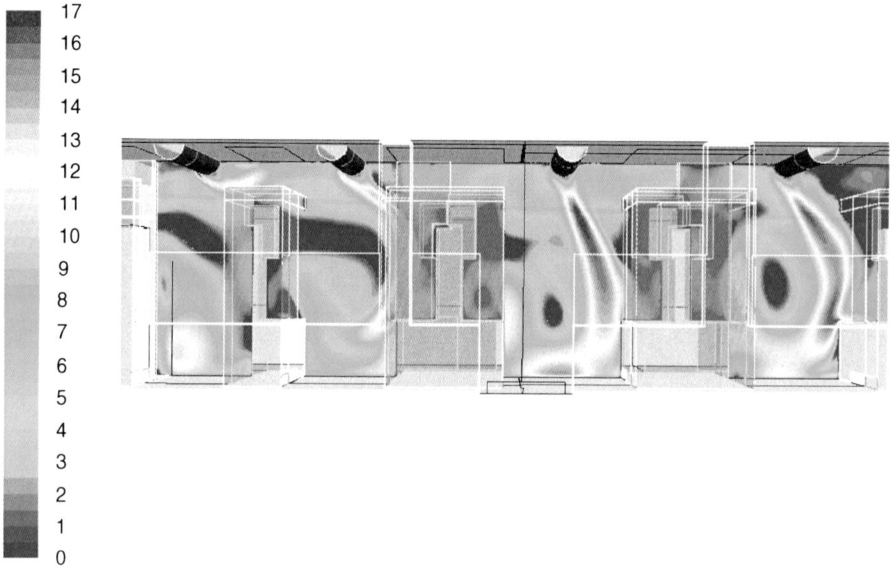

Figure 10.2 Semi-cylindrical textile diffusers, directly installed under the bare ceiling. The draft risk according to DIN ISO 7730 is shown. The value forecasts the percentage of persons perceiving this situation as draft. Values from 15 to 20% are classified as average comfort. All open jets emitted by the air outlets are diverted to the right; the cause is warm air rising from the condensers of a refrigerator group on the left-hand wall.

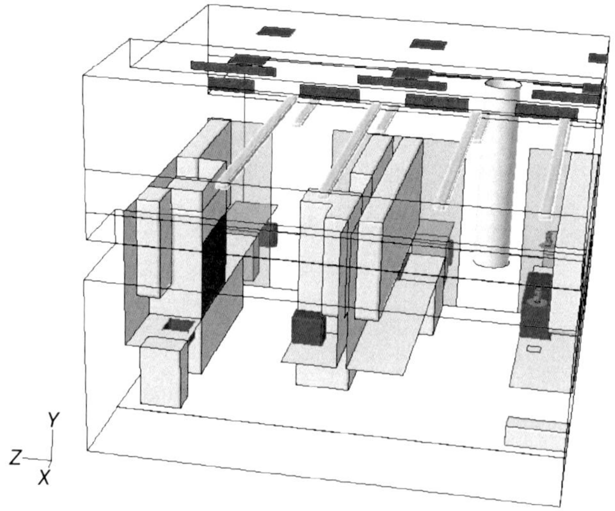

Figure 10.3 Geometry models of the laboratory room. Viewing direction from the corridor side toward the façade. The model contains the following elements: Furnishing including digesters (gray), heat sources (violet), separation between laboratory and writing area (light-blue), lighting (yellow), supply air diffusers (dark-blue), and exhaust air diffusers (red).

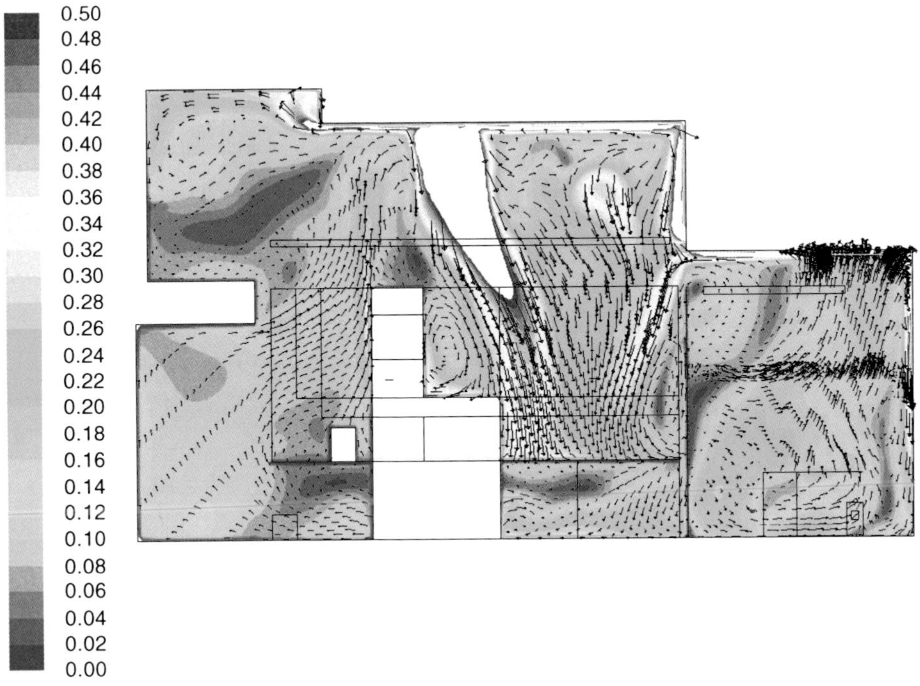

Figure 10.4 The illustration shows the velocity in a vertical cross section. The latter runs vertically to the laboratory aisle exactly through the front edge of a table, which is indicated as a dark-blue horizontal line. The increased velocity in the wedge-shaped area is the result of the interaction of two neighboring swirl diffusers installed flush with the ceiling. At head height, velocities of up to $0.4\,m\,s^{-1}$ occur, which means increased draft risk.

air into the room via fields with micro-perforation arranged in a structured manner (Figures 10.8 and 10.9); it can be realized, for example, as textile diffuser, but may also consist of other materials. At the same time, due to the large surface, it has the effect of a cooling sail, that is, heat exchange between the diffuser surface and the room takes place via radiation and free convection as well.

In the close range, supply air from the micro-perforation mixes intensely with indoor air (induction); in a certain distance from the air diffuser, the flow is predominantly thermally driven and corresponds to the displacement ventilation with air supply above the occupied zone described at the end of Section 10.3.2.

In the illustrated case, the digesters are switched off and the exhaust air is discharged exclusively through the ceiling diffuser. However, the concept also shows good results, when the exhaust air is discharged via digesters only.

In terms of heat removal, this system with an air exchange rate of 6.4/h under boundary conditions of the simulation achieves the same result as an ideal mixed ventilation with an air exchange rate of 8/h. Regarding the removal of

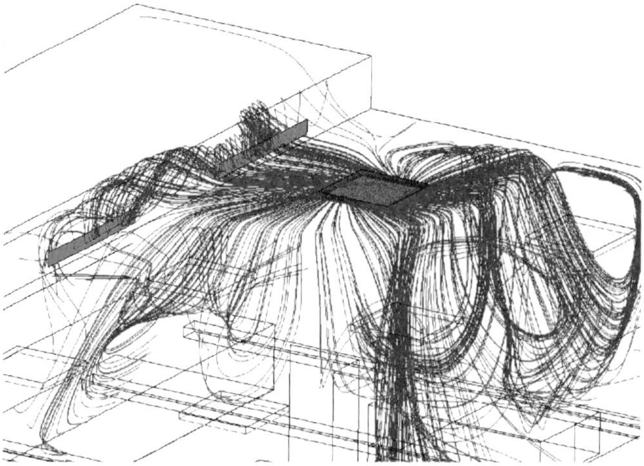

Figure 10.5 Streamlines (red) starting from one of the swirl diffusers (blue). A partial ventilation short circuit exists: Approximately 1/4th of the supply air is directly exhausted by the neighboring exhaust-air diffusers (green) so that this part of the air only contributes little to the removal of heat and substances.

hazardous substances, it is even superior to the comparable case mentioned earlier (Figure 10.10). The reduction of the air volume by 20% can be translated into energy savings of at least the same magnitude.

Other simulation variants have shown that it is possible to favorably combine the system at higher cooling loads with (passive) cooling bars to the right and to the left of the air diffuser.

The ventilation system shown in the simulation can be classified as a hybrid system with mixed ventilation in the occupied zone and layer ventilation above it. It unites the advantages of both systems:

- Efficient removal of heat and hazardous substances similar to that of layer ventilation.
- Low air velocities, which is important not only for thermal comfort but also for the efficient retention capacity of the digesters.
- Homogeneous distribution of temperature and hazardous substances in the occupied zone, similar to mixed ventilation.

10.4.4 Case Example 4: Laboratory for Laser Physics (Fritz-Haber-Institute Berlin)

This simulation refers to a laboratory for laser physics at the Fritz-Haber-Institute in Berlin. The central topic is the temperature maintenance in the experimental area. There, the temperature constancy up to 0.1 K is required in the presence of approximately 5 kW cooling load in the room. The ventilation system must not

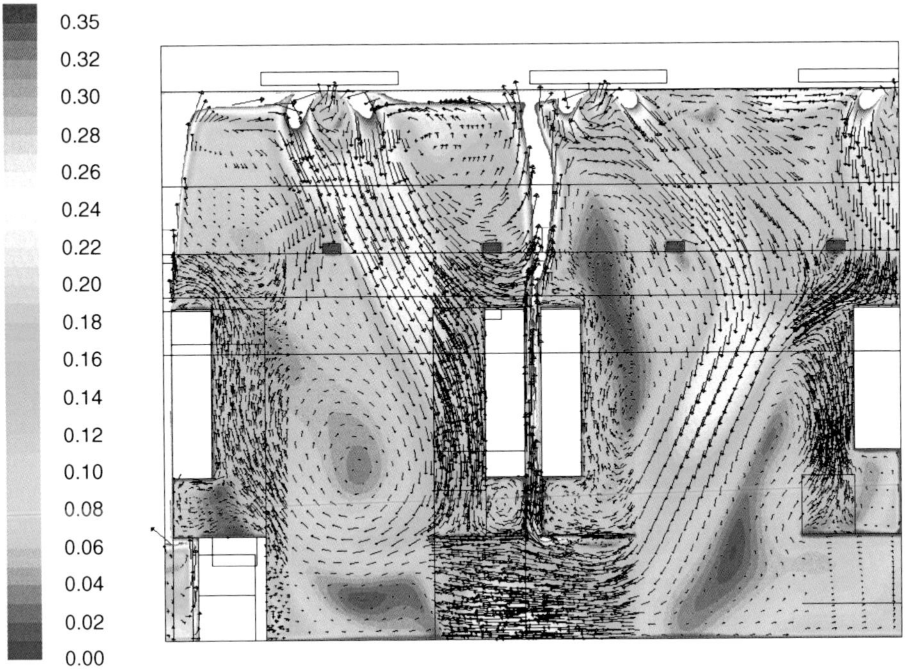

Figure 10.6 The positions of the displacement diffusers (Procondif) are determined by the white rectangles above the flow pattern. At head height, the velocity is below 0.25 m s^{-1} and the draft problem is smaller than with swirl diffusers. The flow profile, bell-shaped in the ideal case, proves to be sensitive toward disturbance flows, in this case generated by thermal buoyancy. At the backsides of the wall cupboards on the left and at the center of the picture, warm air rises which diverts the supply-air jets to the right. The supply-air jet rightmost in the picture is diverted by the furnishing toward the occupied zone.

restrict the flexible utilization of the room. Further requirements are placed on low air velocities and the discharge of released dusts. The air flow system originally provided proved to be unsatisfactory in a first simulation. Having considered several variants, it was possible to develop a system that fulfills all requirements. In this system, the supply air is introduced at insufficient temperature via a large, centrally arranged textile diffuser ("CG-Distributor") on the ceiling (Figures 10.11 and 10.12); all around the walls, a strip-shaped exhaust air outlet is arranged which is separated from the supply-air diffuser by air aprons. A part of the heat emitters is enclosed and exhaustion takes place directly. It was important for the performance of the task to conduct the warm air emitted by the remaining heat sources in a controlled manner to the boundary areas of the room, where it can rise and be extracted without causing disturbances in the experimental area. One third of the exhaust air is discharged via the double floor in order for larger, sedimentation-prone dust particles to be collected as well.

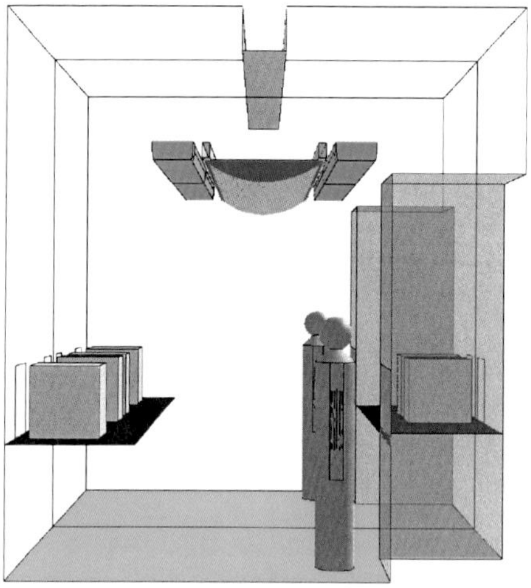

Figure 10.7 Geometry model of a model laboratory room with the following components: Two digesters (gray) with front sash (green) and opened gap (red); two persons (brown); several heat-emitting devices (gray) on the worktables (black); above the laboratory aisle, the extensive air outlet (blue), optional cooling bars (green) to the left and the right, which, however, have been removed for the variant shown here. Above the air outlet, the exhaust-air duct with slot-shaped exhaust-air passages arranged on both sides.

10.5 Energy-Efficient Systems Engineering

10.5.1 Fans

The selection of fans, their ideal integration into the duct system, and their control during operation has great influence on the energy consumption of laboratory buildings. It has already been pointed out that air ventilation by fans accounts for 30–50% of the primary energy demand of lab buildings. The energy consumed by air ventilation is also important, as according to the energy savings decree EnEV, it will be part of the evaluation parameters of a building.

Air ducts should be dimensioned as large as possible, since pressure loss is proportional to the power of 2 of the flow velocity and the fan capacity is proportional to the power of 3. In central devices, velocity should not exceed $1.5-1.8\,\mathrm{m\,s^{-1}}$. For this purpose, shafts, intermediate ceilings, and ventilation control center should be sufficiently dimensioned. Routings as short as possible and the respective arrangement of the ventilation control center also contribute to the optimization of the energy consumption.

The fan efficiency should be at least 75–80%. State of art for partial load operation is a speed control of the fans via frequency converter. Double fans can result in more efficient operation in the lower partial load mode.

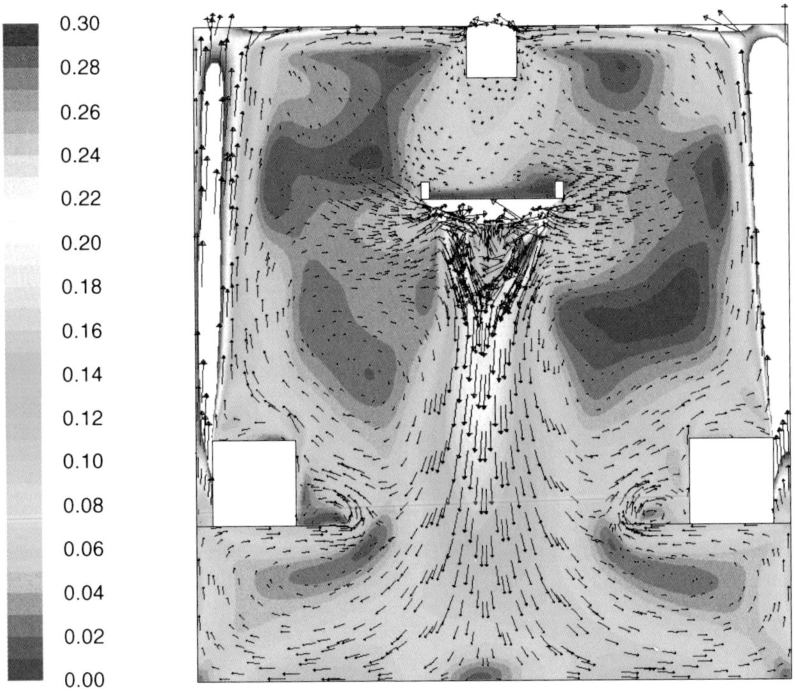

Figure 10.8 Flow velocities (in meter per second) in a vertical section transversely to the laboratory aisle. The flow shows a certain time fluctuation; the picture shows the values averaged over time. The flow velocity at head height below the air outlet amounts to a maximum of 0.22 m s^{-1}, the draft risk is correspondingly low. To the right and to the left in the picture, in the area of heat sources, a thermal buoyant flow with velocities of >0.3 m s^{-1} can be seen.

Air flows can be reduced by about 60–80% in ancillary operating hours in the absence of people, which accounts for 64% of the total year in case of a lab utilization of 12 h from Monday to Friday.

10.5.2 Heat Recovery

The air supplied to the rooms by the ventilation system must be processed outdoor air according to DIN 1946-7. Secondary air would be generally permissible, if it were possible to exclude dangerous concentrations of hazardous substances, which is usually not the case. Therefore, high outdoor air volume flows are necessary in laboratory rooms, which, for energy reasons, render highly efficient recovery of heat and cold indispensable.

The transfer of hazardous substances from exhaust air to supply air in the case of operational disruptions or defects can only be reliably excluded when circulatory composite systems are used. Here, a brine circuit connects the air-brine

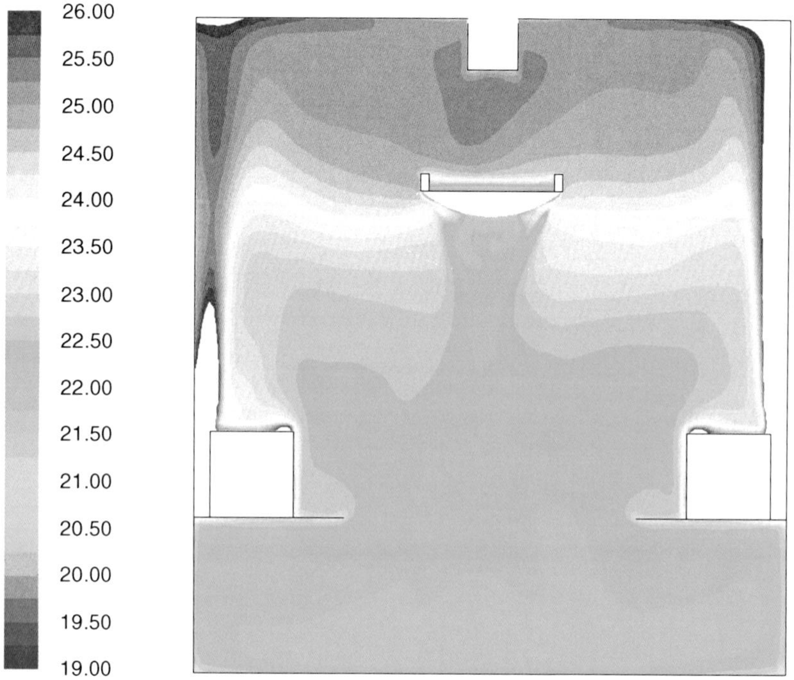

Figure 10.9 Sectional plane, as in Figure 10.8. The temperature image (in °C) shows a clear division of the room, in which a hot-air cushion with temperature stratification is formed above the air outlet, while below, that is, in the occupied zone, a lower, largely homogeneous temperature exists.

heat exchangers in the supply and extract unit. A heat pump in the brine circuit increases the efficiency of the system. Efficient cold recovery is achieved by adiabatic exhaust air humidification.

Thus, the possibility of many small, decentralized exhaust air systems on the roof, previously chosen to minimize the horizontal ventilation ducts on the floors, is excluded. Heat recovery requires centralized extract-air systems. However, for very special areas self-sufficient, separated individual systems may be required or at least practical. With the help of highly efficient circulatory composite systems with water heat exchangers and adiabatic extract-air cooling, the energy demand has been drastically reduced, in particular for heating, but also for cooling. Under winter conditions, performance efficiencies of about 75% are achievable, under summer conditions about 30%, 20% in particular in case of thunderstorms.

The operating cost of heat recovery is relatively low and the systems are economically highly effective. The annual recovery of heat ranges between 90 and 95%; a generally valid statement for cooling is not possible, since this depends strongly on the internal cooling loads and the additionally cold-water based cooling systems.

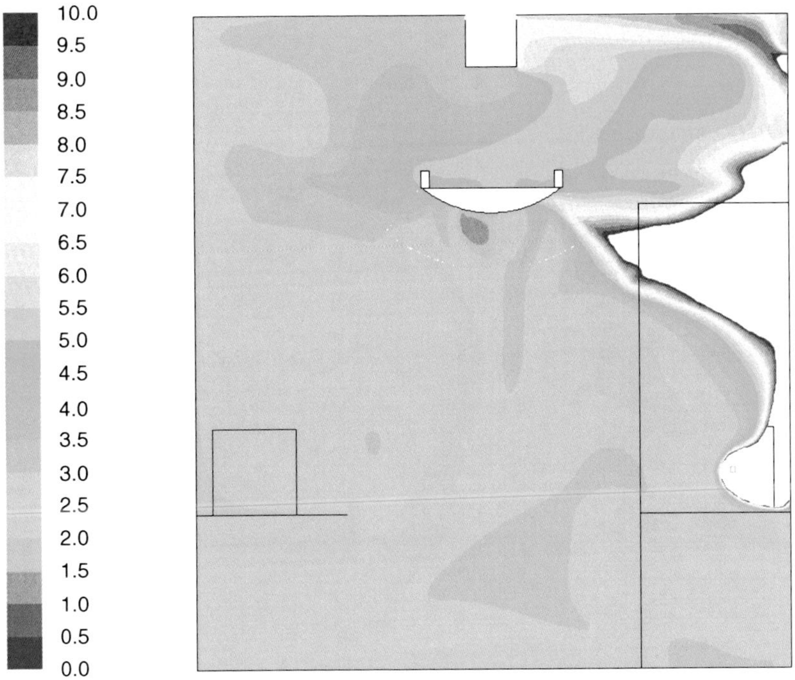

Figure 10.10 The tracer gas source on the right-hand table (green-edged square) releases 2 l min^{-1} of tracer gas (nitrogen + approx. 8.5 volume parts per thousand of sulfur hexafluoride (SF6)). The buoyant flow of neighboring heat sources leads a large part of the tracer gas upwards where it is effectively collected and removed from the room. Another part of the gas is recirculated into the occupied zone. The average SF6-concentration at face height amounts to 2.9 ppm. In comparison: In case of an ideal mixed ventilation with the higher air-change rate of 8/h, the SF6-concentration in the room would uniformly be 3.5 ppm.

10.5.3
Humidity Treatment of Supply Air

Humidification and dehumidification of supply are very energy-intensive processes.

DIN 1946-7 requires:

- Relative humidity at least 30%, occasional lower deviations are acceptable.
- Relative humidity 65% at the most
- Absolute humidity $<11.5\,g\,kg^{-1}$

More stringent requirements as to humidity constancy should be realized in a decentralized way and be limited to as few rooms as possible.

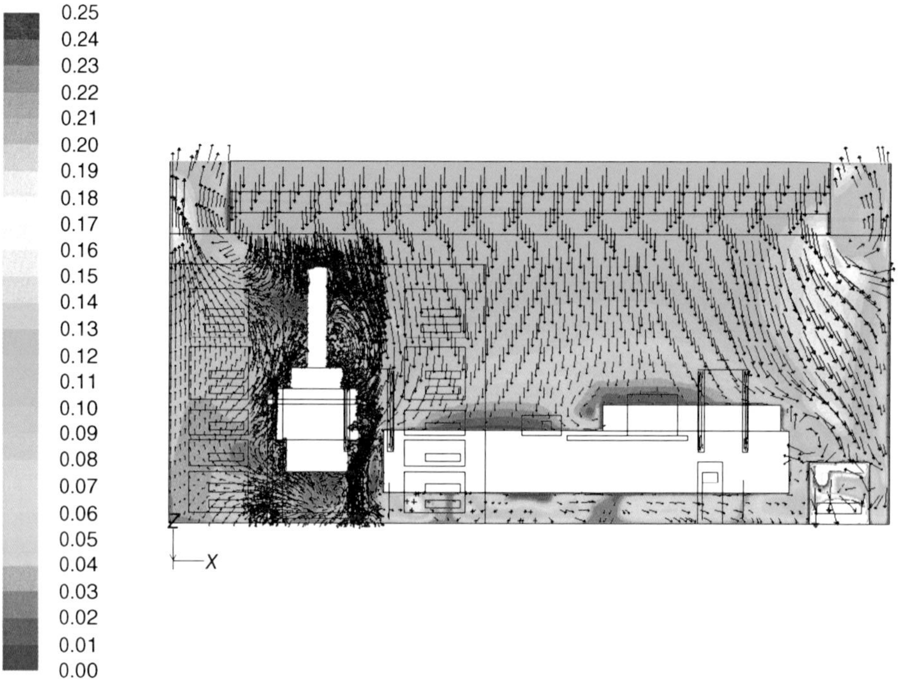

Figure 10.11 The velocity of the free incident flow above the table is 0.12 m s^{-1}; it decelerates toward the stagnation point. The buoyancy tendency above the warm laser (to the right on the experimental bench) can be suppressed by this flow.

10.6
Installation Concepts for Laboratory Buildings from the Point of View of Ventilation and Air-Conditioning Planning

10.6.1
Arrangement of the Central Ventilation Unit in the Building

The arrangement of the central ventilation unit has a fundamental influence on the building structure. The following alternatives are available:

- Arrangement in a central roof system
- Arrangement in an engineering basement (below the lowest useful floor)
- Arrangement in an engineering mezzanine (required for buildings with more than 10 floors)
- For very high buildings, a combination of these alternatives.

Later, the advantages and drawbacks of the solutions mentioned will be discussed.

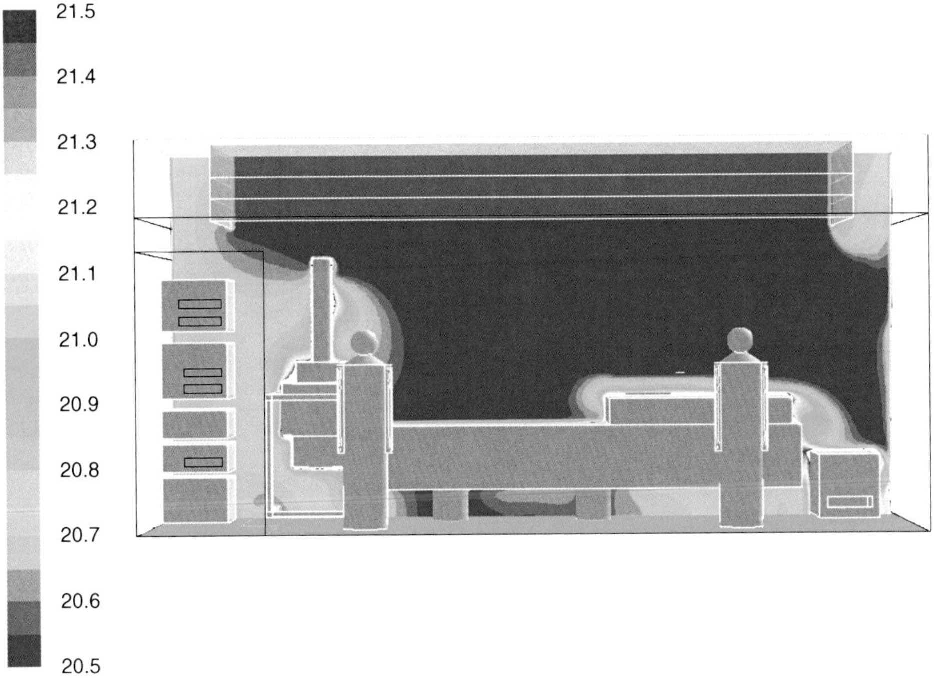

Figure 10.12 The temperature deviation from the set point is below 0.05 K in the experimental area (to the right of the person on the left).

- Central ventilation unit in the attic floor
  In practice, the central ventilation unit as central roof system has proved to be the most advantageous solution. Usually, the building envelope is a lightweight-construction. The clearance height is about 5–6 m. If possible, the central roof system extends across the entire building length, however, at least across all shafts. A close allocation of the central units to the respective shafts contributes to the reduction of the duct lengths. If required by building regulations or architectural concepts, the peripheral setback of the central roof system will usually be possible.
- Central ventilation unit in the engineering basement
  If, for instance, due to a permissible building height the central roof system allows one useful floor less only, a central ventilation unit in the basement would be an alternative. It would be arranged next to the other supply units (heating, cooling, sanitary, media, fire protection, electric).
- Central ventilation unit in the engineering mezzanine
  For high buildings with more than nine floors, an engineering mezzanine is required.
- Supply air unit in the basement floor, exhaust air unit on the roof.
  If circulation-connected heat recovery systems are used, the supply and extract air system can be arranged in separate rooms. In laboratory buildings, supply

air units are sometimes arranged in the engineering basement and extract air units in the roof area in order to achieve a duct layout in opposite direction and thus a minimization of the shaft area.

10.6.2
Central Units

For security of supply reasons, at least two central units are to be provided; depending on the size of the buildings, correspondingly more. Flow velocities in the central units of about $1.5-1.8\,m\,s^{-1}$ are a good compromise regarding installation size (investment costs) and energy efficiency (operating costs).

Complete redundancy would require the remaining units to be able to supply the total air volume in case of a unit's failure (in consideration of the simultaneity factor of the consumer). For economic reasons, however, complete redundancy is hardly ever required; but, in coordination with the user, it will be determined, which consumers, in case of average, need complete air supply, and which can be operated with reduced air volume.

10.6.3
Vertical Access

In the building concept, particular importance is attached to the arrangement of the vertical shafts, since they define the structure of the building along with staircases, lifts, supply shafts, rooms for electrical/IT sub-distribution panels. Clear specifications from the technology planners are required regarding the number and the size of the shafts. The shaft arrangement depends on the floor plan of the building and on the access concept.

When dimensioning the shafts, the focus should be on the accessibility during installation work and subsequent modifications and retrofits, as well as on the operability of the fittings. Each shaft should be accessible from each floor via a door. The shaft itself shall be provided with a servicing level. On each level, the assigned supply areas are accessible via the shafts.

In addition to the air ducts, today even the complete media supply (heating, cooling, vapor, sanitary, and gases) is accommodated in a common utility shaft. In practice, the accommodation of the electrical supply in a separate shaft has proven its worth. In most cases, both shafts are arranged directly next to each other.

Usually, the shafts along with the central ventilation unit are defined as a common fire area; in this case, all air shaft passages into the floors require fire dampers. Alternatively, special concepts providing full sprinkler systems and comprehensive fire alarm systems are possible; this compensation measure makes fire dampers redundant.

10.6.4 Horizontal Access

In general, the overall coordination of all technical trades for the installation on the floor levels lies with the climate control engineer. Basic parameters are the determination of the clearance height and the selection of the ceiling system (ceiling suspension, visible installation, and system ceiling).

Reasons for ceiling suspension are the facts that visually attractive solutions can be achieved and hygienic requirements can be fulfilled if high-quality pharmaceutical ceilings are used. A disadvantage is the higher rebuilding expenses in the case of modifications and adjustments. Areas with increased hygienic requirements will usually be furnished with a ceiling suspension or at least with the possibility of subsequent retrofitting.

The renouncement of a ceiling suspension provides a cost advantage while modifications and adjustments are easily feasible. In this case, however, all installations, such as pipelines, fittings, insulations, attachments, and so on, must meet the visual and hygienic requirements, which cannot always be achieved completely.

A ceiling system enables prefabrication and quick and collision-free installation. Modifications and adjustments can be carried out quickly and easily. Visually appealing solutions can be achieved.

11
Electrical Installations

Oliver Engel

In this chapter, central building control systems for laboratories, power supply, data networks, and laboratory lighting in the industrial sector are described and their operating instructions, application possibilities as well as case studies are provided.

11.1
Power Supply

The electrical energy supply for lighting, ventilation, and fume cupboards requires the provision of separated electric circuits, see DIN VDE 0789-100.

The necessity for separated circuits results from the different requirements of different consumers. Lighting in the laboratory, for example, must be independent from other circuits, since the failure of a fuse for the fume cupboards should not lead to a dark laboratory. Also, the actuation of the lab emergency shutdown must not cause the failure of the lighting. The entire lab equipment must be shut off via this emergency shutdown as required by the laboratory guideline.

11.1.1
General Distribution

There are several possibilities to supply power to a laboratory. The conventional method is the supply via laboratory distribution systems with an integrated lab emergency shutdown. The laboratory distribution system can be arranged in an electrical service room, but it is to be supplied in the lab as well. The layout and concept depend on the demand. Different installation sizes with corresponding currents can be used to design the extensive distribution in the lab. In labs with instruments that require the securing of continuous operation (as e.g., analysis instruments or deep freezers), emergency power supply in the form of an independent grid is established. The kind depends on the number and the power of

The Sustainable Laboratory Handbook: Design, Equipment, Operation, First Edition.
Edited by Egbert Dittrich.

the consumers. Here, the preferred variant is an emergency power network via separate distributors to supply consumers, such as the lighting of hallways and data systems technology. In case of a failure of the mains, if a secured network does not exist, the most common variant is the supply of the emergency power network via an emergency diesel generator.

When switching on the emergency diesel generator, a short mains interruption occurs (starting of the emergency diesel generator in case of mains failure). For maintenance work and tests, however, this can be avoided by a mains synchronization mechanism. The installation site must be well selected, since the diesel generator in operation (test run or maintenance) causes noise and vibrations. Also, the exhaust gas pipe must be routed in a way that it can be properly operated with regard to fire protection and ventilation. The installation room of the emergency diesel generator requires supply and exhaust air and a diesel tank to be filled from outside.

For safety-related consumers such as emergency lighting, fire alarm system, or staff alerting systems, supply by an emergency diesel generator is not sufficient. Additional separate supply is required.

11.1.2 Shutdowns

11.1.2.1 Emergency Shutdown

In laboratories, emergency switches for the disconnection of the power supply voltage must be available. Each person permanently or temporarily working in the lab rooms must know the exact position of these emergency switches to be able to react immediately in case of danger. Tests with voltage applied to accessible parts may only be carried out at sockets secured by the emergency switch.

The restart of danger-prone power supplies may be possible only after the reset of the emergency switch. The reset must not trigger the restart of these power supplies.

The emergency shutdown causes the shutdown of the entire laboratory. Furthermore, local shutdowns of special experimental setups could be necessary. Local shut-downs can be realized by the locking of doors or sashes.

11.1.3 Consumers

11.1.3.1 Plug Connections

Basically, only CEE round plug connections according to VDE 0620–0625 are permissible as three-phase plug connections. However, each laboratory should use only one system to avoid the use of adapters.

11.1.3.2 Switches and Sockets

Switches and sockets at laboratory benches should be installed above the work surface or, if arranged underneath the table top, they should be recessed to such an extent that they do not represent a source of danger in case of leaking or splashing liquids. Sockets of fume cupboards shall be arranged outside of the fume cupboards. In case sockets are required in the work room of the fume cupboard, they must be clearly allocated and switchable from the outside.

11.1.3.3 Motors

Due to their usually long operating times, electrical drives are significant power consumers. Therefore, it seems reasonable to attach great importance to the energy efficiency of drives. On the one hand, the drives themselves should work as efficiently as possible; on the other hand, it is extremely important for the drive to be well adapted to the respective drive task.

Excessive safety margins lead to over dimensioning and thus to additional losses.

Modern motors provide good efficiencies in the nominal load range from 80 to about 100%, which facilitates the selection.

11.1.3.4 Rotational Speed Control with Frequency Converter

A frequency converter is used to continuously vary stator voltage and frequency of an asynchronous motor, thus turning a standard motor into a variable speed drive system. When driving a pump, it is used to control the flow rate over a wide range by means of rotational speed changes.

With suitable motors, frequency converters can be retrofitted. In consequence, a pump with fixed rotational speed can be converted into a variable speed pump. This control type achieves the highest energy savings, but requires higher investment costs as well. Moreover, maintenance costs of the motors will also be reduced.

11.1.3.5 Pumps

In general, pumps should be dimensioned in stable operating mode, that is, without changing operating conditions, exactly to the required output. Each over dimensioning has negative effects on the energy consumption. If, however, the hydraulic systems are modified subsequently or if the power requirements are reduced, and so on, volume flows will need to be adjusted to the new conditions.

11.1.3.6 Vacuum Pumps

Vacuum pumps are subdivided into three categories: backing pump, high vacuum pumps, and ultra-high vacuum pumps.

1) Backing pumps are mostly used for vacuum distillations, in food production laboratories, and in the development of light bulbs.
   Pressure range: atmospheres up to 10^{-3} mbar.
2) High vacuum pumps are used in laboratories of semiconductor and picture tube technology.
   Pressure range: from 10^{-3} to 10^{-7} mbar.
3) Ultra-high vacuum pumps in the pressure range of less than 10^{-7} mbar.

They are used in laboratories of surface physics and particle acceleration technology.

11.1.4
Routes

The planning of cable routes and supply lines in laboratories and in the areas leading to the lab is mostly very complex, since their planning is headed by ventilation engineering and media supply. When bus bars are planned, attention should be paid to minimizing the number of changes of direction. The cable routes must also be designed to house the cables of ventilation engineering and data systems technology.

Here, the coordination of the trades involved in the planning of cable routes, power buses, media, sprinkler systems, and ventilation is very important, with ventilation engineering having the highest priority because it requires the largest areas. 3D-planning is recommended in particular when planning for existing buildings, since in laboratories many trades are involved in the area of ceilings or even suspended ceilings. Whenever laboratories are planned, intersections or junctions arise. This requires the particular coordination of the individual trades (Table 11.1).

When laying out the routes, it is recommended to provide space for subsequent expansion. This mainly applies to data systems technology because when power buses are used, only tap-off units need to be installed. Due to the introduction of the bus technology in the EI&C technology, cabling is kept within limits. Conventional cabling with DDC technology requires the respective adjustment of the planned routes.

Regarding the feed lines to the laboratories, attention must be paid to the LMAR in the corridor areas (escape and emergency routes). Due to the increased assembly costs and with regard to subsequent expansions, it is reasonable to avoid this area or to reduce it to a minimum.

Apart from cable routes, ventilation ducts, and pipelines, the specifications of the operator need also to be considered with regard to grounding and potential equalization (discharge capacity of the floor, potential equalization connections).

When it comes to the crunch: cross-trade planning and coordination is worth the effort!

11.1.4.1 Air Ducts

The ductwork in buildings and industrial facilities is usually installed when the installation of the main constructions/main plants, which entails frequent curvatures and changes of cross-sections, has been completed. In addition, rectangular air ducts are mostly installed, although circular ones are more favorable in terms of energy.

Moreover, after the installation, a fan system must be designed in a way that the required air quantities are achieved everywhere. This design partly comprises the

Table 11.1 Interfaces in the laboratory (auxiliary services according EN 13792).

Consumer list																								
Equipment		**No.**	**Size**	**Weight**	**Electrical requirements**				**Auxiliary material**															
No.	Name	Piece	mm	Kg	400 V	230 V	24 V	KW	WP	WPC	WPH	WNC	WCR	WCF	WDI	WST	WCO	CA	RA	CO_2	N_2	G	V	Comments

application of throttle valves, which entails additional pressure and energy losses. Their minimization requires the correct planning of the ventilation system.

11.1.5
Hazard Analysis

11.1.5.1 Equipment with Special Risks

In laboratories, where mechanical devices or electrical equipment represent a risk, work may be carried out only in the presence of at least one other person.

Only especially trained and instructed employees are allowed to work at the high-voltage test site.

11.1.5.2 Danger Symbols and Sources of Danger

All employees working permanently or temporarily in the lab must know and adhere to the operating instructions and signals as well as to warning lights at machines, measuring devices, apparatuses, and other equipment. To ensure this, employees must become acquainted with the respective operating instructions and observe them when using the equipment.

11.1.5.3 Explosion Danger through Electrostatic Charge and Protection Measures

Explosion dangers through electrostatic charges can arise when working with flammable liquids. Protection measures include, for example, grounding of conductive vessels with a joint grounded connection point (see BGR 132).

11.1.5.4 EMC

Regarding the EMC, interference resistance is the characteristic of a device not to be disturbed by the disturbance variables in its electromagnetic environment. The emitted interference deals with the disturbance variable emitted by a device in the electromagnetic surrounding.

11.1.5.5 Regulations for Access to High-Voltage Laboratories

High-voltage laboratories may be accessed only by employees who have been sufficiently instructed on the hazards.

Other persons may enter the high-voltage laboratories only if accompanied by a trained electrician.

Employees from external companies, for example, cleaning personnel who have occasional or regular access to high-voltage laboratories can be included in this regulation "Access with sufficient instruction," if this seems to be possible with regard to the hazardous situations. In other cases, assistance and supervision need to be ensured. If persons without sufficient instruction, for example, customers, access the laboratory, instructions on behavior must be given previously and the assistance of a trained electrician must be ensured.

11.1.5.6 Noise Protection

Hearing protectors must be worn in noisy areas identified as such by mandatory signs, as well as during all noisy works from 85 dB – even during temporary exposure to noise.

11.1.5.7 Explanation: Trained Electrician

A trained electrician is a person who knows the relevant regulations and who is able to evaluate the work assigned to him/her and to identify possible hazards due to his/her professional training, knowledge, and experiences. The professional qualification is usually proven by the successful completion of a training course, for example, as electrical engineer, electrician, master electrician, journeyman electrician. An electrician, comprehensively trained and qualified for all electrotechnical areas of expertise does not exist. Decisive characteristics of a trained electrician are the knowhow and experience in a certain field or work (apprenticed trade).

11.1.6 Instruction

Once a year, laboratory employees are instructed on safety issues according to the operating manual.

The instruction of new or temporary employees on the electrical fields of activity is carefully carried out by experienced employees and includes a laboratory tour, during which all safety-related facilities are presented. Temporary employees may carry out work only according to the instructions given to them previously. The participation in the instruction is confirmed by signature and archived. All persons working permanently or temporarily in the laboratory confirm by signing that they have taken note of all relevant accident prevention and safety regulations for the working area, that they will adhere to them, and that they will take all precautions to protect both themselves and other persons from accidents.

11.1.6.1 Explanation: Electrotechnically Instructed Person

An electrotechnically instructed person is someone who is instructed by a qualified electrician on the tasks entrusted and on the possible risks in case of improper behavior, and who is also trained and instructed on the necessary protective equipment and protective measures, if necessary.

11.1.7 Behavior in Case of Electrical Accidents

Despite safety precautions and various safety indications, if an industrial accident occurs due to human inadequacies or technical faults, each person is obliged to be responsible for strictly following the occupational first-aid measures.

After electrical accidents the following tasks have to be carried out:

- Immediately disconnect power.
- Determine whether apnoea exists.
- In case of emergency, immediately call a doctor and start rescue measures (breath donation, cardiac massage, adequate positioning in case of unconsciousness, or shock).
- First aid administered by first aiders is no substitution of medical attendance, but only a makeshift until the doctor arrives. Also, in case of a seemingly harmless accident, a doctor should be consulted, as cardiac arrhythmia may still occur up to 24 h later.
- Each accident must be reported to the superior.

11.2 Lightings

11.2.1 Lighting Systems

The selection of the lighting system depends on the environment and the structural conditions. The most widely used lighting systems are surface-mounted luminaries and suspension luminaries. The surface-mounted luminaries are used only in closed ceiling systems. Floor lamps are not used in the laboratory.

11.2.2 Illuminance Level

The illuminance levels are determined in the DIN EN 12464-1 and are 500 lux for laboratories and 1000 lux for laboratories with color testing.

11.2.3 Lighting Control

There are various approaches to lighting control. On the one hand, there is the usual light switch at the doors in the form of a complete on/off-switch or in the form of a serial button with two areas, which can be used to switch, for example, the window side separately. Furthermore, installation relays with central function enable the central on/off-switching as well as switching via automatic timers. The central switching functions can also be carried out by the BMS.

11.2.4 Lighting Regulation

Numerous variants of lighting regulation can be used, which range from central lighting control and individual room control up to the control of individual luminaries. All controls can be coupled with a window shade control. Here, it is useful to carry out a profitability analysis. The individual room control in combination with a window shade control is recommendable.

11.2.5 Emergency Lighting

Emergency lighting is not always absolutely required in laboratories. In any case, a risk analysis should be carried out in cooperation with the user, since the planner does not know the work flows and the respective safety requirements. It has to be aligned with the fire protection certification and the building requirements. Usually, emergency signs and luminaires are mounted at doors and emergency exits. Emergency and evacuation plans need to be observed when arranging the emergency lighting of the escape routes. Here, self-contained luminaires, group battery systems, and central battery installations can be chosen. The choice of the system depends on the number of luminaires and the maintenance expenses. A maintenance contract is recommendable.

11.3 Data Networks

11.3.1 Data Systems Technology

When planning a laboratory, it must be taken into consideration that laboratory devices are increasingly networked with each other and that occasionally even multiple data connections per device (different physical nets) need to be provided.

Furthermore, data connection requirements must be defined in time as the positions of the connections must already be determined before the laboratory furnishing is ordered.

Subsequent holes or changes can influence the approval of, for example, a lab fume cupboard (sufficient placeholders for later extensions, provide for empty conduits to the connector socket with taut wire).

In clean rooms, windproof connector sockets must be used (pressure cascades) and environmental conditions may dictate protection against dust and humidity. Different networks (physically or logically separated) are often marked by various colors.

Lockable connection covers are frequently used for mobile lab equipment or measurement technology.

Place and environment of data distribution must also be defined in good time, since length restrictions for copper wiring have to be observed and active components in the distribution (environment of data distribution) entail the provision of sufficient cooling. Active components require lockable installation (data security) and permanent accessibility for the service staff. In case of safety-relevant equipment, it is necessary for the active components of the distribution and for various terminals to be supplied with electricity without interruptions via a UPS-system (note lab emergency shutdown).

11.3.2
Fire Alarm System

Fire alarm systems are characterized by different ways of detection, such as detection by temperature difference, temperature, optical detection (fumes), aspirating smoke detectors, flame detectors, and manual call points. They are used according to the requirements in laboratories, as for example, in fume cupboards and housings. The aspirating smoke detector is applied where accessibility is not ensured. Here maintenance takes place at a central point. The alarm is subdivided into optical and acoustic alarm on the one hand and the connection to telephone system and data systems technology on the other hand. The system design must correspond to the respective start-up conditions. This also applies to redundancies. Each alarm system requires a dedicated maintenance contract.

11.3.3
Telephone System

The telephone wiring can be combined with data cabling; thus, each data connection can also be used as telephone connection. The telephones can also be installed as wall-mounted telephones or wireless telephones. Wireless telephones require a repeater. Due to the high degree of technical equipment in the lab, the installation of the repeater is possible only after completion, since the positions can be determined only by measuring during operation.

VoIP-phones are increasingly used in new systems. Here, compatibility must be taken into consideration when selecting active network components.

11.3.4
Access Control

Access control ensures controlled access which is necessary for secrecy, security, quality assurance, registration, and archiving.

Security gate control requires additional locking and traffic signal controls. In this context, it should be noted that an emergency opening or a positive opening in compliance with escape and emergency routes must be ensured in case of average.

11.3.5
Miscellaneous

When planning laboratories, further user-dependent systems may be required as, for example, electro-acoustic systems, intercommunication systems, gas monitoring, gas deficiency monitors, and gas emergency shutdown. These systems are partly desired by the operator or demanded in safety meetings.

11.4 Central Building Control System

The specifications of the employment protection provisions have to be taken into consideration beforehand when equipment or work processes are planned, since the correction of failures during subsequent operation will be costly.

11.4.1 Nodal Points

The central building control system consists of a multitude of nodal points. They can be switched on as inputs or outputs in the form of binary or analog signals. Analog inputs are temperature, air-volume, and humidity measurements, or also sash positions of fume cupboards. Binary inputs are fault messages from the multitude of devices in the lab, but also from the ventilation system itself.

Analog outputs are necessary for the control of air volumes, temperature, or humidity. Binary outputs are used for the switching or signaling of operating states.

11.4.1.1 Planning and Coordination across Trades and Disciplines

In order to ensure the exchange of information and the consideration of interdependences, the overall concept must include all areas from building automation to heating and ventilation and air-conditioning systems as well as low-voltage power supply up to lighting.

11.4.1.2 Signaling Devices and Warnings

If required for safety and the protection of health of people working in the labs, each machine must be equipped with signs and/or instruction plates for its use, setup and maintenance. They have to be chosen, designed, and manufactured to be clearly visible and permanent.

11.4.2 Regulation

Laboratories can be regulated by two different types of control: single room control or central control. A combination of these two types is also possible. The single room control transmits the required temperature, humidity, and air volume to the central unit. It uses the total values of the single room controls to determine the total demand, which it then makes available to the system. In laboratories, larger heat flows emitted by devices must be collected and discharged directly at the point of release, in case they could lead to higher room temperature. Attention should be paid to the fact that the supply air is supplied free of draft. The most efficient combination of motor and ventilator is the direct coupling only on one shaft. Consequently, unnecessary elements between motor and ventilators should be avoided.

Asynchronous motor + gearbox versus HT-direct
The following measures can be taken:

- no gearbox required
- optimized efficiency
- low operating costs
- small dimensions, weight advantages of up to 50%,
- smaller foundations
- lower maintenance costs and higher reliability
- less noise
- higher dynamics, stiffness
- aligned system, such as, for example, with Sinamics G150, S150, S120
- due to increased winding insulation, no sine wave filter required in the converter.

11.4.2.1 Air Volumes

The operator of the laboratory is responsible for adequate ventilation. This includes that equipment for technical ventilation, for example, ventilators, air ducts, guiding plates, supply, and exhaust openings will not become ineffective over the years and the whole ventilation system is subjected to functional testing at reasonable intervals. See Appendix Standards and Regulations.

11.4.3 Operating Modes

11.4.3.1 Operation

The control or operating mode selected must be superior to all other control and operating functions, except for the EMERGENCY STOP.

In case the device is designed and built in a way that several control and operating modes with different protection measures and/or work processes are possible, it must be equipped with a selector switch for control and operating mode, which is lockable in each position.

Each position of the selector switch must be clearly identifiable and may correspond to only one control or operating mode.

11.4.4 Monitoring

The monitoring of laboratories is partly demanded by the operator, but also by the GMP (good manufacturing practice) and GLP (good laboratory practice). Many climate and hygiene-related data, such as temperature, humidity, air volume, differential pressure, and particle density, but also personal data, such as access, work processes, and credentials can be recorded. GLP is a quality assurance system dealing with organizational procedure and framework requirements. In order to substantiate them, the respective data are recorded.

The GMP is the guideline that defines quality assurance in production workflows and production environment for pharmaceuticals, active ingredients, and medical products as well as for food and feeding stuff. Quality assurance plays a central role in the manufacturing of pharmaceutical products, since here quality deviations may have a direct impact on the health of consumers. A GMP-compatible quality management system will ensure product quality and fulfill the requirements of the health authorities necessary for the marketing.

Link to the professional association Precision and Electrical Engineering: *http://www.bgetem.de/start/index.html.*

12
Service Systems via Ceiling

Hansjürg Lüdi

12.1
General Discussion

In today's building construction (and renovation) environment, most structures designed, that include requirements for laboratory space (or any general space for that matter) are done so without a precise understanding of the actual usage that will occur at the time of final construction, not to mention the decades of reconfiguration after construction. Service systems via the ceiling provide a mechanism that maximizes flexibility, both with respect to the physical fluctuation in lab space required and the precise delivery of services to anywhere within any open space concept, with minimal effort and no disruption in operations. That is to say, delivery of utilities and services to any point within the open space design can be easily designed even when the final specific location for usage is unknown.

Many owners and designers of laboratories are facing these significant obstacles when they try to design a laboratory with unknown, specific utilization. A lack of known, detailed requirements provides insufficient information with respect to usage, necessary for proper final designs, thereby forestalling proper planning of laboratory buildings. Additionally, open architecture concepts allowing highly flexible, quick modifications require a new approach for the engineering of all building services (Figure 12.1).

Pressures to accelerate construction and the pace of real estate resale of laboratory buildings, as well as change of users or tenants and reconfiguration of workflow, create significant consequences for engineering laboratory media services, in addition to constant modification of the requirements for laboratory equipment and furniture. The resultant costs are both substantial and unacceptable, while being completely avoidable through the utilization of a service system via the ceiling.

Media services distributed from the ceiling, containing all laboratory utility and service requirements, have been engineered and designed since 2004 throughout the western Europe. During this same period, the concept of open architecture has penetrated lab design, and open space designed laboratories have been achieved. The redesign of small-room lab units into spacey rooms with areas of hundreds of

The Sustainable Laboratory Handbook: Design, Equipment, Operation, First Edition.
Edited by Egbert Dittrich.

Figure 12.1 Open architecture laboratory.

square meters produced a large number of problems with respect to the seamless integration of all technical building services.

In addition, increasing requirements of flexibility in laboratories, in terms of variable; lab sizes, work tools and utilities, thermal loads, and emission controls did not allow for the continuation of orthodox planning. Hence, with respect to contemporary needs, the historical design process of single utility components developed by separate lab planners, as well as the coordination of individual 2D CAD plans were counterproductive.

The efficient planning of an open architecture concept can only be achieved utilizing an integrated complete design of all required technical services on a 3D CAD platform. The 3D files may not be the initial step, but of course is the resultant chosen methodology. This design process has to be accomplished constructively, visualized, verified, and approved. A major implication is the complete elimination of all time wasted on cost-intensive coordination meetings during the construction phase.

In addition, everyone talks about sustainability, energy consumption reduction, safety, health, and reconfiguration cost reduction. All these are achievable through the general application of a service ceiling concept and the specific utilization of a new patented technology. Its unique air exchange and lighting components, combined with completely integrated services, makes it the most energy efficient, sustainable, open space design solution available in the market today.

12.2 Flexible Laboratory Room Sizes/Configuration

12.2.1 Planning

A detailed review of local fire protection ordinances always is conducted prior to the design of open architecture labs. Furthermore, specifications of national insurance organizations and their safety departments, as well as other responsible bodies are to be incorporated into the design of comprehensive utility/services distribution for all open architecture laboratory space. Usually, the ideal recommended individual lab space is between 300 and 800 m^2. It is required that open architecture space is equivalent with any specific restrictions within local fire protection codes. Later is a summary review of the basic special requirements for a typical design.

12.2.2 Height

The ideal room height is 3.5–3.7 m and thus a resultant floor height of about 4 m. The media ceiling clearance from the floor should not be less than 2.8 m because of the normal height of the fume hoods. Therefore, the building engineering services will be provided the required construction space of approximately a 0.7 m. That could become a design issue for the horizontal main fresh air and extract ducts depending on the layout of the building. The floor height should not be extended because the optimal levels of sustainability and efficiency have been scientifically determined (Figure 12.2).

12.2.3 Width

The width (center lines) has to be chosen according to the application and usage of the occupants. All EN 12128 specifications and requirements have to be met, as a minimum, under all circumstances. The most efficient width is 3.45 or 3.6 m between the center lines.

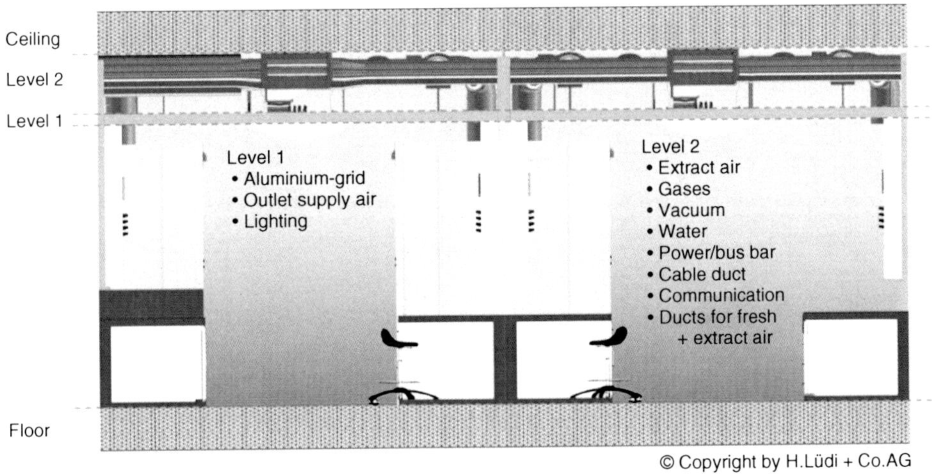

Figure 12.2 Service ceiling system side view.

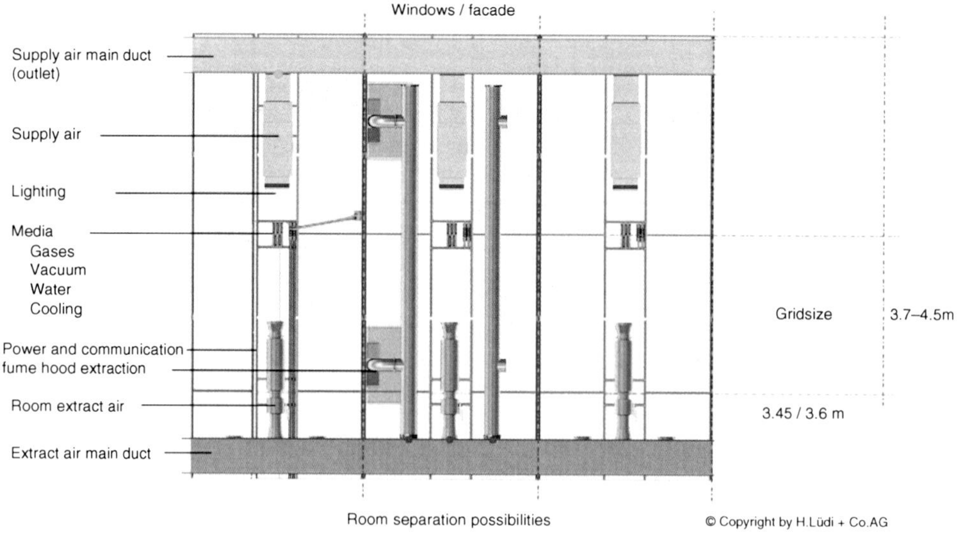

Figure 12.3 Service ceiling system plan view.

The dimension between the center lines should be fixed within the architectural concept of the building. It has to be identical with the grid of the façade and the grid of the structure (columns) (Figure 12.3).

In this respect, partition walls set up at a later stage will allow an easy room separation with connection to the inside of the façade.

12.2.4 Depth

The depth depends primarily on the architectural concept and may vary widely. Coincidently, it is important to properly consider all traffic patterns and escape routes. Generally, from every position in the laboratory, two escape routes must be available. In order to reduce traffic routes, it is recommended to include a second transverse joint in the grid, if depth exceeds approximately 7 m.

12.2.5 Analytic/Composition Areas

Most current advanced designs of lab buildings provide analytical and writing space, utilizing office style furniture, in proximity to windowed areas. These locations should be separated with partially glazed partition walls. Hence, it will be possible for the occupants to work in these areas without normally specified safety equipment such as eye glasses and laboratory coats. Also, it could be allowed to eat and drink in these areas, especially if there is another access route besides the laboratory itself. One of the most redeeming features of this type of space is their easy reconfiguration and assimilation into additional lab space.

12.2.6
Room within Room Solutions

Open area designed lab spaces quite often generate an underlying demand for special rooms, possessing unique requirements. These specific uses such as; clean rooms, rooms incorporating bio-safety features, localizing increased emissions, dark rooms, robotics, cubicles for specialized large fume hoods, and independent temperature control, require complete separation from the balance of the open space lab. This requires a longitudinal and cross separation from the floor to the concrete ceiling or up to the false closed ceiling. The separation of these rooms follows the building grid along the longitudinal (X) direction and the furniture grid along the depth (Y) direction.

12.2.7
Flexible Separation Walls

A grid in the depth or Y direction, should allow for the utilization of standard furniture sizes of; 90, 120, 150, and 180 cm plus the thickness of the separation wall and the usage of standard separation wall units in both the X and Y directions. Door modules with specialized power devices and connection elements such as cross angle or T pieces should be incorporated. Separation walls inside a fire protection area normally have an aluminum lightweight construction without additional fire protection attributes. These separation walls may be screened with a variety of materials, but in wider areas, they should only be glazed allowing visibility into adjoining rooms (Figure 12.4).

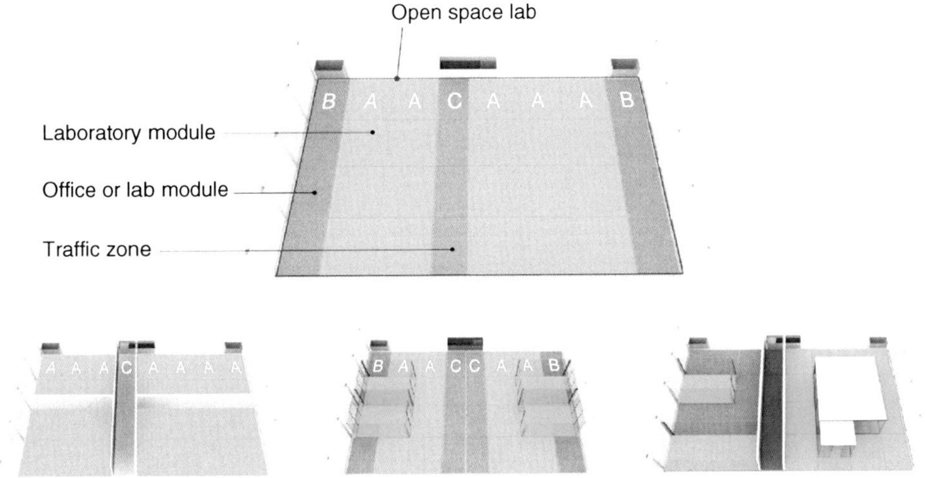

Figure 12.4 Building grid.

12.2.8 Reconfiguration due to Change in Work Content or Process

Restructuring and reconfiguring lab space are occurring at a faster and faster pace. Some of the major reasons for these necessary changes are:

- Laboratory type modification (wet chemistry, life science, pharmacological, etc.).
- New scientific team amending internal processes.
- Lab expansion to accommodate more scientists.
- Major equipment utilization (size, number, energy load, and services).
- Alteration of component services required (emission control and bio-safety).

The potential for any of the aforementioned restructuring requirements has a significant impact on engineering all building services and will pose varying difficulties based upon the initial design. Through the utilization of Service Systems via the Ceiling, engineering all building services optimizes flexibility through the incorporation of modular concepts, while incurring little to no interruption in laboratory or building operations.

Utilization of completely flexible, highly integrated, service ceiling systems require an increased focus in the initial planning of all building services to maximize the benefits these systems can deliver and minimize cost of construction. The designs can be completed independent of specific user requirements, which can be incorporated later with no further modification of the services distribution scheme required. Once the precise internal processes are known, the final placement of major components (service columns and fume hoods) can be made anywhere within the lab, given the unique, equal distribution of airflow and light throughout the open space.

Given the complete flexibility of the service columns, all services can be easily directed to their final destination at the time of initial construction and in fact, reconfigured at a later stage without disruption to the operating laboratory. Specific requirements such as: modified air supply and extract air for a unique subspace, utilizing chilled beams for special thermal loads, directing all services to specific workstations through service columns, and placement of fume hoods can be decided at the time of the final building drawings with no impact on infrastructure of either building or service ceiling designs (Figure 12.5).

All service columns have completely modular component parts and can easily be relocated anywhere throughout the lab. They are so flexible that you can replace a gas module in 15 min or upgrade the column with an additional gas module in 30 min. All of this can be accomplished without any interruption to operations at any other workstation. The columns never touch any of the underlying furniture thereby allowing an open, easily reconfigured space solution.

Figure 12.5 Service columns.

12.3 Major Differentiating Components

12.3.1 Ventilation

Air ventilation is the most important component for a sustainable laboratory and is the largest differentiating factor between the new solution and historical standard designed laboratories. It should be driven by a BMS (Building Management System) system allowing at any time for the adjustment of air exchange rates according to end-user specifications and necessary safety requirements. The most important issues for the design are:

- Variable air volume calculated from the fundamental air exchange rate and air volume needed by the fume hoods.
- Nighttime and weekend adjustment to air exchange rate.
- Necessity of specific air exchange levels for special usage.
- Minimization of draft exposure.
- Controlling exposition of toxic emissions.
- Creating internal airflow environment allowing fume hoods to operate properly.

The DIN EN 1946 recommends an air exchange rate of approximately eight times per hour as a standard for the design of airflow demand and represents a reasonable level. A detailed risk assessment is required for precise fixing of the final designed level. For example, if a specific risk can be eliminated, four times exchange rate per hour may be reasonable. On the other hand, flexibility needs to be designed into some systems to allow for up to 16 times exchange rates for provision to specific lab areas requiring special treatment (Figure 12.6).

Figure 12.6 Supply air pattern.

In the contemporary service ceiling lab, the specific ventilation component delivers intake air evenly throughout the light, utilizing a patent pending process, and it is diffused into the lab space through a proprietary outlet pattern in the light foil. This consistent and even distribution of airflow throughout the lab achieves the positive result of sustainable air ventilation. Two counter-rotating air circulation patterns with high ascending force on the sides are generated. These ascending forces, produced from the induced airflow pattern and the hot air emitted from electrical equipment on the benches, transport potentially harmful, leaking gases toward the ceiling to be exhausted with all other hot air through the extract air system. In every instance, we recommend using computational fluid dynamics to simulate and calculate properly the room climate because even small changes with respect to equipment positioning or process parameters may cause a significant alteration of the lab climate.

The combination of our unique air intake and distribution system, combined with our air extract process above the grid, allows for an even airflow throughout the space, with very low draft risk and much lower exposition to toxic emissions. Additionally, a common problem of improper operation of fume hood exhaust is completely eliminated. Fume hoods can operate efficiently and safely at any location throughout the lab without risk of reverse drafts caused by typical diffusing designs. This allows for an easy reduction, addition, or relocation of fume hoods within the lab space. The complete air ventilation component incorporated within the service ceiling lab allows for the reduction of harmful particles in the air while saving significant energy consumption.

12.3.2 Lighting

New LED lighting system comes in two different designs, ECO and diamond, and represents the most advanced energy saving, multifaceted, lighting source available for laboratory utilization today (Figure 12.7).

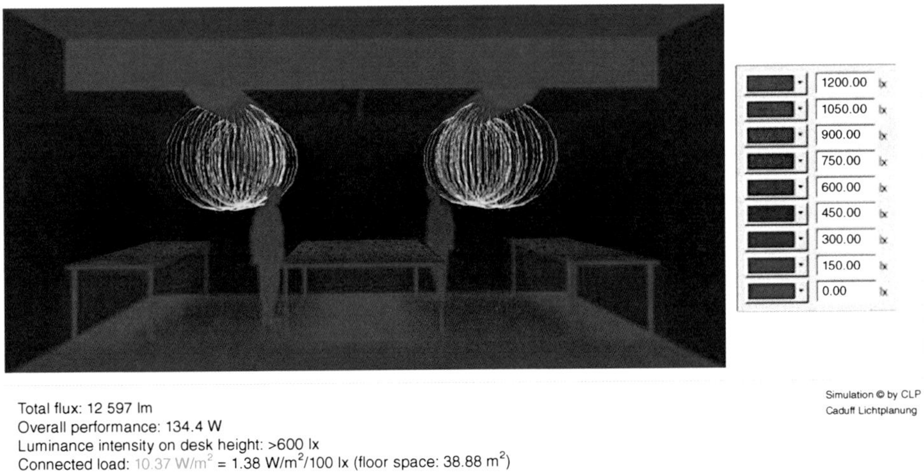

Figure 12.7 Illuminance.

It produces characteristics of almost a daylight environment in a practically shadow-less workplace. It is easily adjustable with respect to intensity and color, and remarkably simple to maintain and clean. The light foil quickly unclicks from the grid and provides easy access to lighting sources and hangs vertically for steaming if necessary.

12.3.3 Other Services

All other services such as water, electric, data, communication, and gases are completely designed and integrated into the system so there is absolutely no interference or complications during construction. All these services are delivered through connecting blocks, down through the columns to the workstations. Each connecting block and column work independently and therefore, when modified in any way, require no disruption to any other lab operation.

12.3.4 Prefabrication and Installation of Service Ceiling

The highly engineered service ceiling solution is rigorously designed in 3D CAD software, developed throughout the past 15 years to deliver the highest of precision and integration of the complete lab infrastructure system. Heavy-duty anchors are fastened into the ceiling structure of the core building construction to which the service ceiling is later attached. It is constructed at chest height in the open and clean lab space by first assembling the aluminum grid structure and then positioning all service components. Once assembled, special lifting jacks

Figure 12.8 Assembling of grid.

hoist the system into place utilizing laser leveling techniques to secure in the previously installed anchors.

One of the most important features is the lack of any conflicts between services competing for the same space. All components are completely integrated and any adjustment to the design and layout automatically insures the error-free routing of all services. The open grid system and the coordinated layout of all services not only provide easy access for maintenance and reconfiguration but also significantly reduces ongoing laboratory operating costs (Figure 12.8).

The high precision design allows, due to complete parallel installation of all trades a team of six specialists, to install 500 m^2 of lab space, in 1 week. That is to say, given a clean empty room of 500 m^2, the system can be installed and the lab infrastructure could be completely operable in 1 week. The installation process alone is a considerable savings in the construction process because it is not only efficient and quick but it also allows for late changes in the design without significant redesign work. Further, it is the nature of this design that lends itself to significant reduction in reconfiguration costs and operating lab downtime during any repositioning initiative.

12.3.5 3D CAD Design versus 2D Planning

The proprietary design of prefabricated service ceiling systems utilizes a modular concept. All component subsystems and services (fresh and extract air, power, data, communication, water, etc.) are incorporated and integrated within the 3D

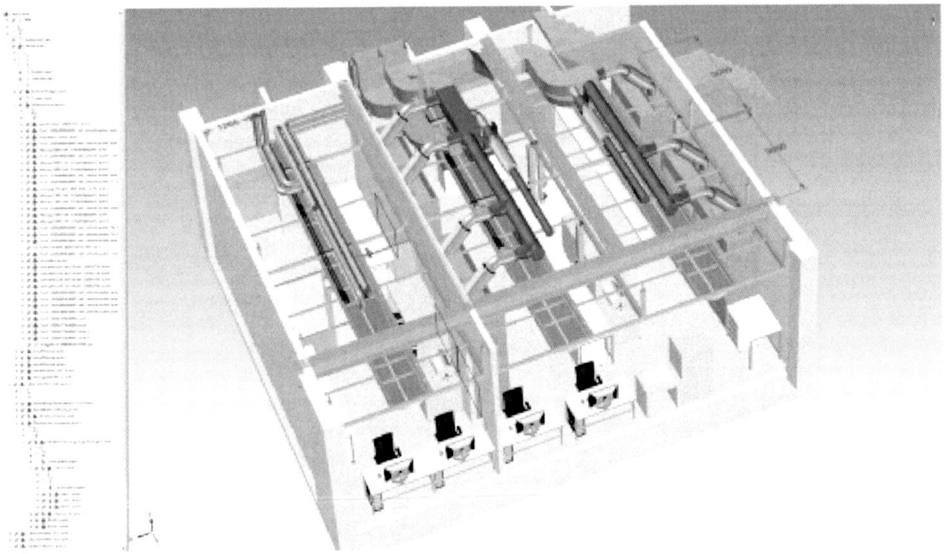

Figure 12.9 3D file.

CAD design as parameterized modules. The highly advanced 3D CAD software package, developed over the past decade, simultaneously integrates all component subsystems without conflict, generates bills of material, and communicates with vendor ERP systems. Designing within a 3D framework allows for the easy review of even the most intricate detailed parts as well as assuring the unobstructed flow of all elements throughout the complete design (Figure 12.9).

The 3D CAD files are produced as a 3D PDF and presented to the customer for review and approval. The releases can be viewed with the help of a free Adobe viewer. This allows entire laboratories or complicated buildings to be visualized in detail without using CAD workstations. The system simultaneously generates all required 2D views and perspectives. All displays and data can be printed out or provided to involved entities as 2D AUTOCAD files without requiring 3D software. Once a master grid has been designed, the adjacent grids are easily linked with replicable applications of specialized techniques within the 3D CAD internal software package. Any previously generated 2D views, when modifications occur, are automatically updated. This seamless reproducible process allows for easy change and adjustment at any stage of the project.

This simulated engineering/design construction system is successful if and only if, the following conditions are met:

- All component parts are accurately measured and uploaded as 3D elements.
- Special parts are measured as precise interference contours.
- Collision analysis is completed resulting in no interference.
- All component parts must be designed to be mountable.

- Assembled subsystems are easily accessible for maintenance.
- The technical performance of all component parts is known in advance.

When the complete 3D CAD files are finished and available, this process eliminates the standard, highly iterative coordination that is usually required on the job site for constant modifications. The service ceiling can be designed at the initial time of the building design and given its complete flexibility, can be easily adjusted at the time of the final building design as long as basic building infrastructure remains the same.

The first truly sustainable lab, providing owners a significantly lower cost of ownership due to its unique design and technical superiority, additionally provides a residual value producing the best economic solution available.

13
Laboratory Logistics

Ines Merten

Efficient logistics, just-in-time deliveries, universal availability – the catchphrases of modern enterprises – are increasingly being mentioned in the context of laboratory buildings. Similar to the situation in hospitals, for example, laboratory buildings are also places where a large quantity of goods is turned over. On the one hand, there is supply, with consumable material, chemicals, solvents, special media, and sterile goods – also office supplies, lab coats, and post – on the other hand, the various waste materials have to be disposed of and, in some cases, autoclaved; glass equipment must be transported to the laboratory rinsing room and lab coats must be transported to the laundry. The goods turned over in hospitals are well recorded in terms of quantities and their point of need; therefore, the inventory management systems can be created in an easy manner. In the field of laboratories, however, we find a far higher diversity of goods and, particularly at universities and institutes, the defining of regular need turns out to be very difficult. One more task is the observance of statutory provisions for hazardous substances and the handling of cooled and frozen goods. All these aspects have to be taken into account by the supply and waste logistics of the building and the respective distribution and storage systems.

Sustainable lab concepts are supposed to have prompt supply and disposal units, and decentralized storage should be reduced to a minimum – especially in the valuable areas exposed to light.

13.1
Classic Systems

Logistical problems are seldom considered during the operation and planning of most lab buildings. The procurement of large quantities of goods is carried out at regular intervals; the goods are stocked in central storage facilities in the distribution centers in the basement. The distribution in the building is rarely performed in a regulated way and often by the lab technicians themselves. In order to provide for shorter distances to be covered, the lab technicians create extra interim storage facilities between the floors and in the lab rooms at the

The Sustainable Laboratory Handbook: Design, Equipment, Operation, First Edition.
Edited by Egbert Dittrich.

workplace, storage facilities, which are planned to a higher or lower degree. Material distribution is performed in the usual way, resorting to classic elevators – a very staff-intensive way of distribution, which requires a large number of (goods) elevators approved for the transport of chemicals. Storage of the goods on the floors reduces the amount of valuable areas exposed to light. Storage in the lab causes overstocking and leads to confusion about existing goods; this in turn leads to a situation where the lack of urgently needed goods might be detected too late. This does not only apply to durable consumer goods but also to nondurable consumer goods. Glass goods are being stored by the individual laboratories in large quantities and washed and rinsed by the lab technicians themselves. Often, there is a lack of suitable rinsing machines and drying areas so that uniform hygiene standards cannot be met and valuable laboratory workspace gets lost.

13.1.1 Drawbacks of Classic Systems

- Quantities of goods ordered is too large
- Central storage facilities
- Unplanned extra storage facilities/storage in the lab
- Valuable lab space gets lost
- Overstocking, lacking control of inventory
- Hardly any registration of goods
- Several elevators required
- Staff-intensive distribution in the building
- Lab technicians burdened with "extra jobs"
- Lack of hygiene standards.

13.2 Centralization and Implementation of Logistics Systems in the Building

13.2.1 Centralization

The initial step in the implementation of logistics systems is mostly carried out via the centralization of individual areas. In lab buildings, this mainly applies to laboratory rinsing rooms, autoclaves, and media kitchens. The centralization enables the concentration of the distribution in the building, for example, by means of automatic or semi-automatic material flow systems, and thus, helps to save on resources with prompt supply guaranteed at the same time. The following example depicts the completed installation of a universal logistics system in the building created through the reduction and centralization of rinsing rooms.

13.2.2 Vertical Linking of Several Laboratory Rinsing Rooms

The 16 rinsing rooms existing previously had been reduced to 3 under the restoration of a research center. This means that – next to the horizontal distribution – a vertical distribution has also become necessary. The infrastructure given, that is to say the number of elevators available, would have caused an unacceptable situation regarding the movement of goods. Due to the structure of the building, it was not possible to install one elevator to be used exclusively for the rinsing rooms. Now, the solution has been found by using an automatic transportation system connecting the rinsing rooms lying above each other. All floors are given a room for loading used for the deposit and collection of the goods to be rinsed. The automatic transportation system represents a significant enhancement compared to other solutions (elevators). Time lost due to waiting in front of the elevators is reduced to a minimum and there is the possibility of establishing controlled material flow. Automatic or semi-automatic material flow systems require less space, enable prompt supply, and are more efficient in terms of staff deployment (Figure 13.1).

13.2.3 Material Flow Systems

The example of the rinsing rooms displays the optimization of the vertical transport through a material flow system. In order to get an overview of the goods stored so that storage areas can be reduced to a minimum, the system has to be expanded. It is possible to record the goods and to follow their stay in the building by means of central delivery with subsequent consignment. In addition, directed buffer areas provide for an even work load for the staff members through the absorbing of peak loads (Figure 13.2).

13.2.3.1 Consignment and Concentration of the Flow of Goods

Several lab modules are connected through a vertical and horizontal material transportation system in the new construction of a pharmaceutical company. Samples, consumable material and chemicals are delivered and handed over to the central consignment department; they are recorded and prepared corresponding to the target departments. Shipment to the lab modules via a transport system is carried out in uniform carrier boxes. Should time buffering of the samples or goods turn out to be required, there is the possibility of storing transport containers in the buffer storage facility (high rack storage area with lifting bar delivery device) and, when the need arises, to dispatch them to the appropriate place.

Using the transport system, all carrier boxes containing waste and recycling material are also transported to a waste buffer storage facility where they are disposed of by the logistics personnel as necessary by dumping them into the waste containers available after correctly sorting them (Figure 13.3).

Figure 13.1 Loading room beside scullery.

The material flow system works to capacity in a more controlled way, load peaks can be reduced and a higher priority can be assigned to urgent shipments due to the buffer areas. The staff is given the possibility of performing irregular tasks during down times. The concentration of the commodity flow and the pre-consignment of the goods provide continuous product recording. What is more, the degree of stockpiling can be reduced to a minimum (Figure 13.4).

13.3 Consignment and Automatic Storage Facilities

The use of a semi-automatic consignment automat for consumables can be a worthwhile supplement for time-efficient consignment; this may even be a required addition in very large lab buildings. Hazardous substances and cooled materials can be manually supplemented into the consignment boxes. The entire

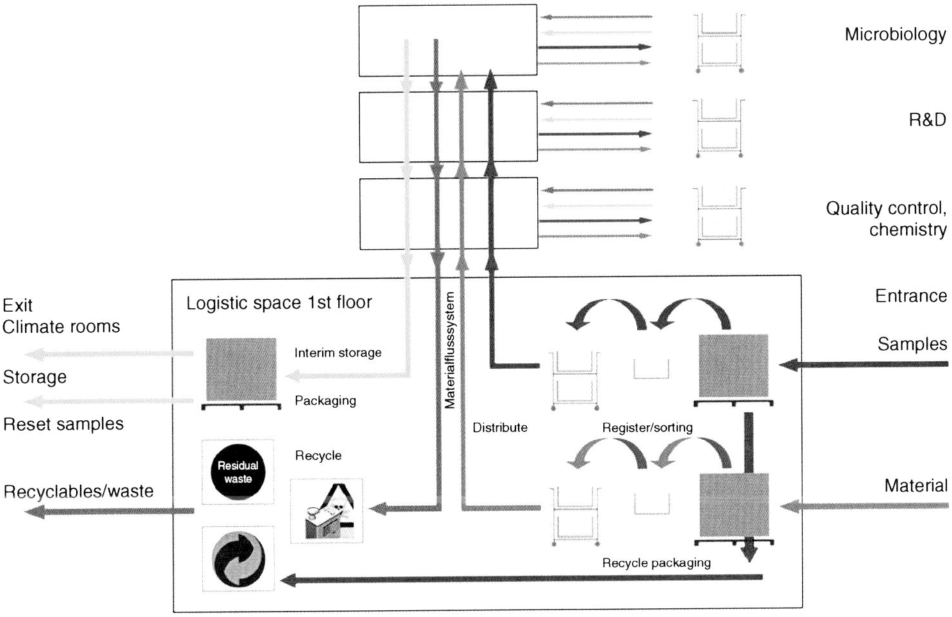

Figure 13.2 Logistics scheme.

Figure 13.3 Delivery of goods within system.

transport of the goods is, thus, carried out in standardized carrier boxes, which can be identified via barcodes or are labeled with another system. The address on the carrier box makes sure that the material is, for example, allocated to the respective work group, which has ordered this box, perhaps via intranet, according to the planning of the experiment. Dispatch for disposal or of goods to be rinsed can be easily performed via programs, which are simple to use.

Automatic container stores can be incorporated into the consignment chain. Goods delivered are unpacked and arranged into the corresponding transport boxes. The entire storage takes place within these transport boxes. An automatic

Figure 13.4 Delivery-detail.

storage and retrieval machine provides for the storage and retrieval of the goods. The storage boxes are promptly made available for consignment by means of conveyor belts and a material transport system. The goods are then consigned to become single orders. The transport back to the container storage facility is effected via conveyor belts again.

13.4 Solvents – Supply and Disposal Systems

Centralization, storage, staff efficiency, and registration of goods are essential topics in the discussion of logistical contexts. A lesser importance is attached to the topic of safety in the storage of hazardous substances and, above all, to the transport of goods within the building. However, the safety of staff members comes first, particularly in lab buildings. Various pharmaceutical companies serve as an example of how not only the staff members' safety can be enhanced through the automation of processes and the concentration of tasks, but also how the staff members can be relieved of various tasks at the same time.

13.4.1 Solvents Disposal Systems in the Pharmaceutical Sector

We find the installation of solvent disposal systems, allowing the collection of solvent waste through the in-house cleaning service, in all new modern lab buildings in the pharmaceutical sector. Possible solvent contact between the employee and filling activities are reduced to a minimum. The laboratories are equipped with safety cabinets for hazardous substances, including collection containers, which automatically store the waste solvents of, for example, HPLC systems. The containers are equipped with fill level control and overfill protection. The filled containers are emptied by the in-house cleaning service by means of a pump, which is attached to an emptying vehicle with a transport container. In order to enable this,

Figure 13.5 Ancient solvent disposal.

a hose connection is created via quick coupling connections. Therefore, the staff members no longer have to resort to the frequently occurring and laborious filling from one container into the other by hand. Similarly, direct contact with solvents no longer occurs. The decanted solvent waste is then transported to the site of an intermediate bulk container, where the waste is pumped from one container into the other in the same way. The contents of nonstationary containers can also be manually extracted by suction using a suction lance. Similar systems enable the emptying and transport of radioactive wastewater, acids, and bases (Figure 13.5).

Concerning the supply of solvents, the kinds of systems should be installed which allow the smallest feasible number of possible contacts for the staff members. Small quantities are filled by pumping from one barrel to another in

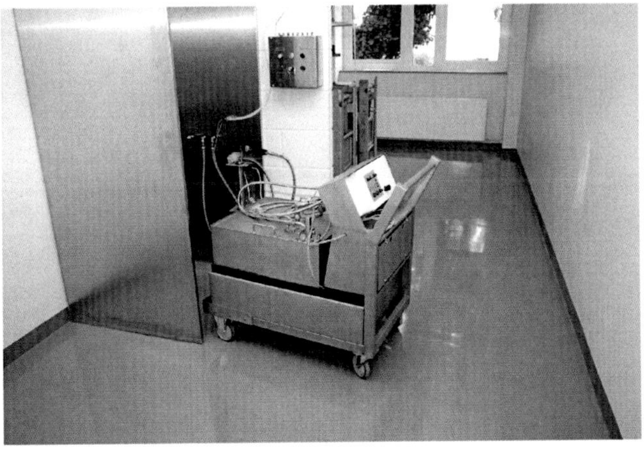

Figure 13.6 Solvant change.

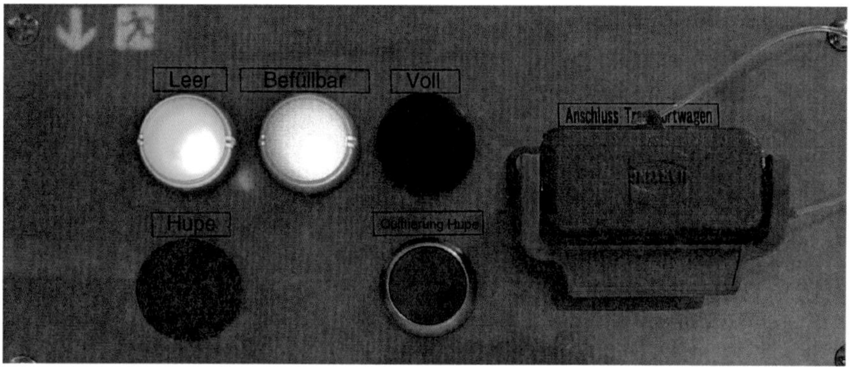

Figure 13.7 Solvant change.

the central supply system; these small quantities will then be distributed to the laboratory rooms. It is possible to install tapping points within individual lab hoods in order to satisfy the needs of increased quantities directly at the point of consumption; the tapping points are supplied via storage containers to be found in the adjoining safety cabinets (Figures 13.6 and 13.7).

13.5
Laboratory Work 2030 – Objective?

> Full of energy, the employee enters the modern lab building in the morning. He finds a fresh, personalised lab coat in his locker right next to a steaming pot of coffee – made just the way he likes it – and the number of today's workplace. All materials required for today's tasks are ready for use at the tidy workplace. Thanks to the efficient supply and disposal of hazardous substances, it is possible to reduce the lab's ventilation to a minimum. The display reports the delivery of a sample to be delivered personally. Our employee makes use of the interruption for a conversation with his colleagues in Australia who are just about to finish their day's work …

This or a similar scenario might be what work looks like in large lab buildings in the near future. However, this side note also shows that the implementation of various logistics systems into the daily routine of modern laboratory buildings does not only require the staff members to start to rethink, but also require a change in the existing working methods. Future projects and the respective material requirements have to be planned in advance in a targeted way, and there is the growing risk that staff members will be burdened with laborious administrative tasks. Personal "stockpiling" through minimization of workspace is no longer possible. Special requirements are hard to implement into standardized processes. The use of automatic systems results in the standardization of the products; for instance, only the pipettes of a single manufacturer are used and personal preferences are not taken into account. Regardless of all current visions, we should never forget that, above all, we have humans working, researching, and developing things in the laboratories. It is for and with these staff members that successful complete systems have to function and be implemented in order to enhance the company's gain in all fields.

13.6 From Small Areas to the Big Picture

A lab technician assesses the efficiency of a total system in the small area of his workspace. The catchphrases are short distances to cover, clear supply and disposal facilities, and rapid availability of the materials.

13.7 Local Transport Systems

Lab trolleys, comprised of a steel frame and two transport boxes, which are inserted on top of each other, can be used in many ways. Regarding the size, the trolley is aligned on a base frame so that the trolley can be stored easily under a laboratory bench and, thus, does not block the aisles and can, nevertheless, be taken to any lab area by the lab technician. The lab technician assembles the materials required in the box and transports it to his workplace. The transport boxes additionally serve as collecting trays, for example, in the transport of the chemicals to the laboratory. After the experiment, the dirty goods to be rinsed can be collected in the trolley and transported to the central rinsing room in the evening. The boxes are suitable for transport systems and can also be easily cleaned in the dishwashing machine (Figures 13.8 and 13.9).

Central points are equipped with similar trolleys for the collection of recyclable materials. These trolleys will also be with placed under the lab benches or in frequently empty spots in the building, which cannot be used in other ways. Bins of different colors make sure that materials are allocated in an unambiguous manner.

In a similar way, the niches in the entrance area can be used for the establishment of lab coat lockers with wastebaskets.

13.8 Supply and Disposal of Chemicals at the Workplace

One possibility for the collection of hazardous substances is the installation of automated disposal systems. Laboratories, however, deal with a large number of different types of solvents and hazardous substances – all requiring separate collection. A chemical hood is the most important workplace to deal with hazardous substances; this is the place where the main portion of the work with solvents is carried out. A useful combination of safety cabinets and lab hoods

Figure 13.8 Lab trolley under-bench-unit.

creates hazardous substances centers in the labs. It is always best to have the laboratory hoods standing some distance from the entrance doors in order to avoid disturbances in the air flow. The free space is used to establish a hazardous substances safety cabinet for the supply of solvents. It is also possible to provide larger quantities, which supply the tapping points within the lab hood via barrels. The space under the chemical hood is used to establish hazardous substances safety cabinets for the disposal of solvents. Similar setups for the supply and disposal of acids and bases are established if required. The automated collection of waste solvents directly at the machine-equipped workplaces has already been referred to earlier.

13.9 Perspective

Dealing with logistical issues in laboratory areas is still a very young and fresh discipline. These issues are often detected for the first time through the centralization of selected areas. Material flow systems for vertical and horizontal transport

Figure 13.9 Lab trolleys and boxes in park position.

increasingly put things right in this field. The reception of goods is enhanced by consignment systems and storage facilities are equipped with automatic operating devices. The safety level of staff members has increased and they are relieved of additional tasks by the installation of automatic supply and disposal systems. In spite of all this support these systems can offer, we must not forget the people and their daily work.

The task of the future will be to continue to implement those kinds of systems which support the staff members, reduce the distances covered and the costs to a minimum without restricting the open minds of the researchers through excessive standardization, and thus, destroying the time won by burdening the staff with additional administrative tasks. This objective is often achieved by simple, cost-effective solutions, which are incorporated into the total system in a sensible manner.

14
Animal Housing

Ina-Maria Müller-Stahn

Biological sciences or bioscience (biological, physiological, or medical research) is a research direction gaining more and more importance in the field of research. Wikipedia defines the term *Biological Sciences* (Greek βιος (bios) = "Life"), also called *life sciences*, as those research directions and training courses dealing with the processes or structures of life-forms, or processes or structures with which the life-forms are involved. In addition to biology, bioscience also comprises related fields, such as medicine, biomedicine, biochemistry, molecular biology, biophysics, bioinformatics, or biodiversity research.

The spectrum of methods in bioscience ranges from the classic molecular-biological procedures to the entire range of chemical analytical procedures. However, not all the questions which arise can be answered *in vitro* by experimental/analytical procedures; essential basic principles, for example, of biochemical processes in the body or the metabolism of specific substances, can only be researched *in vivo* by animal testing due to the direct reference to the life-form. The management of test animals, in particular of small rodents such as mice and rats, is mandatory for such testing methods. The management of special animals, such as fish and insects, and of larger animals, such as dogs and primates, requires separate descriptions which, due to lack of space, cannot be given here.

14.1
General Points

During the past few years, significant efforts have been made in the management of test animals in order to guarantee animal welfare, on the one hand, and to reduce the number of animal experiments, on the other hand. The resultant requirements have raised the standards of animal husbandry.

The standardization of the holding conditions of animals serves to keep the biological variance at a level as low as possible. This makes sure that the number of experiments required to achieve statistically meaningful results is reduced. In spite of these standardization efforts made for the sake of animal-friendly

The Sustainable Laboratory Handbook: Design, Equipment, Operation, First Edition.
Edited by Egbert Dittrich.

husbandry, the management of animals is put into effect along with materials for nest-building and occupation (enrichment), even if this might result in the biological variance increasing again.

The animals are kept under different types of hygiene levels depending on the requirements of the experiments; the hygiene levels most often used are as follows:

- SPF (specific pathogen free),
- Conventional management, and
- Quarantine.

The requirements existing for these hygiene zones will be briefly explained in the following.

14.2
Planning of an Animal Facility

Although there are big differences regarding the requirements at various hygiene levels, there is some common ground regarding the planning and establishment of animal husbandry. The structured design of the rooms, allowing for nonintersecting logistics of clean and dirty material to the highest possible degree, is the basis for any proper kind of animal husbandry. The ideal case would be a cycle (depicted in Figure 14.1), making sure that hygiene standards can be easily met. The zone of animal husbandry is accessed via a reception area where street clothes are exchanged for special zone protective wear. These special clothes allow people to access rooms on the other side of the *hygiene barriers.* Passing a further checkpoint for both personnel and material is required in order to get into the next, higher hygiene zone (see Figure 14.2).

Clean material is introduced into the animal facility via the barrier, whereas dirty material is passed on to wash-up area/cage processing center unit via a transfer lock.

Supply of material is effected via a barrier consisting of a large autoclave, walk-in decontamination lock, and material pass-through. Staff members gain access from the surrounding zone through the barrier via another lock. Basically, an airlock can be equipped with a swing-over bench, an air shower or a compulsory shower, or it can be designed in the form of a two-chamber air-lock.

A pressure stage concept makes sure that the rooms are clean.

The rooms also have to be adjusted to the operator's, respectively, the future users', operating methods so that future operation will be supported by spatial geometry and structure. Such basic conditions have to be determined at an early stage of planning because of their impact on the layout and the technology required.

The animal holding rooms have to be developed in a technical and systematic manner: Each room should have a separate connection to the ventilation system so that it is possible to disinfect individual rooms through fumigation with hydrogen peroxide. The connection to the animal rooms via a separate technology corridor

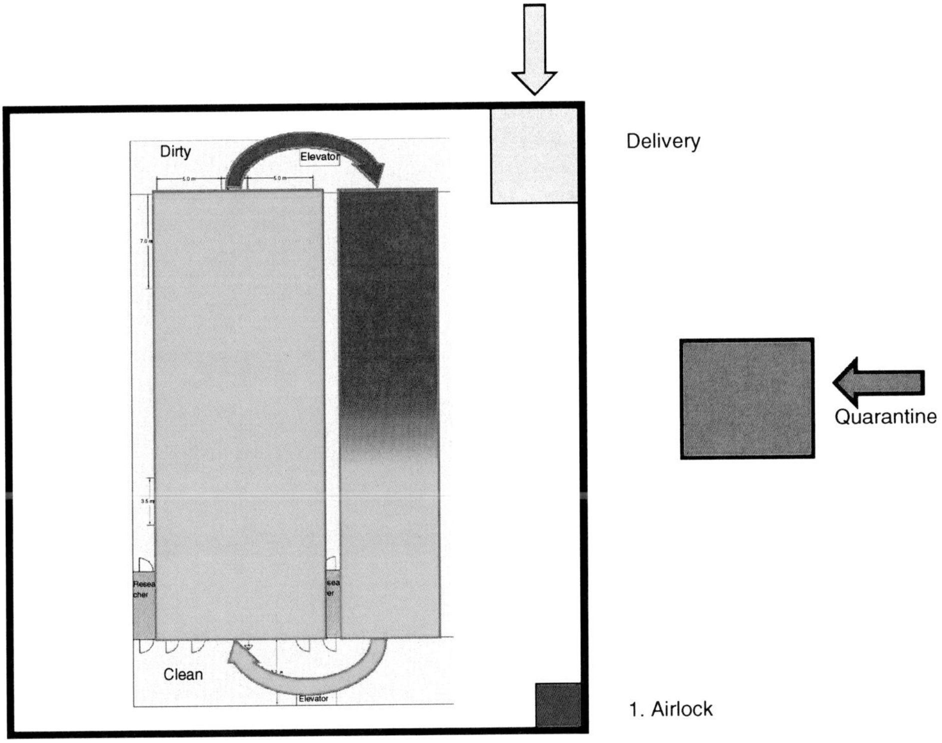

Figure 14.1 Logistics in an animal facility. Principle: establishment of animal husbandry system with surrounding area, delivery, cleaning, and module for animal husbandry. Ideally, the quarantine area is located outside.

is the best possible solution (see Figure 14.3, here with exhaust air filtration) in all areas of the building suitable for this. Thus, only the parts of the technical installation actually required are put into the animal holding rooms because a large portion is installed in the technology corridor. In such a case, maintenance work can also be carried out from the outside. Due to the ventilation connection via the technology corridor, it is possible to carry out fumigation from this corridor as well, so that the other rooms can be continuously operated without any disturbances occurring (Figure 14.4).

14.3 SPF Management of Animals

Animals with SPF status have to be kept within a barrier in order to achieve and maintain the hygiene status. The barrier is comprised of airlocks for people and material airlocks, which seal off the animal facility from the surrounding area in an airtight way. The introduction of material is executed via a sterilization stage

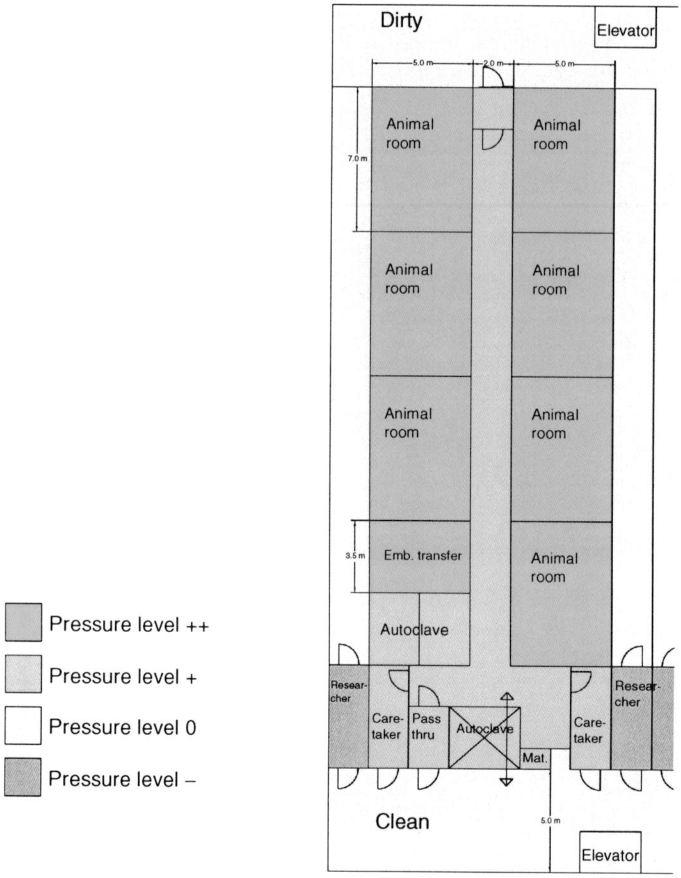

Figure 14.2 Schematic structure of small rodents facility.

(autoclave) or via disinfection, for example, through fumigation with hydrogen peroxide or peracetic acid (PAA).

Animal caretakers and, if the need arises, scientific personnel are introduced to the SPF area via locker rooms. Coordination with the future operator prior to planning is essential because the designated/required procedure of changing clothes must be reflected in these rooms.

The compulsory showers formerly used – staff members had to undergo a shower program via a shower lock running for a predefined period of time before being allowed to access the barrier – are only implemented in individual cases today, because of the skin problems arising due to the frequent showers having to be taken by the employees. The employees suffer from a higher susceptibility to skin diseases and the shower procedure is not generally well received by the employees. Therefore, in today's projects, staff members are required rather to completely change their clothes. They put on area-specific wear (Figure 14.5) and, after swinging over via the swing-over bench, the staff members put on special

Figure 14.3 Technique corridor for individually linked rooms with exhaust air filtration.

Figure 14.4 Barrier consisting of walk-in decontamination airlock, drive-through autoclave, and fumigation material pass-through (not in the picture: access to airlocks for occupants).

barrier wear, including the respective shoes. Furthermore, staff members have to wear a face mask (also covering the beard if necessary), hair-cover, and gloves.

Air showers (see Figure 14.6) are increasingly being used to replace compulsory showers. Air showers were first used in physical and pharmaceutical clean rooms in order to remove particles and, thus, the corresponding pathogens clinging to the clothes. The person to be introduced is cleaned with air emanating from the top and from four sides. In order to support the introduction processes, it is possible to position the air showers into the airlock for people in two ways: In an animal

Figure 14.5 Example of area-specific wear. (Source Photo: Pfeiffertextil Company, Switzerland).

Figure 14.6 Example of air shower.

facility, the first option is to establish the air shower into the airlock for people in the installation between the "undressing area" and the "putting on zone-wear area." In this case, the air shower takes over the role of the swing-over bench. Access to the air shower is gained without clothes from the outer compartment of the lock; after the appropriate time under the air shower, the inner part of the lock is entered and the zone-wear is put on. A second option is the installation at the inner side of the animal facility, where the employee enters the lock after completely changing clothes, including putting on zone-wear; prior to accessing the animal facility, and then the employee is "cleaned" by "showering" under the strong airflow.

The positive effects of air showers in clean room areas is the result of the zone-wear used in these areas: A special material with a smooth surface is used (e.g., TYVEK®, a paper fleece-like substance made of thermally sealed polyethylene fiber). Particles can be blown from this material. Cotton is a clothing material frequently used in animal facilities (trousers with smock overall/T-shirt). Cotton can be easily cleaned and it can be autoclaved; it also provides a high measure of wearing comfort to the staff members. Currently, there is no literature available regarding studies on the cleaning of woven cotton clothing. It is even assumed that the hard-pressing of particles into the fabric is one possible effect of cotton; an effect which stops the release of particles. Another possible positive effect under discussion is that members of staff perceive the passing of the lock along with the blowing of their body in a psychological way: They perceive that they are passing a boundary which then results in the changed behavior of staff members.

14.4
Animal Management under SPF Status

On principle, the holding and breeding of animals under SPF status can be implemented in two ways. Either the animals can be kept in secure systems fixed to the wall or in mobile systems with

- cages opening in the direction of the room (conventional holding of animals) or
- individually ventilated cages (IVCs).

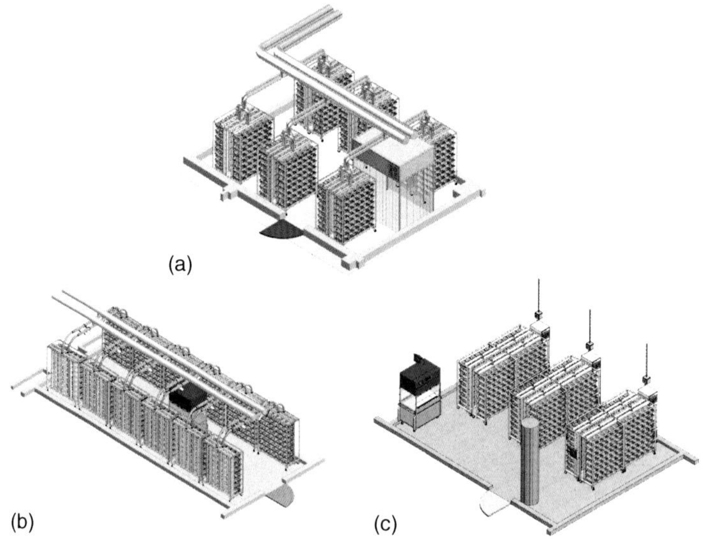

Figure 14.7 (a) Centrally connected double IVC racks with laminar flow module for the transfer of the animals. (b) Centrally connected, single-sided IVC racks placed against the wall. An animal transfer station exists in the central corridor. (c) Double-sided IVC racks with ventilation unit.

The room serves as the barrier under the open management of animals; transfer of animals is effected under a transfer station or a laminar flow module. This type of management is only to be implemented for noninfectious animals. In this case, the respective hygiene requirements for the rooms and the room conditions (air supply) have to be observed in order to protect the animals against infections from the outside.

On the other hand, it is the cage which serves as the barrier when keeping animals in IVC cages. On principle, the requirements in the animal rooms could be reduced in this case. However, the holding rooms for animals usually meet the same requirements (e.g., regarding size) – irrespective of the method of keeping the animals.

Supply of the cages or of the rack the cages are connected to can be ensured in a central manner (see Figure 14.7a,b) by means of a room ventilation system or in a decentralized way by a connected ventilation unit with air supply (see Figure 14.7c).

14.5 Decentralized Connection of IVC

The room air supply using this method of animal husbandry is generally conducted via a HEPA filter type H14, and is, thus, sterile-filtered. Transfer of the animals from a used cage to a cage with fresh litter is carried out using changing stations or laminar flow units, so that there is no risk of contamination at this spot. The material made ready for transfer is piled and stored on transport trolleys in the corridors and transferred to the animal rooms. It is also worth conducting the air supply to the cages; a filter is used in order to avoid contamination of the cages through the air in the room.

14.6 Central Connection

In this case, the IVCs are connected to a central air supply and the racks are supplied equally via a delivery plenum. The air supply is conducted via a HEPA filter type H14 so that there is no contamination from the outside.

Establishment of the ventilation units required for decentralized connection is no longer necessary in the animal holding room, which results in a lower size of area required because the space for the fan units' footprints can be saved.

14.7 Extract Air

As there are allergens (e.g., animal hair and dust) in the extract air of the animal cages, it is disposed of via a central connection for reasons of work safety. One further aspect of this is the reduction of the odor in the animal rooms/the animal facility. If necessary, the exhaust air can be treated again. In the case of infectious animals, for instance, it is possible to conduct the exhaust air over sterile filters. Alternatively, the exhaust air can be conducted over activated carbon filters or a bio-filter in order to reduce the odor nuisance, and then disposed of via the roof.

14.8 Supply through the Barrier

The supply of water and food to the animals must comply with the existing hygiene requirements. Water for the drinking trough is either autoclaved in bottles or is sterile-filtered. Since the autoclaving of water consumes a lot of time, time actually needed for the autoclaving of other material, the sterile-filtration of water via filter cartridges in a designated supply room in the barrier is common practice today.

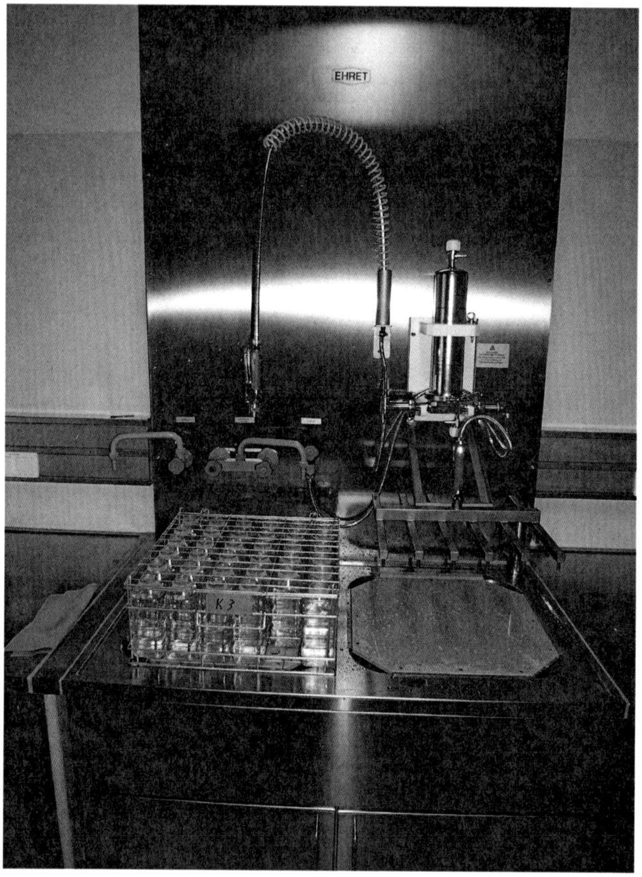

Figure 14.8 Bottle filling with single and multiple filling ("rake").

Water is often acidified in order to achieve a better shelf life of the water. Filling is performed by means of a filling screen; several bottles are filled simultaneously in a bottle case (see Figure 14.8).

Food can be passed onto the rack prior to autoclaving; or food itself – packed in bags – is autoclaved or fumigated and introduced through the barrier. Dosing of the food inside the barrier enables more exact adjustment of the amount of food required for the respective number of animals in the cages so that wastage of food will be eventually reduced.

14.9
Quarantine

In the true sense of the word, quarantine does not represent an independent type of animal management for breeding or experimental purposes. Quarantine rather serves as a stop-over station for animal phyla, which are kept prior to their introduction into a barrier animal facility.

After checking their health status, they are introduced into the barrier-protected animal unit – either directly or via embryo transfer. Cages and material from the quarantine system are autoclaved prior to cleaning and, thus, prior to contact with the other material for animal management.

The determination of the position in the building, particularly for this area, is, therefore, very important. Outsourcing into a separate building, which may even have logistics of its own, can be an appropriate solution for larger animal facilities or in the case of a research campus.

14.10
Open Animal Management without Hygiene Requirements

Under this type of animal management, the animals, which require a low level of hygiene requirements, are mostly kept in open cages. Supply of the animals is put into effect in the same way as in the other hygiene levels, with no autoclaving required as a necessary step for introduction into the animal holding unit.

As a rule, the animal caretakers and the scientists gain access to the area through a changing room where street clothes are exchanged for the common, functional area-specific wear, made of cotton, and area-specific shoes.

Drinking water can either be filled in the wash-up area or, similar to barrier keeping, in a corresponding supply room in the animal facility. Water can be treated as and when required prior to filling, for example, water can be acidified in order to enhance shelf life. Food, bedding, and water are normally not autoclaved, that is, the cages are filled with bedding in the wash-up area and transported into the animal facility by means of transport trolleys. In the animal room – prior to the mouse transfer (from dirty to clean cage) – the amount of food, which is adjusted to the occupancy rate of the cage, is placed in the food tray of the cage.

14.11
Experimental Animal Facility

Experimental animal facility does not represent an independent hygiene level but, as a rule, depends on the requirements and the status of the animal management system allocated. As an example, there is no hygiene standard required for some kinds of experiments; in addition, the contamination of the animals cannot occur or contamination is not relevant due to the short duration of some experiments.

It is often the task of experimental animal management to provide animals for research and diagnostic tests, animals which have been bred for this purpose in an SPF animal management system. Breeding facilities should only be entered by a limited number of people in order to protect valuable animal phyla. Scientists, however, must be given the chance to carry out the experiments themselves and also to observe the animals in the experiments; this is why they need to have access to the animal management unit. Thus, the animals in an experiment are kept in a separate unit, which must meet the identical hygiene requirements as exist in the breeding area, and with the same requirements regarding locking processes. SPF barriers and experimental barriers differ in terms of the larger number of people having access to this area. As humans or human errors represent the biggest risk of contamination, the separation of breeding and experiment makes sure that healthily bred animals will be available for remediation in case of contamination.

14.12
Sustainability – An Issue in an Animal Facility?

Due to the stipulated standard room conditions, animal management facilities require large air change rates. A large amount of energy has to be used in order to achieve and maintain the correct temperature and humidity (heating/cooling/dehumidifying/humidifying of the air supply). Large amounts of steam are required for washers and autoclaves to clean the cages and materials used, as well as for heat activation in the autoclave. Special zone-wear mandatory in the hygiene areas must be cleaned and autoclaved more often than lab wear – usually after single use.

The use of energy-efficient equipment can help to establish resource-conserving operations for the machine-aided processes of cleaning and heat activation. The manufacturers of such equipment should be motivated to develop energy-efficient equipment through the respective demand in the market. Specific measures help to conserve resources when in operation:

- Frequency of Change of Cage
- Degree of Cage Dryness after Cleaning/Drying (e.g., use of only one drying stage).

However, the greatest potential for sustainable animal husbandry is the sensible planning and practical implementation: Through the development of functional modules, which implement the user requirements, the cubage of the animal facility is limited to the necessary minimum.

The planning of a central connection reduces the cubage as well, because utility space for ventilation units will not then be required. Smaller room areas require smaller quantities of air. The amount of ventilation required can be considerably reduced by means of the central connection, because the harmful substances from the cages can be discharged by direct air exchange.

The controversial discussions on animal management in decentralized connection *versus* central connection IVC attest to reduced flexibility of the central connection. However, since every modern animal husbandry unit connects the IVC racks to the exhaust air, the installation planned previously has to be adhered to as well, which means it is similarly "inflexible" due to this fixed installation.

Another benefit is the possible visualization of pressure and rack monitoring, which is related to the animal rooms via the central building control system, a possibility, in principle, also existing for the monitoring of decentralized connections; this, however, would result in higher installation costs and in the need for corresponding software, which may even have to be customized to the specific issues in some cases.

The advantages of central connections are as follows:

- Saving of Space in the Animal Room
- Filter Exchange outside of the Area of the Animal Management Facility
- Redundant Operation of Racks through Room Ventilation System
- Pressure Control and Independent Volume Monitoring
- Single Aerating with H_2O_2
- Separation of Low-odor Room Exhaust Air from Olfactory Polluted Exhaust Air
- Reduced Operating Costs
- Reduction of Air Quantity: approximately 20%
- Reduction of Cooling Power Required: up to 55%
- It is even possible to enhance this energy balance by the setting of different temperatures concerning the air supply to rooms and cages.

(Source: U. Feldmeier; Lecture on: Energetic Comparison. Lecture held at Cluster of Excellence University of Cologne/University Hospital Cologne – Comparison of Central/Decentralized Ventilation of Rooms for Animal Management).

Based on experience, the best option for redevelopment measures, in which the room structure mostly cannot be altered, consists of a decentralized solution. This is due to the lower investment costs occurring (since a ventilation system for the animal rooms already exists); in general, planning and implementation turn out to be less complicated and less time-consuming.

In principle, the question about the size of the animal facility is of significance: each planning should be carried out according to the actual demand. With regard to sustainability, consideration should be given – in cooperation, of course, with the approving authorities – to use incubators in case of smaller units or temporarily high demand for certain experiments.

15
Technical Research Centers – Examples of Highly Sophisticated Laboratory Planning Which Cannot be Schematized

Thomas Lischke and Maike Ring

Technical research centers, in contrast to laboratories, are distinguished by flexible areas which are not bound to a strict laboratory grid. These areas can be designed more freely and are characterized by equipment and processes that are not depicted in a grid.

In contrast to laboratories, reactants and products are handled in a greater volume in technical research centers and all safety aspects must, therefore, be considered to accommodate these dangerous substances and chemicals.

Technical research centers are needed

- for process optimization
- as a pre-stage for the production of smaller batches
- solely for research.

In the planning of technical research centers, the regulations must be clarified first and the laboratory guidelines and other regulations, which must be applied, established. The Federal Immission Control Act requirements, in particular, must be assessed for the approval of a technical research center as certain thresholds may be exceeded due to the volume being handled in such centers.

Technical research center areas may also be functionally joined to production areas: production → technical research center → laboratory or are research-driven laboratories → technical research centers. The first may be viewed more as a downscale for the accompanying examination and optimization of production processes. The second aspect is more research driven and involves large-scale laboratory tests and the conducting of an upscale.

Production-driven technical research centers are usually easier to plan as the requirements of the user can be better defined. With technical research centers which are more research driven, greater flexibility is required and a broad area of applications usually needs to be covered.

In addition, there are also manufacture-driven technical research centers which are, for example, used in the pharmaceuticals industry. In recent years, there has been a trend for offering machine manufacturers, alongside the technical research centers solely relating to technical processes, also those with production and GMP (Good Manufacturing Practice) capacity. These enable pilot tests

The Sustainable Laboratory Handbook: Design, Equipment, Operation, First Edition.
Edited by Egbert Dittrich.

with customers as well as the production of small batches and clinic samples. Both require a greater development standard and outlay for qualification and manufacturing documentation for GMP production.

- Structural boundary between the technical research center and the laboratory
- More flexible grid
- Greater ceiling heights
- Comprehensive media distribution
- No support pillars
- If possible, free spaces (location of shafts)
- Modified escape and emergency routes
- Greater explosion/fire protection requirements
- Possibly also greater requirements due to clean room status, hygiene (e.g., also issues such as impervious facades)
- Door size, transport openings
- Materials handling technology, crane tracks
- Lighting, hall depth
- Development/extension possibilities, for example, floor
- Supporting ability, area load
- Possibly also clean room and GMP requirements.

Furthermore, the usage cycles within the technical research center should also be considered. With regard to the infrastructure, there are also different requirements compared to those of the laboratory.

For example:

- Greater media volume
- Greater media spectrum
- Safety technology, alarm technology
- Cooling/thermal loads
- Waste water volume, waste water treatment
- Contaminated process exhaust fumes
- Flexible media supply systems
- Open installations (with exception of the clean room areas).

In comparison to the laboratory which is strongly characterized by the set-up grid of the laboratory work rows as well as the processes, the technical research center is defined more by the size of the facilities/equipment and processes. The flexibility of the free spaces and corresponding layout of route guidance, escape routes, media arrangement, and spatial division is also oriented to this.

Technical research centers are used in the fields of

- Fermentation technology, biotechnology
- Chemistry, distillation
- Pharmacy, galenics
- Foods, cosmetics, strongly defined by batch facilities
- Physical research with increased requirements in electricity supply, heat, and possibly also radiation protection.

Due to the greater volume being handled, there are also increased requirements in logistics, which differ from those of laboratory operation. Material handling, for example, requires delivery areas, interim storage space which meets regulations, as well as handling aids and this must not be underestimated in normal operation. The use of a crane track is also a meaningful alternative in a technical research center.

Special aspects in technical research centers also include the maintenance and upkeep of the equipment and facilities, and also media feed and safety technology. Furthermore, it can also be meaningful to have remote monitoring, in the form of a control room, for the equipment due to noise development and heat/odor emission, pollution, and safety aspects.

Technical research centers are also characterized by different operating times. Ongoing trials and overnight tests may also be carried out here, and this must be considered in the layout of the equipment and monitoring technology. Such trials and tests may also occur in laboratories although there are increased requirements here due to the greater volume being handled.

In the layout of technical research centers, special attention should also be given to the media supply and estimates should be made, together with the building client, regarding the equipment and required amounts particularly in electricity, cooling water, and media. Particularly important here is the synchronicity to ensure there are no over installations.

Development/extension capacity should be designed more robustly than in laboratories. Here you should allow for more lifting equipment traffic for plant supply/disposal and also plant assembly/disassembly, which means that industrial flooring/tiles are more likely to be used depending on the requirements relating to imperviousness, discharge capacity, and so on.

16
Clean Rooms

Thomas Lischke

In the manufacture of many products today, there are certain requirements concerning low particle loads and/or germ counts to ensure a high degree of quality or fulfill certain laws and regulations, for example, in the manufacture of medicinal and pharmaceutical products.

In such cases, production rooms are designed as clean rooms.

In the pharmaceutical environment, laboratories are needed for quality controls while clean room laboratories are also always required for the microbiological examination. For research and development projects for the manufacture of new medicinal products or for the optimization and development of procedural steps, laboratories are required as clean rooms in order to maintain the same conditions as in the production environment.

Before small batches or hospital samples can be manufactured of technical research center standard, it is necessary to conduct reliable pilot tests in laboratories. As a result of increasing requirements in production, a trend toward clean rooms has been noticed particularly also in laboratories.

Clean rooms can be very diverse and their design is generally orientated toward the requirements they must fulfill through the process or product. Hereby, there are special design characteristics depending on the application. With clean rooms which are used in the fields of microelectronics, optics, materials research, and measurement engineering, particles play a much more significant role than in the pharmaceutical industry or medicinal products where the focus is usually on the germ count.

In addition to the clean room requirements concerning particles and germs in accordance with EN ISO 14644/1 (clean room categories ISO 1 – 9), there are also often further requirements.

In work with biological or genetically modified substances, for example, it is also necessary to observe the respective protection stages in accordance with the German Ordinance on Biological Agents (BioStoffV) and the Genetic Engineering Safety Regulation (GenTSV).

Hereby, the respective requirements may contradict each other. In room ventilation, overpressure is usually built up from the clean room to the other areas.

The Sustainable Laboratory Handbook: Design, Equipment, Operation, First Edition.
Edited by Egbert Dittrich.

However, to protect the environment and employees, it may also be necessary to have low pressure in the room due to the harmful properties of the substances being used. In such a case, a material and personnel lock can be used as a pressure drop or as barrier in the overpressure to the clean room and the adjoining, nonclassified area.

The question of whether a clean room is required or not can usually be answered relatively easily. Either legislature stipulates the use of this as is the case in the pharmaceuticals industry or there are economic reasons which determine this. The products cannot be reproduced otherwise or manufactured in the desired quality.

The clean rooms in laboratories should be designed in accordance with the clean rooms in the associated production.

If the germ count is one of the main criteria, focus should be particularly on air flow as well as good cleanability. Difficult-to-access corners should be avoided not only in the room elements but also in the facilities. The materials used must also be resistant to the cleaning and disinfection agents, which will be used. As it is usually only small substance/material amounts, which are used in laboratories, which are also mostly transported by push trolley, it is possible to use a cost-effective flooring, for example, PVC or synthetic rubber. With the walls, a drywall construction may be used with 2k coating as the risk of damage is low and the effect in case of damage is not as serious as in production areas.

The installation of scratch or impact protection is not usually needed in laboratories.

However, if particle concentration is the most important aspect, the flushing of the room with air, the prevention of turbulent air flow or air rolls are the most important aspects in the planning. Hereby, the air which is blown in via the ceiling is guided to the return air duct through the floor or near the floor for which perforated floors are used, for example, in microelectronics. This guarantees a maximum and even perfusion of the room. A further possibility for air flow is horizontal air flow which also enables a maximum and even perfusion of the room.

Different pressure levels are built up between the various clean room categories to achieve an overflow from the clean room to the other area. This aims to prevent poor quality air entering the clean room. To achieve an overflow, it is only necessary to have a difference in pressure of a few pascals. In the pharmaceuticals industry, Annex 1 (PIC/S GMP (Good Manufacturing Practice) – Guideline for the Manufacture of Sterile Pharmaceuticals) stipulates 12.5 Pa, which ensures that the overflow is maintained even if the pressure fluctuates due to the ventilation system.

With biological risks and the set-up of ventilation barriers, higher pressure levels of about 30 Pa are required.

Metal-free clean rooms represent a further characteristic in the planning of clean laboratories. Metal-free clean rooms are used, for example, in the trace analysis of metals, for example, in sea water and sediment samples. Rare metals, heavy and trace metal elements are isolated out of these samples. As these are at a severe risk of contamination with even only the slightest dust and metal particles,

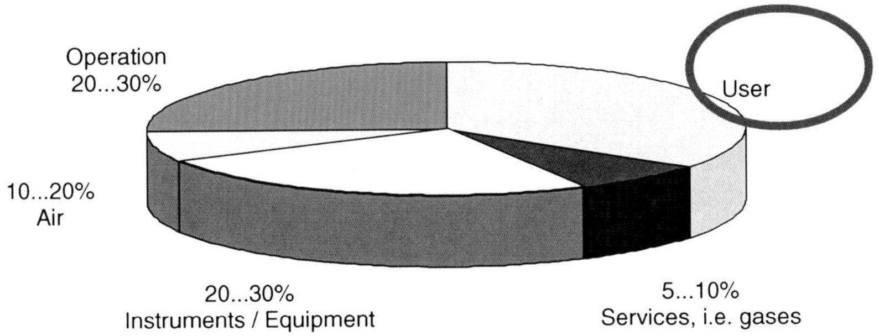

Figure 16.1 Sources of contamination in clean rooms (Dastex).

clean room conditions and a metal-free work environment must be guaranteed. Alongside the necessary clean room conditions, the workplaces should also be kept metal free and each workplace also has its own supply and exhaust air. As work is partly also conducted across several levels, air flow should be horizontal. The finishing materials are largely Trespa, plastic sheeting, polypropylene, as well as PVC for transparent windows.

Besides the air flow in a clean laboratory, humans are also an important influencing factor in the clean room climate (Figure 16.1) with view to the germ count as well as the particles. Each square centimeter of human skin has up to about 30 million bacteria and every hour the human body sheds about 600 000 skin particles. A well thought out clean room clothing and lock concept is, therefore, extremely important to ensure that these particles and germs do not get into the ambient air.

Another possibility to prevent contamination of the clean room through humans is to keep these from sensitive areas. This is not possible in all cases but should be attempted in so far as possible.

A standard definition of the clean room categories is given by EN ISO 14644/1 (clean room categories ISO 1–9), which states the maximum permitted particle concentration per cubic meter air depending on the particle size. This is important so that the user can give the planners and plant construction companies limit values, which are sufficient for the requirements of his process and which can be proven.

Contamination sources in the clean room .

Particles <0.5 μm do not play a role in the pharmaceutical industry although they do in microelectronics. One of the most important documents in the pharmaceutical industry in Europe is the GMP Guideline and its appendices. For clean rooms and limit values this is particularly Annex 1 – "Manufacture of sterile medicinal products" which outlines the limit values for particles as well as microbiology depending on the clean room category. It also states which process steps need to be taken in the respective clean room categories. For want of other requirements, the limit values for the manufacture of nonsterile medicinal products are also based on these. The clean room categories defined here from A to D are extended to include categories to be defined separately, for example, E and F.

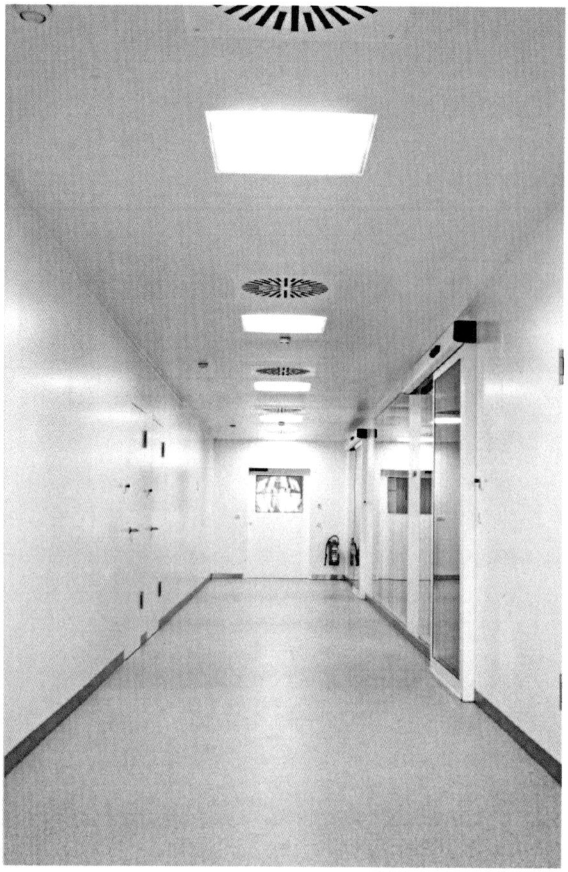

Figure 16.2 View with louvered strip waffle ceiling.

A further important guideline is VDI 2083 which outlines, for example, in part 1, the ascertainment of particulate air cleanliness categories in clean rooms in accordance with various norms ISO 14644-1, FED STD 209E, and the EC-GMP guideline based on measured concentration levels of airborne particles of various standardized sizes in comparison (Figure 16.2).

The building materials of clean room laboratories can generally be oriented to the materials of the production areas as described earlier.

16.1 Wall materials

- Drywall coated with epoxy resin is usually sufficient
- Pharmaceutical industry walls: coated aluminum/sheet steel or stainless steel in shell construction form with hollow cavity for supply of media

- Pharmaceutical industry walls in monoblock wall with media channels
- Floor to ceiling glass wall.

Glass and stainless steel in cases where high/highest standards are required concerning disinfection ability.

Flooring, possibly deflective or conductive with hollow area

- PVC
- Synthetic rubber
- Epoxy resin coating.

When using sheet or tile flooring, use prefabricated corners where possible, which do not have a cut edge on the inner or outer side. Chemical resistance is given with the standard coverings.

Flush doors and windows with fitted bands and without any hard-to-clean cavities with drop down seal to maintain pressure.

16.2 Ceilings

- Drywall 2k coating with access panels
- Clip cassette ceiling with or without jointing
- Modular grid ceilings – rare
- Flush mount installation of filter outlets/filter ceilings
- Flush mount installation of clean room lights.

16.3 Fixtures and fittings

The requirements concerning clean room laboratory fixtures and fittings can range from standard laboratory furniture to stainless steel furnishings. In contrast to standard laboratories with open installation, all cavity areas in the clean room laboratories should be clad or paneled, and barriers should be sealed and jointed up to the suspended ceiling. Barriers for laboratory media are set up via the suspended ceiling and usually integrated in the overhead cupboards. In the conception of fixtures and fittings, attention should be paid to the radius of the fillet of the floor covering.

In the planning of the clean room areas, it is important to define the requirements of the individual process steps and to limit the clean room only to these areas. Due to the high air volume needed, clean rooms are relatively expensive, not only in the investment but also in their operation. In the increase in costs which can be expected in energy, the greatest saving potential lies here.

If the areas which must fulfill certain requirements are confined, work is often carried out with so-called Laminar Flow workbenches or isolators. These offer a

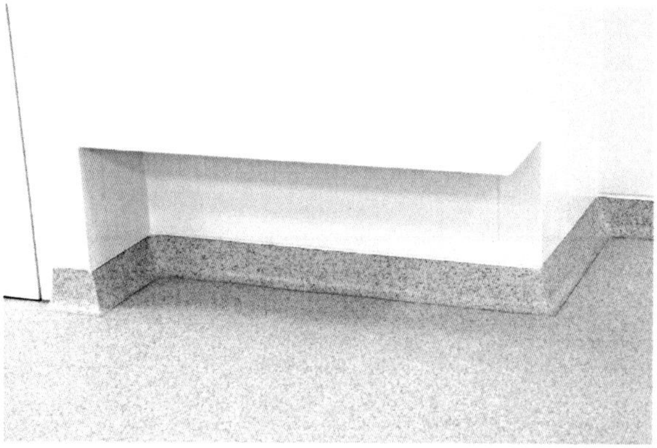

Figure 16.3 Floor extract.

cost-efficient possibility for achieving the desired clean room category in a relatively small space. However, hereby, you should ensure that these workbenches only have the desired protective effect when they are set up in a controlled environment. This clean room also has significantly lower requirements and can, therefore, be installed and operated more cost effectively (Figure 16.3).

17
Safety Laboratories

Michael Staniszewski

The term *safety laboratories* is generally used and not very specific. The expression is used to designate laboratories and function rooms of various disciplines, such as chemistry, biology, and physics, to name just the most well-known and important ones. Today, it mostly refers to biological safety laboratories. However, safety laboratories are also used for the handling of explosives and the development of identity documents. This chapter focuses on the most important safety laboratories intended for the handling of:

- biological material,
- radioactive material as well as
- active and highly active substances.

Safety laboratories are necessary for the handling of highly potent substances or organisms, which have a potentially large impact on human beings and animals. Therefore, the legislature has passed specific laws and regulations to ensure adequate protection of the health of both workers and the public from exposure to those agents and substances.

17.1
General Remark

Please note that the below-mentioned laws and regulations are normally only valid for Germany, except, for example, EU-norms. For other countries, the comparable acts and standards must be considered in detail.

The objective of these provisions is to reduce the contact with hazardous materials to an absolute minimum – irrespective of the group of people affected.

The following paragraphs deals with the three different types of safety laboratories mentioned earlier, including their basic requirements and specifications. In addition, in this context, the simultaneous handling of biological and radioactive materials in safety laboratories needs to be considered (Figure 17.1).

The Sustainable Laboratory Handbook: Design, Equipment, Operation, First Edition.
Edited by Egbert Dittrich.

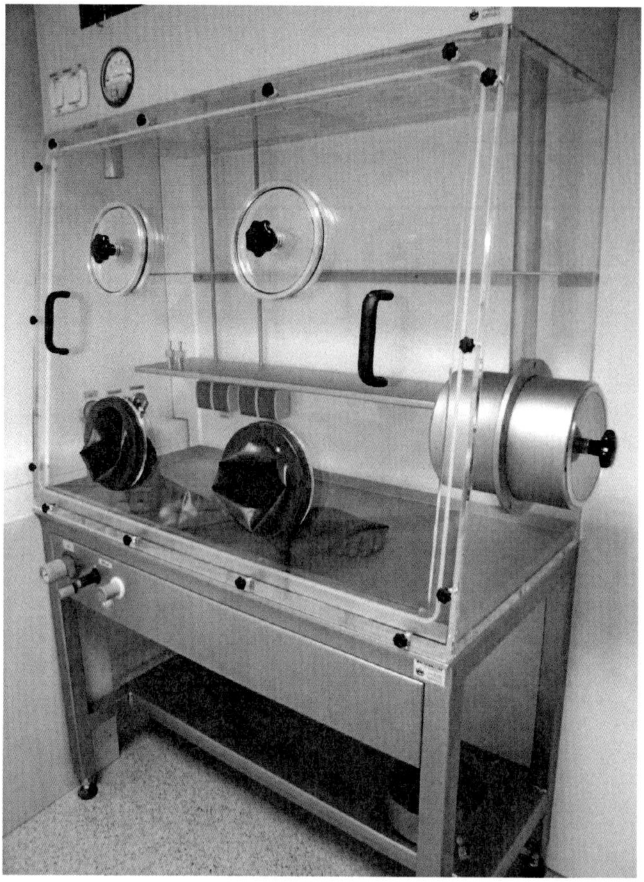

Figure 17.1 Glove box for open handling of hazardous substances in a self-contained working area.

17.2
Types of Safety Laboratories

17.2.1
Biological Safety Laboratory

The activities in biological safety laboratories can be divided into two main categories, both regulated by different laws:

- Protection against Infection Act (IfSG) and
- Genetic Engineering Act (GenTG).

The IfSG and the Ordinance on Biological Working Agents (BioStoffV) are decisive for the requirements in laboratories designated for the handling of organisms classically originating from the fields of human and animal health

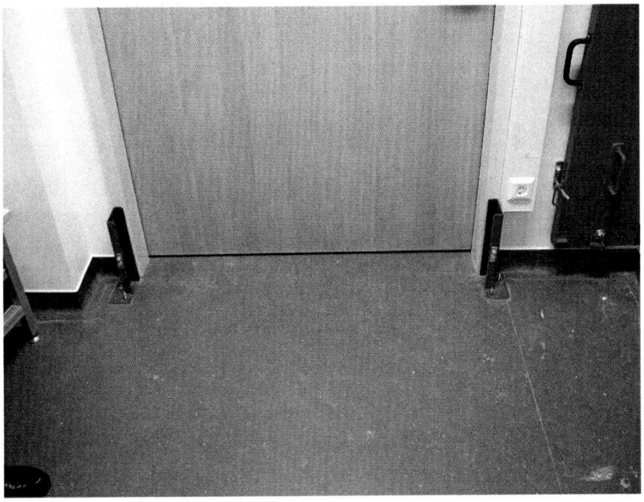

Figure 17.2 Mobile fire water retaining basin in a safety level 3 (BSL3) genetic engineering laboratory.

and health protection. These laboratories deal with pathogens in the form of microorganisms, such as viruses (e.g., hepatitis viruses – liver disease), bacteria (e.g., vibrio cholerae – cholera pathogen), and fungi (e.g., aspergillus species – aspergillosis) or unicellular and multicellular animals (e.g., amoeba and nematodes – diarrhoeal diseases and roundworms, respectively Figure 17.2).

The GenTG in combination with the Genetic Engineering Safety Ordinance (GenTSV) defines the requirements for laboratories using methods of genetic engineering. In this context, we are talking about genetically modified organisms (GMO).

In both cases, the generated waste has to be sterilized in safety level 2 (BSL2 or Germany S2) or higher. In safety level 3 (BSL3 or Germany S3), the facilities for sterilization need to be directly associated with the corresponding laboratories.

Moreover, the Technical Rules for Biological Agents have to be mentioned, which substantiate the general requirements on biological laboratories (Figures 17.3 and 17.4).

Biological safety laboratories generally possess no particularly large fire loads. In Europe, there exist no specific requirements for fire protection measures in this type of laboratory. An exception represents Germany, where the fire brigade issued a paper on the fire protection in genetic engineering facilities of safety level 1–3 (S1–S3). (In Germany, there currently exist no facilities of safety level 4 (S4) according to GenTSV.) The paper was originally developed as an internal document and summarizes the existing regulations of the Building Act, the Fire Protection Act, and the Work Protection Act pointing out their impact on genetic engineering facilities. The creation of the paper was carried out in dependence on the existing regulations for isotope laboratories, even though biological and radioactive materials differ completely with regard to the occurring risks.

Figure 17.3 Wash hand basin with sterilizer integrated into the undersink in a safety level 3 (S3) laboratory.

The requirement to equip safety level 3 (S3) laboratories with an EI90 façade and an automatic extinguishing system raises the costs of construction significantly, without a safety-enhancing effect. Contrary to the intention, the use of automatic extinguishing systems increases the risk of further spread, as the biological material is widely distributed and perfect growth conditions are provided. The inactivation and disposal of extinguishing water contain a large risk of contamination and has not yet been resolved in most cases. Infectious material was handled in safety level 3 (S3) laboratories for over more than a century without the occurrence of serious incidents. Safety measures should respond to potential threats as needed. To implement the appropriate protective arrangements, an interaction of all specialist fields is necessary. The separation of safety level 3 (S3) laboratories by means of EI90 walls from other areas in order to prevent fire from spreading is beyond doubt. However, in any other cases, the effort for safety procedures needs to be carefully weighed against the safety-related benefits.

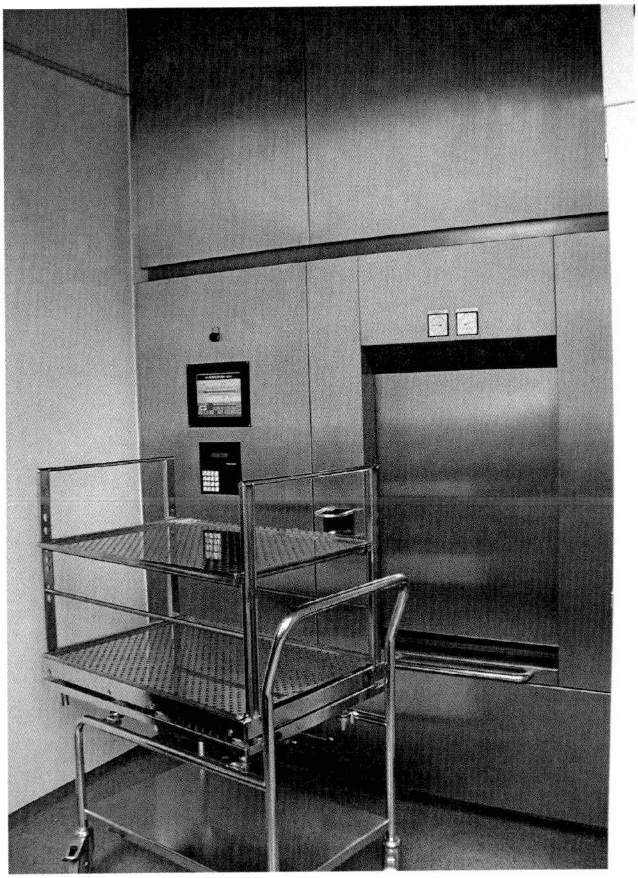

Figure 17.4 Four hundred liter pass-through autoclave with charging trolley in a safety level 3 laboratory. Autoclave with integrated gas-tight separation.

17.2.2
Safety Laboratory for Radioactive Material = Isotope Laboratory

The handling of radioactive substances in isotope laboratories is regulated by the Atomic Energy Act (AtG) and the Radiation Protection Ordinance (StrlSchV). In analogy to biological safety laboratories, the activities in isotope laboratories can be distinguished in different fields. First, we find the classic radiochemistry with the handling of nuclear fuel, on the one hand, and basic research of elementary chemistry and physics, on the other hand (Figure 17.5).

Furthermore, isotope laboratories also play an important role in life science disciplines such as biology, medicine, and pharmaceutics, where various isotopes are used to investigate biological processes. For this purpose, the isotopes are brought directly into the metabolic cycles or used in downstream analytical reactions.

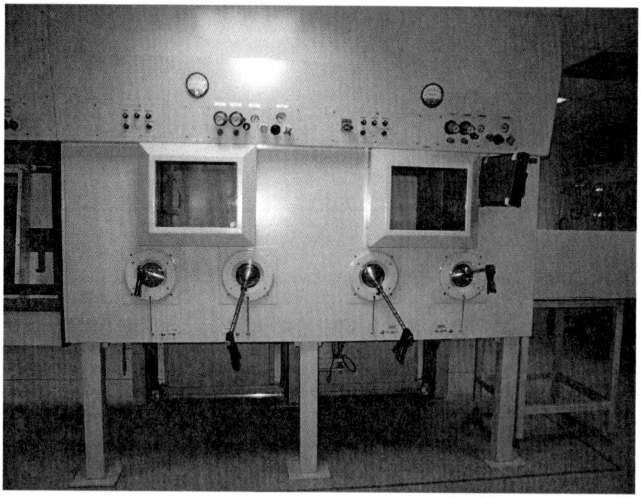

Figure 17.5 Hot cells with manipulators.

Additionally, the DIN standards 24525 Part 1–5 and 25422 have to be mentioned in this context, which provide corresponding specifications.

The regulations apply to a wide range of applications in isotope laboratories and reach their limits regarding specific requirements. Therefore, it seems advisable to have a more detailed look on the different activities, for example, in case of the general specifications for the required load capacity of ceilings, which is 15 kN m^{-2} for safety level 0 up to safety level 4 (Figure 17.6).

- *Molecular biology*: Single Acrylic Glass Shield (figure at top)
- *Radiochemistry*: Small Lead Container (figure at bottom).

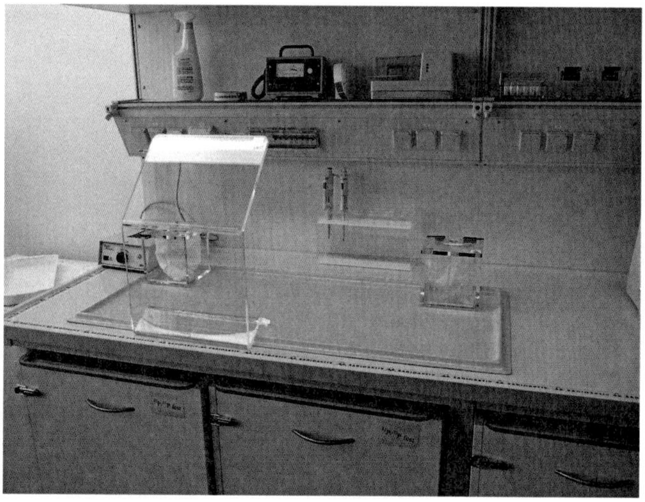

Figure 17.6 Workplaces in an isotope laboratory.

This design certainly applies to laboratories equipped with large lead containers and shielded devices. However, it does not meet the requirements of most laboratories in the life science sector. This becomes particularly clear when taking into account the flat ceiling structure common in modern laboratory construction, which ensures a safe transfer of loads.

Present safety philosophy no longer comprises the large-scale fitting of isotope laboratories with declared control areas. The usual approach today is to concentrate the activities to a limited number of rooms, the so-called core facilities. Thus, synergistic effects can be exploited according to the pursued approach.

17.2.3
Safety Laboratories for Active and Highly Active Substances

The legal situation for the handling of active and highly active substances in safety laboratories is somewhat different. In this field, the decisive regulations are the Chemicals Act (ChemG) – Act on the Protection against Hazardous Substances – and the Hazardous Substances Ordinance (GefStoffV). They actually represent a general body of laws which applies to all safety laboratories described in this article and to the substances these laboratories handle in the broadest sense. Further, specific regulations such as for biological agents and radioactive substances, do not exist in this field (Figure 17.7).

This is the result of the circumstances that active and highly active substances are used as medicinal products. Thus, minimal amounts of these substances can already affect the metabolism of humans and animals. Staff members handling

Figure 17.7 Hazardous substances workplace in form of an extracted dust workplace for the determination of grain size. Equipped with laminar-flow ceiling units and a local extract system.

Figure 17.8 Isolator, operable from both sides.

these substances require a special measure of protection regulated by special laws for medical products (MPG) (Figure 17.8).

Hazardous substances are classified and graded into different levels with regard to their potential risk to humans by means of the regulations mentioned. These levels, in turn, are decisive for the design of the rooms, their fittings, and the preventive measures required.

Isotope laboratories as well as biological laboratories are classified into following levels:

- S1 – S4 according to the Ordinance on BioStoffV
- S1 – S4 according to the GenTSV
- RK0 – RK3 according to the Regulations for Isotope Laboratories.

Easily confusing might be the usage of the abbreviation S1 – S4 in Germany for either all types of biological laboratories (until 2013 also used for Isotope Laboratories), handling BioStoffV or GMO, respectively.

In case of isotope laboratories, an additional point has to be taken into consideration according to the fire protection and protection against theft. In 2012, the abbreviation changed from SK (0–3) to GS1–GS4. The definition of these classification levels is not based on the incorporation probability (RK), but on the specified exemption level:

- direct individual relationship to persons (RK) – generally effect mass dependent (SK).

All classifications – irrespective of whether we talk about biological or radiation protection – have one thing in common: the lowest level of threat is given the lowest number and the highest level of threat is given the highest one.

Meanwhile, depending on this classification, the pharmaceutical companies have introduced their own safety levels for active substances (e.g., Cat 1–4 or BIEL 1–4) (Figure 17.9).

These classifications have been partially adjusted to each other. The factor of daily handling with permanent exposure to the smallest amounts is taken into account and assessed under the substance-specific classification. Thus, working with the same substance may sometimes require working under full protection, in an isolator (e.g., in synthesis labs – kilogram amounts), or under partial protection at hazardous substance workplaces (e.g., weighing station in the analytics laboratory – microgram amounts).

Thereby, exposure to dusty substances represents the maximum threat.

Regarding the laboratories for the handling of active substances, the structures and the protection concepts of isotope and biological safety laboratories have been adopted and adapted (Figure 17.10).

Implementation of Acts and Regulations in the Design of Safety Laboratories

Irrespective of the type of laboratory, there are basic structures that apply to all kinds of safety laboratories.

Figure 17.9 Weighing station for active substances with integrated extract air filter.

Figure 17.10 Cleaning station for active substances with coupling option for an adjacent isolator.

The separation from and to adjoining rooms becomes increasingly strict and restrictive along with the increasing degree of classification. This applies to the actual building structure as well as to technical installations and fittings.

17.3 Building Structures

Concerning the building structures, access may be possible only via locks. The fitting and the design of safety laboratories become more complex along with the increasing safety level, since the personal protective equipment get more extensive as well. This applies to the handling in both directions: the inward and the outward transfer. In particular, the taking off of the protective equipment when leaving the laboratory is associated with special increasing efforts in consideration of possible contaminations.

Thus, we see a wide spectrum of physical structures ranging from direct access under the lowest levels, to single locks, up to locks of several rooms with decontamination showers, and so on, under the highest levels. However, there are variations in the required structures depending on the respective specialist field. A lock in a biological safety laboratory up to safety level 2 (S2), for instance, may only consist of having to change the lab coats in the entrance area of the laboratory and having to install a hygiene basin with the option of disinfection. Irrespective of this, it obviously makes sense to establish locks in the form of a separate room, particularly under the aspect of providing self-contained room units.

Figure 17.11 Floor plan examples of an S3 genetic engineering laboratory and an isotope laboratory.

The provision of internal circulation space and directly allocated infrastructure facilities, such as material locks or autoclaves for sterilization, becomes necessary in addition to the establishment of pure laboratory rooms.

17.3.1
Technical Equipment: Ventilation, Electrics, Media

One main purpose of the technical equipment is to make sure that the compartment is supplied with the technology required. In lower safety levels, the laboratory rooms can still be connected to the general supply. Higher potential

Figure 17.12 Filter cabinet in an isotope laboratory with filter boxes for prefilter, activated carbon filter, and hepa filter.

risks, however, result in an increasing necessity to establish independent systems – above all concerning the ventilation system. Laboratories classified in higher safety levels need to be supplied with technical equipment in a completely separate way.

As an example, the following picture shows genetic engineering and isotope laboratories of safety level 3 (S3) and room category 3 (RK3), respectively (Figure 17.11).

In the depicted case, there is a special situation arising from the double corridor shown in the figure. Both laboratory units are separated from the adjoining rooms by EI90 walls. At the side facing the facade, this separation is realized at the inner wall of the actual laboratory. This results in the formation of a gray transition area, which allows to leave each room independently and to separate activities from each other (Figure 17.12).

The extract air filter systems, which are placed in the internal corridors, bear special significance. The advantage of this structure has to be seen in the integration of the technical equipment in the laboratory unit itself. Thus, no separate technical rooms are needed, which would have similar requirements as safety laboratories. Furthermore, used filters can directly be sterilized using the autoclave before being transferred outward and discharged as prescribed (Figure 17.13).

17.3.2 Fittings

The different fitting elements need to be designed in such a way that surfaces can easily be cleaned and decontaminated. The number of joints necessary due to technical reasons has to be minimized. Suitable facilities, such as safety cabinets and chemical hoods, which are able to hold back und remove the released substances

Figure 17.13 Contamination-free filter exchange by means of changing bag method.

Figure 17.14 Isotope laboratory subconstruction for waste collection and flow decay rack.

(gasiform, aerosol), are required for the open handling of hazardous substances. Suitable containers and facilities have to be used for disposal (Figure 17.14).

The fittings in isotope laboratories of higher safety levels have to be designed, as far as possible, in an incombustible manner.

Part III
Laboratory Casework and Installations

Egbert Dittrich

Laboratory buildings are dominated from casework and installations providing efficiency and safety. Furthermore, finish, design, and material selection of the technical facilities can be influenced of durability against the commonly in laboratories used aggressive substances and chemicals. Due to the increasing orientation to sociocultural determinants, which align the working environment of users on a pleasant, satisfactory, promotional appearance, also choice of color, modern design, and other "soft" feel-good factors are of importance. In this respect, there is a tendency to highly specialized, specific, and differing requirements imaging device parts, but where a parent design is based. Increasingly involved investors, operators, users, planners, and suppliers look at laboratory disciplines as complex-occurring coordinated unit, that is, a functional guarantee is expected, which requires an intrinsic development and planning of the entire trades. In the following chapter, the state of the art, that is, the available solutions will be shown in a structured way.

The particular components of laboratory casework and installations are:

- Lab furniture, fume cupboards, and workbenches
- Gas, water, and electric installations
- Building connected emergency showers and eye washers
- Devices for clean-, safety-, and special rooms and safety cabinets

The Sustainable Laboratory Handbook: Design, Equipment, Operation, First Edition.
Edited by Egbert Dittrich.

18
Laboratory Casework

Egbert Dittrich

Laboratory casework in the German fee regulations for project planners and engineers (HOAI) is handled under the cost-group 400 and thus, in terms of cost but also factually is connected with the technical building equipment in the laboratory. The commonly used term, laboratory furniture is indeed used interchangeably but does not describe the full range of the trade.

Laboratory facilities are highly specialized and differentiated according to application cases – see laboratory typology Heinekamp; although, in regard to ergonomic conditions, the same requirements exist for the size classes of humans and other key figures, which are also used in the manufacture of furniture for work. In respect to a plenty of installed technical services, working with aggressive or harmful substances and increased utilization of laboratory plug-in-devices have also introduced unique features of laboratory casework. It can, therefore, only be emphasized that casework not developed for laboratories is not suitable there.

18.1
Design

The design of the laboratory casework, quite certainly, is a subject of fashion trends or national practice. Essentially, laboratory furniture must meet the following requirements:

- Functional quality,
- Occupational safety, hygiene, and cleaning,
- Aesthetic quality,
- Ergonomic quality,
- Chemical and mechanical resistance,
- Economy,
- Ecological quality.

The Sustainable Laboratory Handbook: Design, Equipment, Operation, First Edition.
Edited by Egbert Dittrich.

Figure 18.1 H-frame.

The change in laboratory work to the monitoring and paperwork with computers demands more and more comfortable and pleasant design. Colors no longer serve convenience only, as related to pollution or produce contrast to the benches, but also provide well-being. After decades of abstinence, there are more accentuated surfaces in wood design again. The designers put emphasis on individual uniform *color schemes* for the entire building, based on performance-enhancing factors, dedicated psychological effects or corporate identity of the owner. Accessories or functional parts, such as handles, in addition to the function become design features. There are recurring design elements.

The width of the laboratory furniture is usually followed by a grid of 300 mm and its multiples, and therefore, comply with the standard laboratory building grid, but there are also often deviations to be seen. The current standards describe standard sizes. As a result of the average growing body sizes in the countries in the next few years, normative changes can be expected.

The selection of superior attributes such as wood, steel, and the suspension construction C-, H-, cantilever frame, and so on, depends on a mixture of individual practice, but also has functional reasons (Figures 18.1–18.4).

First of all, the material decision between steel and wooden furniture must be taken, furthermore, if the suspension frame is zinc or epoxy powder coated and in case of wooden furniture, if the casework furniture shall be made of particle boards cased with melamine.

Following selection criteria are relevant[1)] (Table 18.1):

The comparison shows the conditions for high quality basic materials from, that is, particleboard with largely sealed edges and coated surfaces, so almost no water can penetrate and the melamine coating does not lift. Qualitywise, lower

1) Extraction of "Vergleichsuntersuchung von Laboreinrichtungen aus Holz und Edelstahl", eretec 1996.

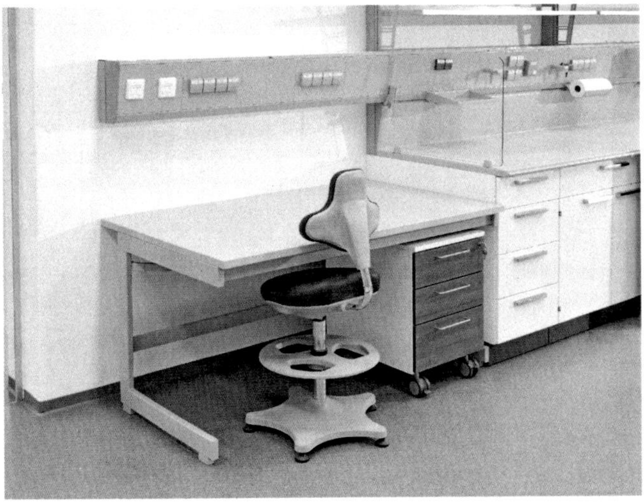

Figure 18.2 C-frame.

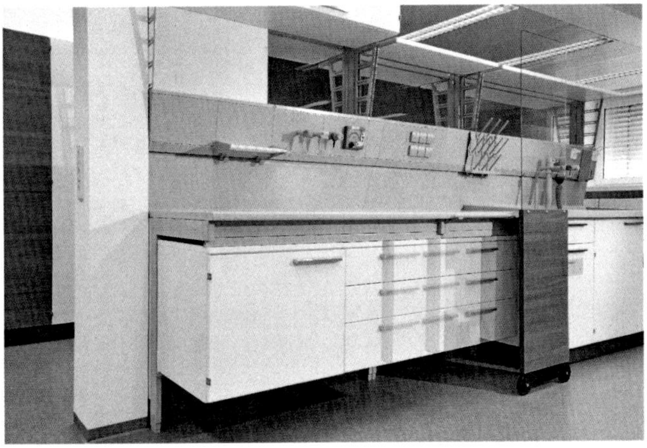

Figure 18.3 Hanging under-bench-units.

wooden furniture, for example, glued chipboard, and glued edge band and not galvanized steel furniture perform much worse and are not suitable for everyday laboratory work.

The strong improvement of melamine coatings and PDF edges in recent years is likely to ensure furniture made of wood to provide functional and qualitative projections. But water penetration is still inevitable.

Continue the selection and country-specific traditions, and development of the laboratory casework plays a determining role.

Figure 18.4 Height adjustable units.

18.2
Functionality and Flexibility

The functionality of a laboratory facility is primarily subject not only to the quality of the planning process but also to the design.

Ergonomic aspects, finishing details, and accessories concept are significant. Especially, the parts that need to be particularly frequently touched, moved, and operated are in the focus of attention. These include handles and handle bars, drawer, doors or service fixtures, and switches. The ability to install on the tables holding devices for glassware and accessories, in general the presence of rail systems for adapting other components ensures function and flexibility in the

Table 18.1 Selective criteria.

Selective criteria	Wood with melamine casing	Steel
Ecological quality	Minor pollution, resources considered, good recycling, thermal use, formaldehyde class E1, minor harmful substances in manufacturing	Recyclable, meanwhile more than 50% waste cycle, high energy use, etchant, and gases when melted
Corrosion performance for low and high concentrated acids like H_2SO_4, HNO_3, HCL, nitrohydrochloric acid, and HF	Heavy to very heavy corrosion of surface	Low to heavy corrosion of surface
Corrosion performance for middle concentrated lyes NaOH, KOH, and NH_3	No corrosion of surface	Low to no corrosion of surface
Corrosion performance for organic solvents	No corrosion of surface	No corrosion of surface
Performance for water	If mechanical damaged swelling and penetration at the end face	If mechanical damaged – rust
Combustibility	Reaction to fire classification EN 13501-1 class A, during first 20 min no extended fire load	Reaction to fire classification EN 13501-1 class B
Performance of low temperature and noise	In comparison with steel better working conditions	Depends on construction
Refurbishment	Easy	Not easy
Performance of mechanical forces and vandalism	Very solid no deformation	Less resistance, bulges

future. All important elements should be attached in the ergonomic gripping area of the user.

Mobile fitting parts are essential to provide flexibility. Modular components ensure quick adjustments, realignments, and even replacement of defective components.

18.3 Trends

In research laboratories, especially in life science interdisciplinary work groups achieve their results as teams. This requires the possibilities, under certain circumstances, of several dozen researchers in a team to communicate with no time interruptions. This claim can only be realized in large rooms and distinct areas of communication. Furthermore, it must be possible to change the laboratory equipment with frequent change of the research groups and the research content. Furthermore, there must be an option to change the casework depending on frequent change of the teams of scientists and the scientific goals.

Sustainability requirements articulate in environmental safety, low follow-up costs, and demands of users. Laboratory casework and installations generating consumptions permanently (materials, energy supply, supply air) are affected mainly.

19
Work Benches, Sinks, Storage, Supply- and Disposal Systems

Egbert Dittrich

19.1
Benches

Laboratory benches in comparison with commercial work benches are a subject of higher demands in respect of stability and corrosion resistance. Bench frames are preferably made from precision square steel profile with reinforced cross-section. You should proof the load of 200 kg m^{-2}. A complete homogeneous powder coating protects against external influences and provides a homogeneous appearance. The working height is 750 or 900 mm. Working height of height-adjustable benches can be adjusted in steps of 25 mm between sitting height (700 mm) and standing height (950 mm). In addition, the bench base can be leveled.

There are bench frames for different needs. Major structural types are C-foot, H-foot, and cantilever bench frames according to need and purpose. The grid width corresponds again to the standards. Form flush leveling feet for C and H frames can accommodate up to 23 mm adjustment travel (optional up to 50 mm). For ease of cleaning is recommended to place constructive leveling about 30 mm above the floor.

19.1.1
Major Bench Frames

- *H-frame* (roll bench also)
  - Provides good stability for attached benches, roll benches, and analysis work benches for sitting or standing work.
  - Under bench cabinets are possible to retract, hang, or move sidewise grid independent. This allows seating niches everywhere.
- C-frame
  - Are very stable and carry a load of 200 kg m^{-2}. Provides enough knee- and leg room together with rolling and suspended under bench units.
- *Cantilever-frame* provides the most knee room and airiness due to construction and design installed with service spine or wall mounted.

The Sustainable Laboratory Handbook: Design, Equipment, Operation, First Edition.
Edited by Egbert Dittrich.

- Racks
  - Racks are vertical stacked units for storing heavy loads and plug-in-units.

In order to improve the individual design options, it is useful to assemble under bench units movable in of C or H frames. If no under bench units are needed, pipe and wire installations can be covered with movable and height adjustable faceplates. The modularity and user-based design should be encouraged to attack and height-adjustable work benches. The wide range of different materials of the table tops allows a very economic adjustment to the respective needs.

The following surfaces are commonly used:

- Phenolic resin
- Melamine resin
  - with polypropylene-edge (PP)
  - with post-forming element PP on flat pressing plate according to DIN 68761
  - Epoxy
  - Stainless steel on flat pressing plate according to DIN 68761
  - Technical Ceramics, acid resistant glazed
  - Glass (safety glass) on plywood according to DIN 68705
  - Ceramic
  - Composite stoneware on plywood according to DIN 68705
  - Solid Surface.

19.1.2 Laboratory Work Bench Material

Selection Criteria for Laboratory Work Benches (Table 19.1).

19.2 Sinks

Despite the regression of water for process purposes in the laboratory, laboratory sinks are core units, especially since they are assigned to multiple jobs, that is, be used by multiple users.

Laboratory Sinks Shall Not Apply to the Disposal of Chemicals

Laboratory sinks fulfill the following tasks:

- supply and disposal of water of different types (WNC, WNH, WDI, etc.)
- Cleaning of equipment,
- Disposal of water,
- Hand hygiene,
- Location for eyewash and disposal of eye wash water.

Table 19.1 Selection criteria.

Properties Types	Critical substances	Damaging substances	Advantage	Disadvantage	Application	Kilogram per square meter thickness in millimeter
Melamine coated, postforming	Acids < 10%	Conc. HCl, HNO_3, H_2SO_4 heated	Even	Moisture sensitive edges, middle chemical resistance	Rolling benches, side benches, window benches, benches for plug-in-units, benches in dry areas	19.6 30
Polypropylene	C_mH_n, CCl_4, $C_6H_8O_7$, $C_2H_2O_4$, diesel	O_3, conc. HNO_3, CHCl, C_6H_6, fuel	No edges, even, lightsoft	Easily scratched surface, low thermal resistance	Areas with high chemical load, HF, radionuklide applications, if no edges required	20.3 30
HPL EN 438	Acids < 10%	Conc. HCl, HNO_3, H_2SO_4 heated	Even, moisture resistant, easy disposal	Thin surface layer	Wet rooms, physical labs, middle loads	26.4 19
HPL (Trespa) EN 438	Acids < 10%	Conc. HCl, HNO_3, $H_2SO_{4\text{heated}}$	Antibacterial, high density, moisture resistant, even, easy disposal	Thin surface layer	Chem., microbiology, genetic engineering lab	26.4 19
Epoxy resin	Different solvents, dilute acids	HF, conc., heated mineral acids	No edges, even, solid, easy disposal, mechanical load	Easily scratched surface, low thermal resistance, limited for acids	Universal	32 19

(continued overleaf)

Table 19.1 (Continued)

Properties Types	Critical substances	Damaging substances	Advantage	Disadvantage	Application	Kilogram per square meter thickness in millimeter
Stainless steel	Cd, $C_3H_6O_3$, $C_2H_2O_4$	Caloric and bromine combinations Verbindungen, CH_2O_2	No edges, solvent resistant, heat resistant	Halogene sensitive	For high loads biology, microbiology, pharmacy, radionuklide, pathology, in Bereichen with H_2SO_4 disinfection	27.5 30
Ceramic	None	HF	Best chemical resistance, stable, easy disposal	Uneven, limited thermodynamic loads	Areas with highest chemical and mechanical loads	56 26
Composite ceramic	None	HF	Even, best chemical resistance, stable, easy disposal, lighter than ceramic	Limited thermodynamic loads	Areas with highest chemical and mechanical loads	40 30
Glass	None	HF	Veen, high chemical resistance	Shock sensitive	Areas with highest chemical loads	38 30
Solid surface $(Al(OH)_3 \times H_2O +$ polyester + acrylic)	Some solvents, dilute acids	HF	Completely without edges, good haptics, even, antibacterial, high chemical resistance, easy repair, high volume of recycling material	Low thermodynamic loads	Areas with high chemical, mechanical, and hygienic requirements	20 13

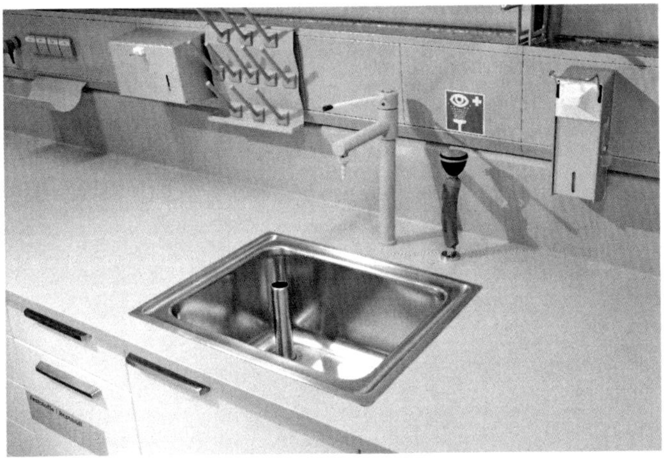

Figure 19.1 Stainless steel sink.

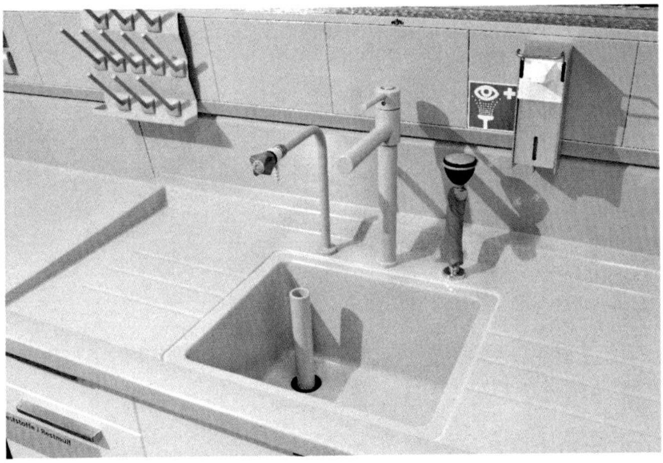

Figure 19.2 Ceramic sink.

The materials match to previous table for lab benches except glass. In various applications, the selected material of bench top and sink can diverge. The sink types in detail are:

- Bench-mounted sink
- Front-mounted Sink
- Drip cup, hanging in service spine
- Drip cup mounted in bench
- Mobile Sink (Figures 19.1–19.3 and Table 19.2).

Figure 19.3 Cup sink.

19.3
Under Bench Units, Cabinets, Storage Cabinets

Storage cabinets are used for storage of equipment and chemicals and are designed for the demands of the work environment in a laboratory. Frequent use of doors and drawers requires stable structures and especially impact-resistant edges. Double steel frames with concealed roller guide of drawers have proved. The classification of drawers is subject to the individual needs of users. For cabinets and work benches, there is a need to allow for optimal leveling the 4 ft.

Table 19.2 Materials for lab tops and sinks.

Work top material	Sink material	Mounting	Application
Ceramic	Ceramic	Sink flush mounted, edge	Work with chemicals on bench
Melamine, HPL Glass	Polypropylene	Sink mounted from above, sink with bulge	Work with less chemicals on bench
Melamine, HPL, Trespa, glass	Stainless steel	Sink mounted from above, sink with bulge	Higher hygienic requirements
Polypropylene	Polypropylene	Sink welded from bottom up, sink with bulge	With high chemical requirements
Stainless steel	Stainless steel	Sink flush mounted, welded	Pathogens, high hygienic requirements
Composite ceramic	Composite ceramic	Sink flush mounted, edge	Work with chemicals on bench
Epoxy resin	Epoxy resin	Sink flush mounted, no edge	Universal
Solid surface	Solid surface	Sink flush mounted, no edge	Hygienic requirements

Following configurations are applicable:

- Under bench units on
  - Pedestal
  - Rolls
  - Pedestal and rolls
  - Suspended to support frame
  - For fume cupboards
- Wall cabinets
- Laboratory cupboards
- Top mounted cabinets
- Pull-out cabinets
- Medical cabinets
- Safety cabinets for acids and bases (see Chap. 22).

Storage cabinets have the following options for the front:

- Swing doors
  - glass
  - partial-glass
  - massive
- Glass-sliding doors
- Drawers (full pull)
- High drawers
- Folding doors
- Open shelf.

Doors can be folded up to 270° to the front of the next unit. Sliding and folding doors need less space when opening.

On the upper level of the laboratory, cabinets and the drawer cabinets and air extract can be adjoined. Acid and base cabinets always have an on-circuit for permanent extract. Under bench units on pedestals are open at the top and provide additional storage space under worktops and laboratory sinks. They are 750 and 900 mm high.

Under-bench cabinets on pedestals have the following options:

- Swing doors
- Swing doors and drawers
- Drawers
- High drawers (für waste disposal)
- Open as shelf
- With or without air extract.

Under-bench cabinets on rolls are covered from the top and can be dislocated anywhere in the laboratory. This ensures easy access to valves and lines behind the substructures. They are available for work height of 750 and 900 mm working height. Under bench units on pedestals and rolls have a cover at the front.

Under-bench cabinets on rolls have the following options:

- Swing doors
- Swing doors and drawers
- Drawers.

Drawers are provided with a safety device against tilt over by allowing only one drawer to open at once.

Under bench cabinets with rolls are featuring two string rolls to fix in front and two string rolls in the back without fixing.

Suspended under bench units have a closed corpus and are propped up in a rail on the table frame and hung with two hinges in the table frame of the lab tables. They have no contact with the laboratory floor. The suspended under-bench cabinets are available for 750 and 900 mm working height. Suspended units can be moved laterally across the table grid. Hinged Cabinets are available with the following options:

- Swing doors
- Swing doors and drawers
- Drawers.

The suspended cabinet can be unmounted by loosening two screws on both sides and move the two fittings from the front top rail of the laboratory work bench and tip forward. Then it can be lifted out of the rear of the lower rail of the workbench. Before unmounting the cabinet off its base, the following requirements must be met:

- Units must be empty completely
- Drawers completely unhinged
- Modular shelves took off.

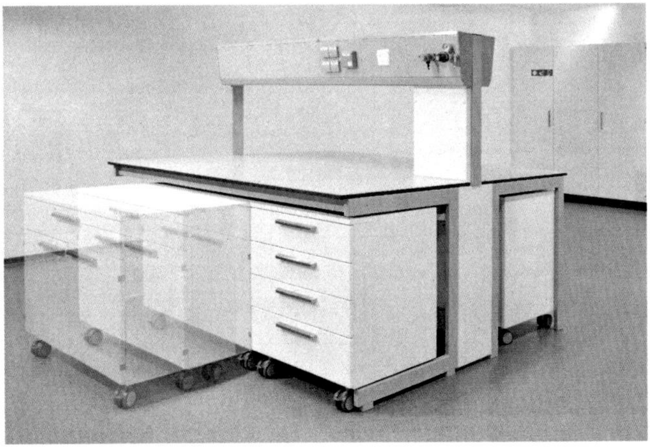

Figure 19.4 Under-bench unit with rolls.

Planners and users are advised to clearly define the demands based on previous calls for bids. The variety of described units allows an optimizing of the furniture units according to requirements such as flexibility, modularity, clear floor, space, and cleaning (Figure 19.4).

19.4 Supply and Disposal Systems

Due to *high safety-related criteria,* the provision of supply and disposal systems of liquid and solid hazardous materials is indispensable even for small quantities. The supply system enables the safe storage and delivery of combustible liquids at laboratory workstation. To prevent damage to the supply system, the compatibility must be ensured with the materials of the supply system. When dealing with combustible liquids, the information is noted in the respective operating instructions according to Section 20 Hazard Substances Ordinance.

19.4.1 Supply System for Combustible Liquids

For the supply system for combustible liquids, special safety cabinets are required. The safety cabinet must be connected to a venting system and an earthing line with potential equalization. For storage and supply of combustible liquids, suitable drums are in the safety cabinet. The barrels are applied via a pressure regulator with a defined pressure of 0.2 bar (0.5 bar safety valve). Due to the pressure the combustible liquid is transported from the barrels to the extraction point.

Supply is handled in the following ways:

- Cyclic supply with different combustible liquids
- Continuous supply with automatic switching to second barrel.

19.4.2
Cyclic Supply

Cyclic supply provides two barrels with combustible liquids, which are placed in the safety cabinet. The barrels are connected by separate lines to the taps. As long as combustible liquid is in the barrel, it can be extracted via the extraction line. If no more fluid can be extracted, the drum must be replaced.

19.4.3
Continuous Supply

With continuous supply, a maximum of two barrels of the same combustible liquid can be stored in the safety cabinet. The drums are connected to a monitoring system and connected via a common line with the tapping point. The monitoring system detects the time when during the extract a barrel is empty and automatically switches to the second barrel.

19.4.4
Monitoring System for Continuous Supply

The two barrels are connected via a diaphragm valve with the line to the extraction point. When extracting, one diaphragm valve is always open. A sensor monitors whether or not liquid is extracted. Once a barrel is empty, the control system switches to the second barrel. When a barrel is blank, it is visually indicated by a signal lamp on the monitoring module. To ensure a continuous supply, the empty barrel must be replaced. Although security issues and laboratory operation are presented in Volume 2, here we point out certain behaviors that are associated with the selected supply system and must be enabled by this. When dealing with the supply system for combustible liquids, the following precautions must be observed:

- Regulations for combustible liquids must be considered.
- Date of manufacture of the barrels must be considered.
- Always connect barrels with earthing line.
- Storage of barrel with combustible liquids only in the safety cabinet.
- Follow safety data sheet
- Consider the monitoring unit
- Close the gas supply without extracting (night mode).
- Thread-adapter at the barrels hand-screwed.
- Clear air supply and extract of the safety cabinets.
- Open electro modules only if power is off.
- Substances in supply system only with ignition point >135 °C
- Make sure that the safety cabinet is extracted with an air exchange rate of 10.
- Make sure that the extract point is operated with the air volume adequate to the particular ventilation requirement data sheet by a monitored extract system (ventilation requirement data sheets refer to an opening of the sash of 500 mm

and the suggested tracer gas maximum) and that no heated surfaces (mainly Bunsen burners) are installed or are in use in the area of the extract.

- Plastic surfaces must be cleaned only with damp cleaning agents.
- The supply system for combustible liquids must be handled only by instructed and authorized staff.
- The staff must be instructed about explosion prevention.
- Use proper tanks of max. 1 l nominal volume.
- Sloped combustible liquids must be wiped at once and the mop must be disposed properly (explosion safe).
- Remove all plugs in the containment before supply and disposal and switch of socket from energy. (Consider power light and continuous check their function.)
- Do not operate supply systems for combustible liquids in the following areas:
  - Areas where lightning strikes can occur
  - Areas where electromagnetic waves (radio waves, HF-generators, flashing arcs, laser, etc.) exceed the ignition from power lines of 80 μJ.
  - Areas with ionized radiation with an ignition point beyond 135 °C of the explosive atmosphere.
  - Areas with ultrasonic sources that could ignite the treated substance with ultrasound.

19.4.5 Disposal of Combustible Liquids

The disposal system allows the safe interim storage of residual quantities of flammable liquids at laboratory workstation. To avoid damage to the waste-disposal system and the transfer system, the compatibility must be ensured with the materials of the disposal system and the decanting system. The following materials are not permitted in the waste-disposal system for flammable liquids:

- Acids and bases
- Gas bottles
- Radioactive substances
- Microorganism
- Genome changing substances.

For the disposal system for combustible liquids, special safety under-bench units are used. The safety-under-bench-units are connected to a *venting system* and an earthing wire with potential equalization. To take in the combustible liquids, suitable canisters are located in the safety-under-bench-unit. The number of cans in the safety cabinet depends on the type of filling and the dimensions of the canisters.

19.4.6 Electronic or Mechanical Level Indicators

Electronic or mechanical level indicators show when the canister must be replaced or emptied.

In order to fill up, the following options are applicable:

- Fill up by funnel in the safety-under-bench-unit
- Fill up by funnel in the containment of the fume cupboard.

19.4.7 Connection for Liquid Chromatograph

Connection for liquid chromatograph – as alternative to filling by funnel liquid chromatograph (HPLC, high performance liquid chromatography) can also be connected in the containment of the fume cupboard. In place of the funnel, corresponding shanks are disposed of on the worktop.

19.4.8 Filling by Funnel in the Under Bench Safety Unit

One or two canisters are filled directly by a screwed funnel with mechanical level indicator. The canisters are in a safety cabinet with high extract under a fume cupboard. The canisters and funnels are connected by a grounding wire to the safety under-bench frame. In the opening of the funnel, a flame-protection screen made of stainless steel is inserted.

19.4.9 Mechanical Gauge

The funnels are attached with a union nut to the canisters and provided with a mechanical level indicator. When the level of liquid increases in the cans, the signal bar is lifted from the float. When the maximum level is reached, the signal rod exceeds the edge of the funnel. The lid is raised and thus, the canister has to be replaced.

19.4.10 Filling by Funnel in the Containment of the Fume Cupboard

One or two cans are filled by one funnel, each located on the worktop or side panel in the fume cupboard. The funnels are connected via a fill line with the canisters. The canisters are located in the safety under-bench unit and connected with an earthing wire to the frame of the unit. In the opening of the funnel, a flame-protection screen made of stainless steel is inserted (Figure 19.5 – 19.7).

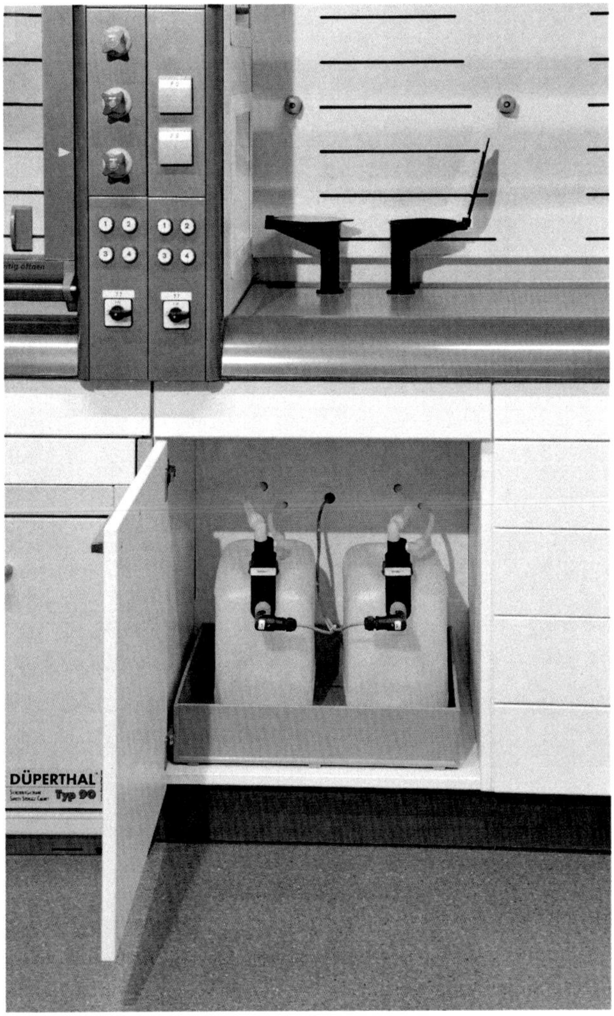

Figure 19.5 Disposal of combustible liquids.

19.5 Service Carrying Frames[1]

The installation of services in laboratories historically resulted in a significant restriction of flexibility, that is, the ability to make quick conversions. The services must be connected for safety reasons in a force-fitting manner to the building, that is, service installation in general are obstructive for refurbishments or technical changes of the building engineering.

1) Services = pipe supply for liquids and gases.

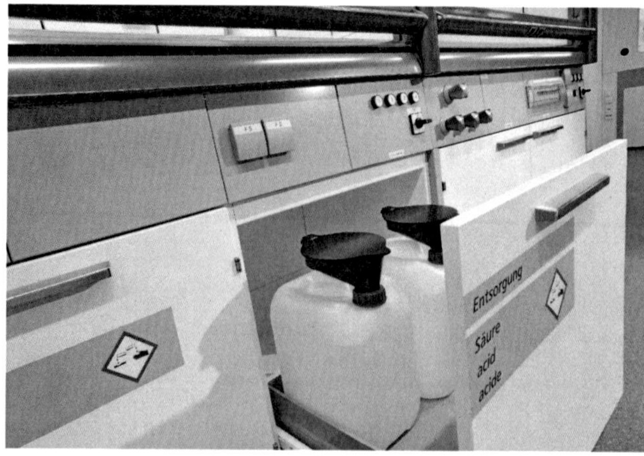

Figure 19.6 Disposal in under-bench unit.

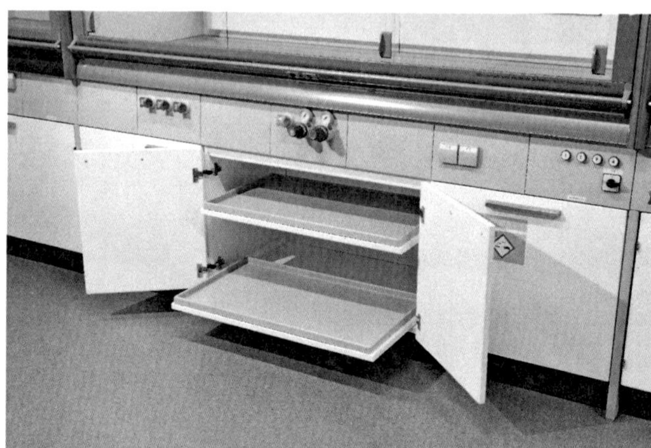

Figure 19.7 Disposal and storage.

There are different service-carrying frames (service carrier):

- Service spine
- Service boom
- Service column
- Service channel
- Service wing
- Service ceiling and movable service columns (see Chapter 12).

The selection of the service carrier is dependent on the future task for the laboratory, the number of services, the expected changes, and the degree of conversion feasibility in the context of laboratory types.

19.5.1
Service Spine

The service spine is a *casket-like* metal structure carrying lab work benches flush mounted with their backs, allowing a compact, very clear installation. The interface to the main supply is located in the bottom (now uncommon), on a wall or overhead. Hence, conversions are expensive but doable. The installations are covered with caskets or covered by large sheets in terms of high hygiene requirements. That prevents fast changes. The service spine is firmly bolted to the floor and is about 10 cm deep. The service fixtures and laboratory taps are mounted above the work surface in an ergonomical position, but restrict the full use of the depth of a work bench. As a basis for the design of laboratory workstations, the service spine provides a very *economical solution.* As a standalone unit, the service spine in combination with freely selectable bench frames either becomes a wall or becomes a double work bench. The service spine also serves as a support structure for cupboards and shelves. The wall unit is mounted on a rail at the service spine. The height of the service spine can be changed via a stand extension to obtain the sufficient height for the cabinet.

There are rail systems, which serve to receive useful accessories such as shelves, reagent racks, support rods, and towel rail. The so attached "helper" might move across a grid and safely fix.

Modular screwless *service panels* can be replaced if required with some effort. Main pipes, for example, for water and compressed air are rapidly expanded with plug-in coupling systems and mounted – with some acceptable disturbance of the laboratory operation. Service supplies housed in service spine, simultaneously deliver services for sinks in the work benches or on the front face.

Configurations

- Service spine for wall bench with carrying service under work bench or console
- Service spine for wall bench with carrying service from ceiling
- Service spine for double work bench free standing with carrying service under work bench or console
- Service spine for double work bench free standing with carrying service from ceiling (Figure 19.8).

19.5.2
Service Boom

Service booms are horizontal ceiling-mounted, space-saving, installation elements, which are fixedly attached to the ceiling and release the floor for a flexible design. Occasionally, such elements are also height adjustable. They are suitable for laboratories with high instrumentation; therefore, many sockets and relatively little plumbing services. For the supply of the work bench, the controls are ergonomically unfavorable (Figure 19.9).

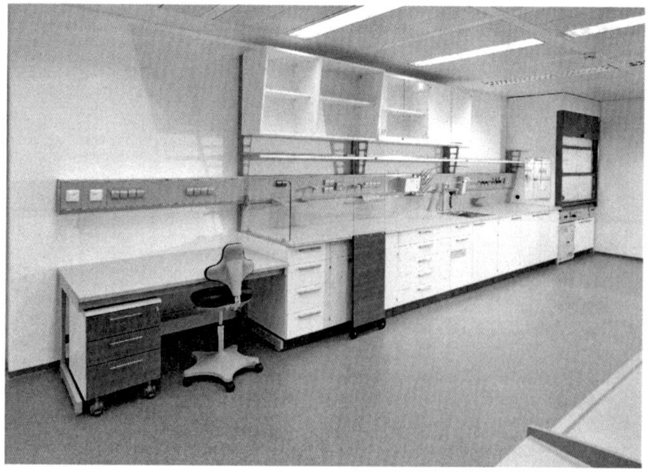

Figure 19.8 Wall-mounted service spine.

19.5.3
Service Columns

Service columns, if they are fixed installed, provide similar functions as service booms, but allow the vertical structure to place in the ergonomic grip range of users. They offer the possibility for transparent room design as compact service supply. Equipped with interchangeable panels and rail systems, the service column can be mounted either directly on the laboratory ceiling or at the service ceiling (see Chap. II.5). The function of the service column is extended considerably when attached to the service ceiling and allows the user during the use of equipment on work benches by dragging the column to make room. Service columns as well as service booms are provided with service fixtures and modules can be upgraded quickly (Figure 19.10).

19.5.4
Wall-Mounted Service Channel

As an alternative to the service spine, the service wall channel can be mounted at varying heights directly on walls or in connection to a wall-side service spine. Likewise, equipped with panel technology and accessory rail for variable mounting, the service wall-side channel is almost identical to the cell media (Figure 19.11).

19.5.5
Service Wing

The service wing is a complete supply system on the ceiling. There are all the major services, such as plumbing, electrical, IT, energy-saving lighting, partial extract, and also the integrated sewerage. If it is not about the freedom to align the rooms

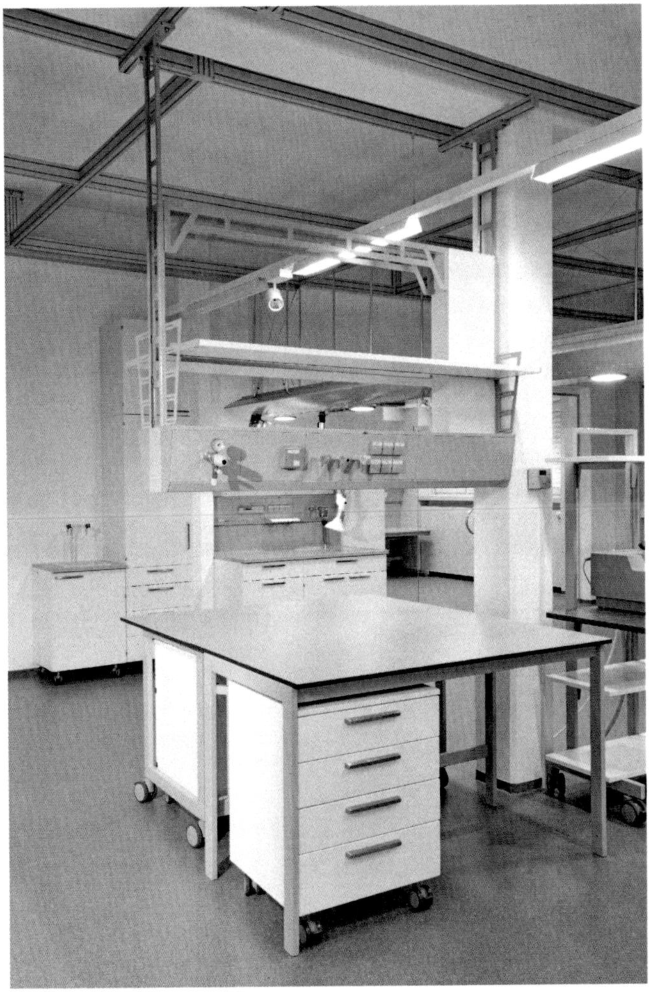

Figure 19.9 Service boom.

in terms of their division and size, the service wing provides a maximum flexibility and freedom of movement.

19.5.5.1 Configurations of the Service Wing

The service wing is available in four different configurations. For every configuration, a matching tee is available in the design of the laboratory services, and a variable energy supply is possible in the laboratory. The following are the expansion stages of the service wing:

- *Stage 1*: Basic electrical duct with panels for electric supply
- *Stage 2*: Basic electrical duct with lights on the wing edges

Figure 19.10 Service columns.

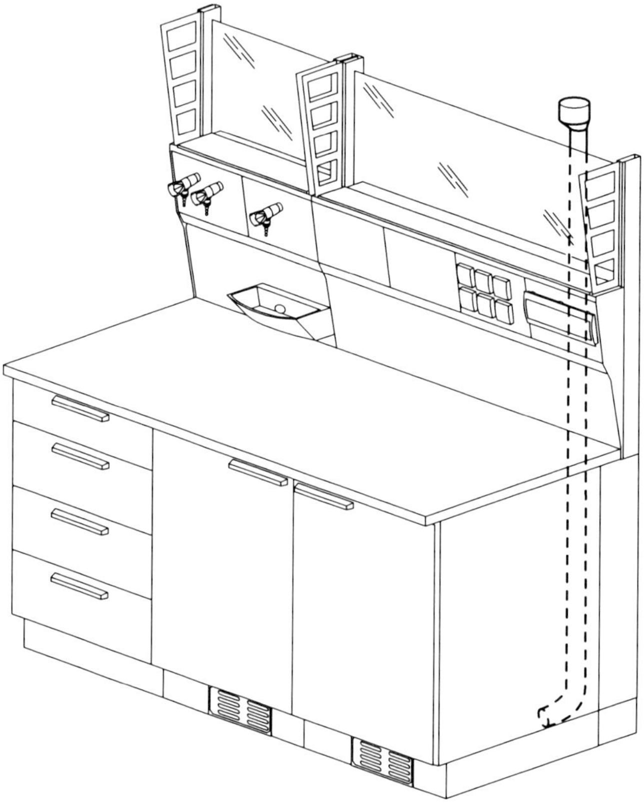

Figure 19.11 Wall-mounted service channel and bench unit.

- *Stage 3*: Electrical basic channel with lights on the wing edges, plumbing services, and ventilation shaft
- *Stage 4*: Basic electrical channel with wing edges without lights, with plumbing services, and ventilation shaft.

On the underside of the wing, element panels for the electric supply are assembled. Within stages 3 and 4, ventilation elements and appropriate supply channels for service fixture are mounted. Stages 2 and 3 integrate direct and indirect lighting. The panels for the electric supply are plugged into the electric base channel. Up to nine sockets for AC or up to four sockets for RC can be attached. Connections for workplace lighting, phones, monitors, data, or speakers can also be installed on the panels.

With the help of the removable service panels, fittings and connections are positioned as desired. A rail system takes on useful accessories such as plug-in-unit carriers, media stations, and tripod holder. These can be slipped scanning across and fixed in any position. Particularly noteworthy is the quick and easy installation, by the industrial pre-assembly. It makes sense to use energy-saving lamps with daylight-dependent control. With the help of T-elements, or out-crossings, different lengths of wing segments open up all areas of the laboratory. In order to avoid oscillations of the service wing, the suspension in the longitudinal and transverse direction is clamped. If the suspension tails-off, it can be restrained.

Usually there is only a general supply and disposal spot.

Service fixtures and control elements that are located above the head level are typical for the service wing system as well as the extensions down to the lab benches. In this respect, service wings are very suitable for labs with many plug-in-devices and educational labs. Work benches, racks, mobile sinks, or mobile fume cupboards with permanent changing locations underneath are available for use. All other units are not assembled inevitably.

To carry out permanent work underneath the service wing might require a service station. The service station is clamped at the work benches and steel frames. Connected with the service wing, the service station offers a considerable free design of the work place. It can also be supplied with other service carriers assembled on the ceiling. The supply is enabled by pipes and fixed connections. The pipes are connected with quick connect fittings to the service wing. Thus ,the pipes offer a flexible service supply underneath the service wing (Figure 19.12).

19.5.6
Bench-Mounted Service Duct

The bench-mounted service duct is a permanently installed, double-sided, service spine, which provides an integrated service unit above the counter top and penetration of all services above the counter top. The bench-mounted service duct, furnishes the services and electricity supply on laboratory work benches and is suitable for double work benches, whose opposite working surfaces should not be separated (cf. Purpose BGI/GUV-I 850-0).

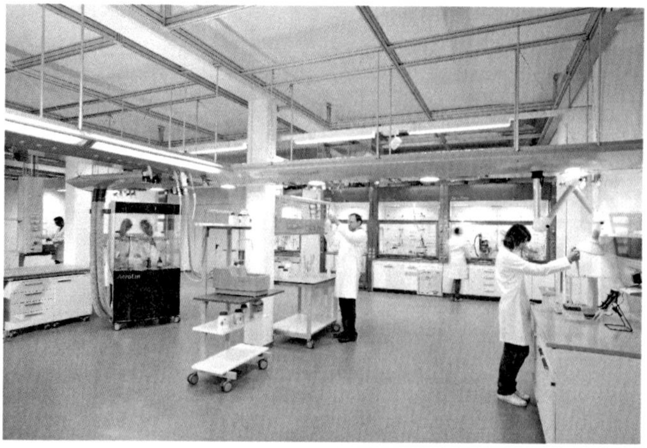

Figure 19.12 Service wing laboratory.

The bench-mounted service duct can serve as a support structure for wall cabinets. Depending on the height of the cabinet, post extensions are attached to the laboratory ceiling. In front of the service duct on both sides, countertops or underbench units are assembled on steel frames.

19.5.7
Service Ceiling

Service ceiling is a construction that takes up a structured and repetitive design by means of a metallic supporting frame in all installations in the laboratory.

Service ceiling does not only signify the implementation of the philosophy to understand, in principle, how to provide the trades with all the services by the clear ceiling area; it is also a planning system and order that allows for quick changes of room layouts and the service supply as required. The service ceiling matches with its dimensions the building grids and supplies laboratory bays. This has the consequence of being able to separate each frame at any time in space by other frames.

By means of the integrated planning and execution of trades, the interface problem is defused and simultaneously a particularly economical execution of the installation by minimizing cable lengths and materials is realized. Industrial prefabrication of service ceiling elements causes extremely short construction times and no waste at the site.

With the help of the service ceiling user-independent planning and execution can be implemented in a very cost-effective and rapid way, because the individual components of the installations have been designed for expansion, extension, and alteration within the shortest possible time. Thus, the service ceiling is suitable, especially for buildings with rapid changes of users or while changing the type of laboratory to provide without compromising effective technical installations in each case.

This implies that all possibilities are not permanently allocated for ubiquitous use, but at the same time if necessary with the help of couplings, fittings, and quick docking facilities to refurbish or realize change in operation. Interruption of work caused by the use of foreign craftsmen is extremely limited and virtually causes no time-outs.

This ensures that there is always an optimal and ergonomically maximized laboratory configured, and the productivity and efficiency of the users are not limited by suboptimal laboratory technology. The service ceiling promotes the satisfaction and well-being of laboratory teams by means of high optical transparency and equivalent expansion of all trades at the same time. Focused particularly on the well-being and safety, integral supply air constructions allow the subsidence of air change rates while improving security by optimal mixing of the air in the inhaled range of users. The service ceiling further contributes to improved space efficiency because space requirements for the technical trades are reduced with an optimized planning of supply ducts and the absence of service spines or service carriers on the ground. Buildings with service ceiling meet the demands of sustainability in special degree; on the one hand, because the life of technical trades come closer to the life time of the buildings, less material must be used throughout the life cycle of the building and on the other hand, the service ceiling teamwork, interaction, and economical work permits. Integral components, such as service columns, are quickly and easily interchangeable and their maintenance is easy.

The service ceiling is considered to be an adequate support structure for service columns that can be easily moved and it can take more loads such as, cupboards or on-site devices, provided the statics of the construction and the building are properly designed. Demands of sustainable – integrated – planning processes are entirely met by the service ceiling. It should be noted that special planners with regard to the design of the single trades are not relieved of an iterative planning. Since the service ceiling is a product designed based on mechanical engineering

Figure 19.13 Service ceiling.

criteria, building allowances must be considered. Changes prove to be particularly advantageous so that the standard in machine design using CAD systems also allows extensions and shipment of prefabricated parts in the future and the possible constructive illustration of the system is always the same as reality construction (Figure 19.13).

20
Fume Cupboards and Ventilated Units

Egbert Dittrich

The demands of accident and professional associations and relevant legislations unambiguously of course require that work involving hazardous and noxious substances be banned in fume cupboards in which the air consumption increases accordingly. Any measure to reduce the air consumption of fume cupboards has a positive effect on the economy. Therefore, in the selection of fume cupboards – next to health and safety issues – the minimum flow rate is important. But again, safety of the occupants is of the highest preference.

It is important to mention at this point that type tests according to EN14175 ff are clearly made in compliance with ideal conditions. This means even regular upcoming disturbances are not taken into account and fume cupboards with applicable very low face velocity provide outbreaks if any minor aerobic turbulence occurs.

20.1
Technical Data and Selection Criteria

General fume cupboards have permanently installed extract vents, which are connected via the extract manifold to a venting system. Sucking the air creates a vacuum in the containment. *By-pass openings* above and below the sash generate airflow when the sash is closed. Aerodynamic inflow profiles ensure optimal air intake. Gases, vapors, aerosols, and dusts in the containment are mixed with the air and extracted. The fume cupboard is opened via the vertically sliding sash. The sash has two or three opposing, *glass cross slides.* The cross slides allow access to the containment and serve as a body guard. The sash cannot be closed completely because of aerodynamic reasons. Slots for air supply are provided if the sash is closed. Bursts of solid or liquid substances affected by these slots follow an angle of max. 45° in order to protect the user.

A front sliding restrictor unit prevents opening of the sash on the largest variable opening. Before the sash can be opened further, a one-hand release on the handle bar must be pressed. The opening of the sash on the largest variable opening in addition is indicated by a visual and audible alarm. Sash supports are subject to

The Sustainable Laboratory Handbook: Design, Equipment, Operation, First Edition.
Edited by Egbert Dittrich.

Figure 20.1 Fume cupboards.

wear and tear and it is important to ensure that both low-wear materials – for example, cam belts – and also mechanical fall arresters are installed (Figure 20.1).

A sash controller allows a motor-driven closure of the sash by a motion detector. A sensor detects obstacles that protrude from the hood interior. The DIN EN 14175 Part 2 calls for continuous monitoring of fume cupboards vent, a technique to warn laboratory personnel in case of failure by optical and acoustic signals. The optical signal must not be deleted. The *function display* is an electronic monitoring system that continuously measures the extract-air volume flow. At eye level informs the side posts integrated soft-touch front panel control display of the operating state. If possible, no turbulence should be generated by the operating element.

The function indicator alerts both in an audible and a visual manner when the set threshold for the extract air is below. When there is lack of air, an alarm message in color red appears and warnings, for example, when there is excess of the *maximum sash opening*, signaled in color orange. The acoustic alarm can be acknowledged by pressing a button. The on/off function of the display can be made optional free by the user. To determine the pressure signal, the differential pressure in the *extract manifold* of the fume cupboard is used. The function display area operates independently of pressure variations and independent of the opening of the sash. In the night mode, the amount of air that is next released can be monitored.

By slightly pressing the sash, the opening and closing processes of the front sash are *motorized* and continued. The sash electronic closes the hood sash when not in use. With the help of a motion sensor, the front sector of the fume cupboard is monitored. No perceived motion is detected in front of the fume cupboard and within a predetermined time period, the sash is closed. By using a sash controller, the specification of the Technical Rule 526, which indicates that fume cupboards currently not working must be closed, is automatically put into practice. The closing delay time after the release of the sensors can be adjusted between 30 s and 15 min.

Figure 20.2 Operating unit.

Modern fume cupboards with supportive air flow technology are used in closed operation of 200 $m^3\,h^{-1}\,m^{-1}$ and with open sash to 500 mm approx $m^3\,h^{-1}\,m^{-1}$.[1)]

Meanwhile, it appears to be sure that special measures such as *aerodynamic reflectors* and wider side posts provide the same results. These values can be achieved only in conjunction with aerodynamic-dimensioned profiles of the side post and the sash in general. The stability of the air flow is better at wide side posts fume cupboards.

Integrated in the rear and side walls or side posts of the fume cupboards are replaceable service panels that provide plumbing and electrical services. Furthermore, integrated cup sinks offer more space to use the containment.

Users must be aware that the information provided by the manufacturers is idealized in terms of air technique of fume cupboards, that is, with appropriate placements in the containment, strong changes of the *outbreak* are revealed. The use of supportive air flow technology neutralizes this effect to some degree (Figures 20.2–20.5).

High containments and a free view of the experimental setups and procedures with the help of glazing above the sash are advantageous at high and expansive experimental setups. Support rods 12 and 13 mm diameter must be firmly and securely fixed.

Energy-saving lights switchable from side posts illuminate the containment. Depending on the application, the tabletop is made of *ceramic, epoxy resin, polypropylene, or stainless steel.* Fume cupboards are mounted at best with self-supporting under-bench units or on a steel frame. Thus, it is possible to equip the fume cupboard with pedestal under-bench units, roll containers, solvent cabinets, and so on.

There are several *flow-related measures* that contribute to the reduction of air consumption, for example:

1) Calculation value.

Figure 20.3 Sensor system for horizontal sliding sash.

Figure 20.4 Central Duct with Control Elements.

- Integration of the fume cupboards in the *metrological ventilation concept* of the building
- Aerodynamically adapted flow profiles at the leading edge of the work bench and side post
- Supportive flow technology
- Automatic sash with motion detectors
- Air guiding function of the sash.

The following criteria are remarkable to maintain the safety and ergonomic aspects:

- No outbreak of gases or solids in closed operation
- No outbreak of gases in open operation

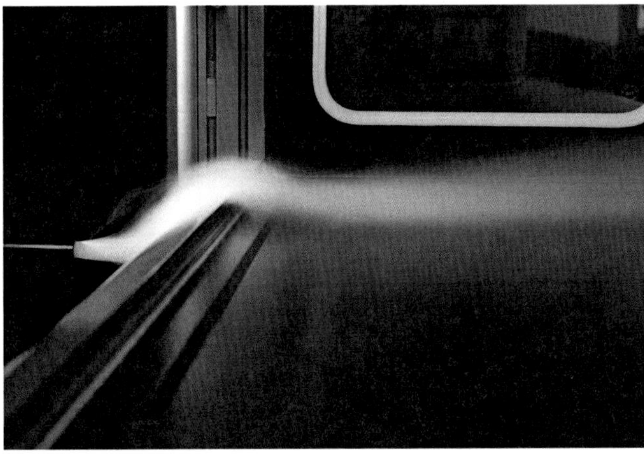

Figure 20.5 Central Unit AC.

- Safety barrier closes automatically when obstacles occur
- A clear view of the entire interior
- All functions of the control panel at eye level
- Explosion opening
- Fall protection of the sash
- Maximum internal volume.

Pros of fume cupboards with wide side posts to accommodate the service fixtures:

- Fittings at eye level
- No manipulation of the nozzles on the rear wall
- Easier maintenance of fittings
- Better air flow comparable with air flow support units.

Cons:

- Low loss of space
- Penetration at the sash opening limited.

Fume cupboards with narrow side posts and under-bench mounted service fixtures.

- Pros:
  - Maximizes space and penetration
- Cons:
  - Read service fixtures severe
  - Air flow suboptimal.

Basically, the requirement applies that only fume cupboards are used, which are reported for the corresponding substances. A standard fume cupboard provides, for example, no safety when working with carcinogens; in this case, only cytostatics workbenches are suitable.

As part of the guidelines for the employment of people with disabilities fume cupboards are offered for use with wheelchairs.

Special application fume cupboards are devices that are working with particularly corrosive and toxic substances. To prevent an outbreak of such substances into the open air in front of the sash, they are often equipped with noxious gas scrubbers installed either in the base or above the fume cupboard. Depending on the nature of the gases, neutralizing acids and bases are added in order to provide a separation degree in the scrubber of at least 80–95%. The neutralized wastewater is disposed of properly and according to the regulations.

Of course, in order to hold highly corrosive substances in fume cupboards, a special selection of lining materials or the bench top material is required. For example, mineral table tops are used when prohibitive substances like hydrofluoric acids have to be stored.

Radio-nuclide fume cupboards allow working with radioactive substances and have lead-shielding coats.

All fume cupboard types in different widths, which correspond to the usual laboratory grid (multiples of 300 mm) are available.

For particularly *bulky structures*, there are *walk-in hoods* and fume cupboards with side post-mounted service fixtures.

The following units are more ventilated:

- Hazardous and solvent safety cabinets, see Chapter 6,
- Cabinets and floor extraction
- Stationary enclosures
- Local extraction, fixed, and flexible.

The extraction of dust, gas mixtures, or thermal loads in the laboratory can have many causes. Potential leaks from containers require the storage of solvents, certain chemicals, and so on, in safety cabinets (see Chapter 6), which must be extracted. Current and long-term experiments in fume cupboards prevent a total shutdown of the ventilation, or at best allow the night-reduction of air exchange.

Furthermore, there are extractors in safety-under-bench-units and on the ground, made necessary by the evolution of gases that are heavier than air; this includes areas with expected high CO_2 developments.

For better handling, some mobile extraction arms are available, for example, to extract thermal loads by open flames or devices.

The following matrix shows the variety of fume cupboards and their specific properties. In addition to the fume cupboards in accordance with EN 14175, more stationary or mobile housings are installed.

The user must consider this device in use that they are not suitable as a substitute for working with vapors, aerosols, or other substances hazardous to good health, but represent an excellent solution for the dissipation of thermal loads and the subsequent heat recovery. In this respect, housings especially recommended for the operation of plug-in-units and other power-operated equipment.

20.2
Fume Cupboards and Sustainability

Due to the air consumption of approx. 150–400 $m^3 h^{-1}$, depending on the sash position, fume cupboards are the major energy consumers in laboratories and it appears to be legitimate to consider these units worthwhile in terms of safe energy consumption and cost-effectiveness. On the other hand, if operators specify fume cupboards with the lowest air volume, they shall be aware that the risk of hazardous outbreaks is increasing. No one knows the previously used set-ups that users install in the containment. Furthermore, disruptive factors of external influences must be considered during operation. The conclusion clearly is to withdraw air consumption in favor of safety. Safety of the occupants beats energy savings in general and this philosophy is inherent only with sustainability. The recent EGNATON[2] certification of laboratory equipment requires beside conformity of EN 14175 a concentration of 0.1 ppm sf6 at the outer grid and the lowest reasonable hold-up time in the containment (Table 20.1).

20.3
Ventilation Control and Monitoring

Laboratory equipment and ventilation of the entire laboratory building cannot be separated from an economic point of view and because of the abundance of extracted devices from each other. An intelligent laboratory control significantly reduces *the operating costs of the ventilation system* and ensures maximum work safety. Furthermore, as a result of associated comfort, efficiency is provided.

Fume cupboards must be integrated as an important part of the laboratory ventilation into the *building ventilation concept.* A measurement and control unit detects reliably and at any time, the terms and conditions of the fume cupboard and precisely regulates within seconds the air flow.

If necessary, may be at any time manually, a higher or lower volume air flow is switched.

The feasibility study clearly speaks about the requirements for the *laboratory control.* Through the efficient use of the ventilation system in accordance with reduced energy use, the investments made in establishing laboratory control pays for itself in 1–2 years.

The central processing unit is a microprocessor-based control and represents the core of the control components. The reference value is set via the sash position. The processor controls this via a specific control response (adaptive or predictive) quickly and accurately. The microprocessor detects the required damper position, has a max. adjusting speed of 2 s for 90° and is provided with a position control. Set point changes are hereby corrected within 3 s. In addition, the matching

2) European Association of Sustainable Laboratories.

Table 20.1 Fume cupboard types and requirements.

Typs \ Properties	Acc. EN14175	Acc. DIN 25466	Acc. DIN 12924	Extract of haz.-mat.	Avoid explosive gases	Applic. For thermal loads and aggressive substances	Extract of radio-nuclide substances and radiation protection	Extract of rhermal loads and dust	Supportive w technology	Air scrubber	Air filter	Splashprotection	Solid-body protection	Service nozzle rear-panel -mounted, fixture under bench	Service nozzele in side panel mounted, ixtures side post mounted	Fixtures under bench mounted, nozzles table mounted	Fixtures and nozzles containment mounted	Visibility from all sides	Opposite lying sashes	Suitable for low rooms
FC bench mounted	+			+	+			+	+	+	+	+	+	+	+				+	+
FC walk ins	+			+	+			+	+	+	+	+	+		+					
FC low level	+			+	+			+	+	+	+	+	+		+					
FC EN 14175-7	+		+	+	+	+	+	+	+	+	+	+	+	+	+					+
FC radio-nuclide		+			+															+
FC filter units	+			+	+			+	+	+	+	+	+	+	+					
FC mobile				+	+			+				+		+		+	+	+		+
FC for sitting use	+			+	+			+	+	+	+	+	+		+					+
FC pharmaceutical	+			+	+			+		+	+	+	+	+	+	+				+
housing units								+												+

calculation for the shield factor is determined by means of a characteristic field, which results from the flap position and the differential pressure.

According to EN 14175, an optical and acoustic alarm is required in case the set point falls short. Also, a visual and audible warning occurs if the maximum sash opening area is detected.

The damper should be placed in the collecting duct. At room heights below 3.30 m, butterfly valves must be used as a duct regulator.

When using this technique, the air support beam must be shut down by the control, when the extract air is below the predetermined amount.

If the supportive air flow technology fails, this must visually and audibly be indicated and the extract value is automatically increased to the value of a standard fume cupboard (Figure 20.6).

Fume cupboards and controls in the current form provide a laboratory operating unit whose precisely coordinated systems are responsible for maximum reliability.

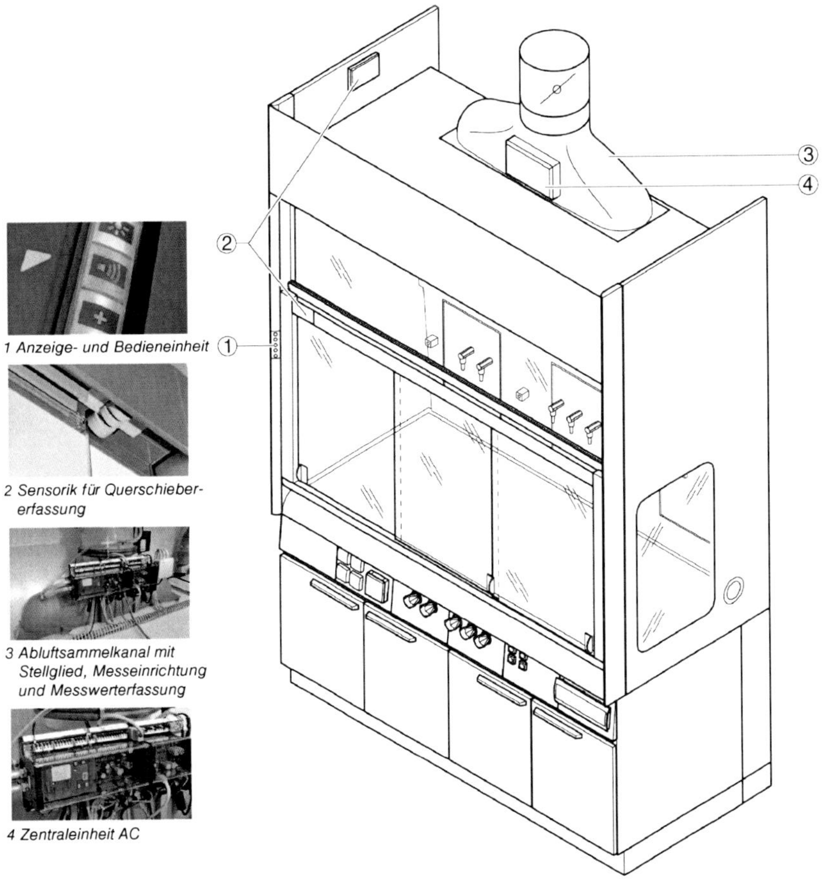

Figure 20.6 Fume cupboard and components.

As a complete safety device, fume cupboards and variable air volume control are together type tested according to EN 14175 part 6. They save precious time while coordinating different trades and mediating in accidents, thereby ensuring legal certainty from a single source.

The variable shield factor and the specific operation of the measuring device enable a volume flow hub in the ration 1:15. In the night mode, an air volume reduction of the fume cupboard of $100\,m^3\,h^{-1}$ is feasible. Also, a measurement accuracy of $\pm5\%$ on the current actual flow value is guaranteed. This is necessary to ensure compliance with the directed air flow in the laboratory, even at low flow rates.

A module cyclically detects the extract air volumes of the extracted individual units in the laboratory, in order to assess the total quantity of extracted air. For four different operating states of laboratory space, a minimum air change can be observed in each case. If the minimum air exchange is not achieved by the minimum values of the air vents, the corresponding minimum value is determined and forwarded to the fume cupboards or extract air volume flow controller through the module. In case the minimum air exchange rate is exceeded by opening a fume cupboard, other fume cupboards or the extract air volume flow controller is reduced to the minimum air value with the help of air control technology. Further, if the minimum air exchange rate is exceeded, the room supply increases.

It is possible to regulate temperature and pressure through the module.

A given *simultaneity* (per laboratory max. amount of extract air) for the usage of fume cupboards extract can be monitored. In case the preset maximum extract air volume exceeds, signaling at the fume cupboards occurs in the laboratory. The control unit triggers via the internal bus system room supply and extract air volume flow controller. Data exchange between laboratory control and the DDC[3] or BMS can be achieved with the following interfaces:

- Modbus RTU
- LON bus
- Profibus
- Ethernet
- BACnet
- analog I/O.

20.4
Fume Cupboard Monitoring, -Control and Room Control

Fume cupboards are to be used in the laboratory to protect personnel. In order to indicate to users whether they can work on the fume cupboard or not, a visual and audible alarm must be carried out according EN 14175 in case of failure. The acoustic alarm can be switched off by the user. The visual alarm is active until the problem is solved.

3) Direct Digital Control.

Fume cupboards can be technically operated with respect to ventilation based on the following two principles.

1) *With constant extract air volume:*
   This means that regardless of the sash position, the same air volume is extracted, according to the construction type examination determined EN 14175 Part 3, and is controlled by the function control. There is no airflow control through the monitoring function.
2) *With variable air volume:*
   This means that a variable amount of air is extracted in accordance with the sash position. The amount of air surveillance and control is affected by an *air-flow controller.* In the closed sash position, 200 $m^3\,h^{-1}$ of air is extracted and at a sash opening of 500 mm at least the extract volume is determined according to type testing EN 14175 parts 3 and 6. The amount of extract air can be increased in steps or as a straight-line depending on the sash position. In the event of an emergency, the user must have the ability via the control-field of the fume cupboard system, even if the sash is closed, to raise the extract air volume to the maximum value. For night operation, the air flow can be reduced to 100 $m^3\,h^{-1}$.
   To have only one person responsible for ensuring safety, it makes sense to use fume cupboards that together with the monitoring control are prototype tested according to EN 14175 part 3 with a constant amount of air or under part 6 with variable air volume.

20.5 Laboratory Control

A *laboratory control system* is nowadays based on state-of-the-art technology and essential for a laboratory building. With this control system, the extract amount of laboratory space is adapted to the present conditions and optimized. Thus, one achieves a significant reduction in loading operating costs compared to the costs involved in old laboratory buildings.

Laboratory control can be considered for the dimensioning of the ventilation system. This means that all ventilation components (duct network, fans, heat exchanger, filter units) can be made smaller. Evidence for the establishment of a simultaneity-factor is based on the following factors:

- Are all laboratories used simultaneously?
- How many fume cupboards are used by an occupant or how many fume cupboards are rarely opened because the current activity does not require them?

Realistically, a diversity of 100% is expected in training laboratories.

You have to start by creating a room air balance to get an overview of the required extract air quantities in the laboratory. In this room air balance, the laboratory space is evaluated and the minimum air quantity defined to be observed.

In most countries, this is as a recommended value for planning, for example, according to DIN 1946 Part 7 (in Germany) 25 $m^3 h^{-1} m^{-2}$ of floor space. This captures all the air consumers present in the laboratory with their maximum extract air quantities and thus determines the required amount of room supply air. To keep the laboratory space in negative pressure compared to the pressure in the corridor, less air is supplied than extracted. The difference between supply and extract air should always be constant. Thus, a constant supply of airflow (sum of the individual differences of the adjacent laboratory rooms) is provided even in the corridor area. Similarly, the strategy for the laboratory control is defined by following the options given below:

1) The entire extract air, except the 24 h-consumer is extracted via the in-room installed fume cupboards. Here, the air value of fume cupboards with closed sashes is set in such a way that the minimum extract volume is achieved for the laboratory space in total. If a fume cupboard is opened, the extract air value is increased and supply air is tracked accordingly. In this variant, of course, is that the first extract air of the remaining fume cupboards is reduced to 200 $m^3 h^{-1}$, before the supply air is increased by the master function (Figure 20.7).
2) Extract air volume required is extracted via the fume cupboards and room air extract systems. This is often the case when separate extract systems are given for fume cupboards and room extract air by the ventilation concept. In this concept, the room-extract works counter-rotating to the fume cupboards. If fume cupboards are closed, the maximum extract-air-flow through the room-extract is taken. If fume cupboards are opened, the room-extract-air is reduced to the minimum value. If even more fume cupboards are opened,

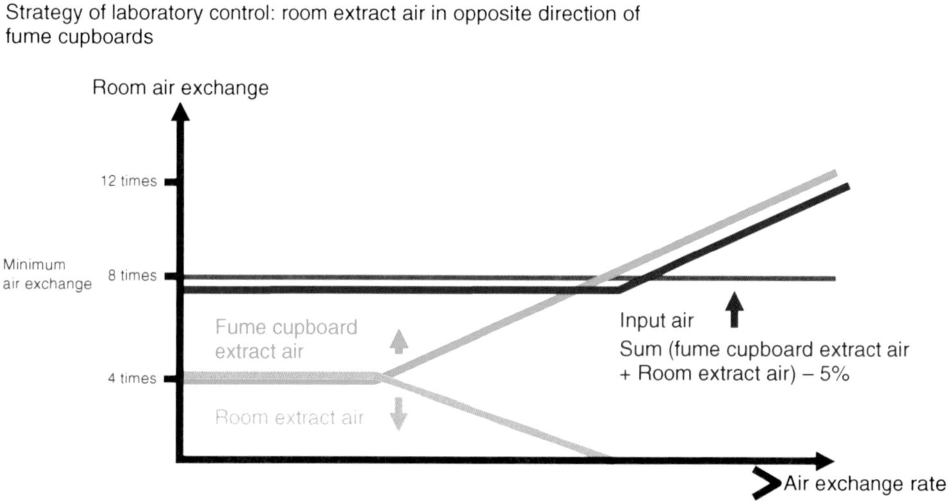

Figure 20.7 Room-extract-air by fume cupboards.

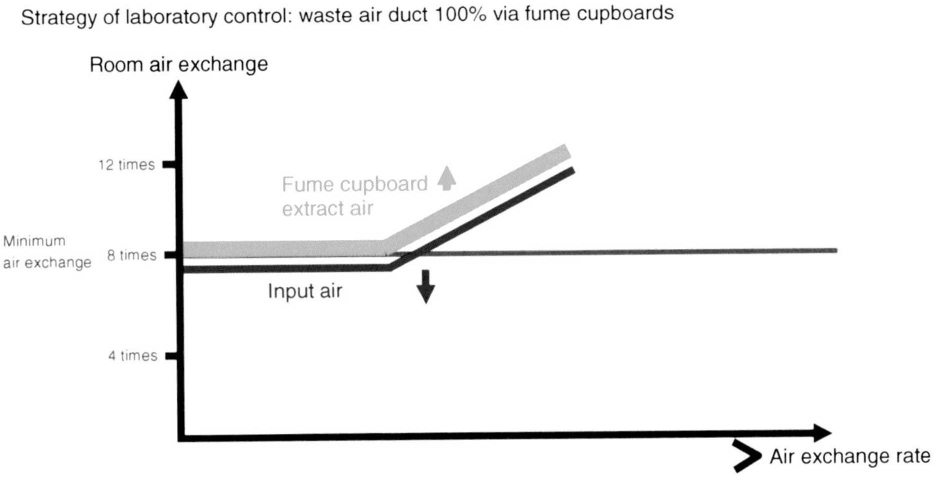

Figure 20.8 Room-extract-air by fume cupboards and counterwise extract.

the room supply is increased. This management follows the master function (Figure 20.8).

Other objects of the master function are as follows:

1) Monitoring the specified simultaneity. If the maximum allowable extraction airflow for the lab is accomplished, the extract value at the last open fume cupboard is frozen and lack of air is indicated. It can also be done as a separate signaling in the laboratory.
2) Implementation of a day/night mode. The approach of the day/night should be coordinated with the plant operators and users. So the day/night mode is switched via a central time program in operations with fixed working hours. For operations with irregular laboratory use such as universities, the day/night switching is often implemented through local switches or with the fume cupboard-controller. That means, if a fume cupboard is opened, the laboratory control recognizes and causes an automatic switch from the night to day mode.
   a. With appropriate requirements for room pressure, the room pressure control is taken over by the master. In addition, the required room pressure sensors as well as room-pressure monitoring units must be available.
   b. Heat load dissipation. If in the laboratory, the desired room temperature is over-reached, a corresponding input signal on the master or the extract-air-volume-flow-controller may increase the air-volume of single fume cupboards and thus, the room air exchange removes the heat loads.
   c. The master is also the interface between the laboratory room control and DDC or BMS.[4] This means that all possible data points of the laboratory room control are present on the master and on standardized bus protocols

4) Building Management Systems.

(e.g., Modbus RTU, Profanbus, Lebus, BackNet, Ethernet) are provided. Also all switching commands of the DDC go to the master and are then passed on to the other components.

The following components belong to a laboratory control system.

1) Variable volume flow controller for user-switchable extract, that is, more hood- or spot extraction.
2) Mechanical volume flow controller for storage cabinets that need to be extracted for 24 h.
3) Extract air volume flow controller
4) Variable extract-flow-controller with integrated air volume control. The flow rate is regulated accordingly the sash opening.
5) Room-supply-volume-flow-controller
6) Data cable for bus communication
7) Master-controller for air management (Optional including room-pressure control)
8) Sash controller.

The components for the laboratory control should come from the supplier of the fume cupboards. This ensures that all components match and the full functionality of laboratory control can be used.

20.6 Sash Controller

A coherent safety and energy-saving concept in a laboratory room is completed only by a sash controller on each fume cupboard. The sash controller ensures that a fume cupboard is automatically closed when the fume cupboard is not at work (according to the requirements of Technical Rule 526 in Germany). By means of closed fume cupboards, safety in the laboratory is increased and operating costs reduced if the fume cupboards are equipped with a variable sash controller.

With help of the sash controller, the opening and closing of the front sash are motorized supported and continued. The sash of fume cupboard is monitored with a motion sensor. No perceived motion is detected in front of the fume cupboard within a specified period, and the sash is closed. By integrating the photocell in the lower edge of the sash, obstacles that are in the sash plane will certainly be detected and the closing procedure is stopped (Figures 20.9–20.11).

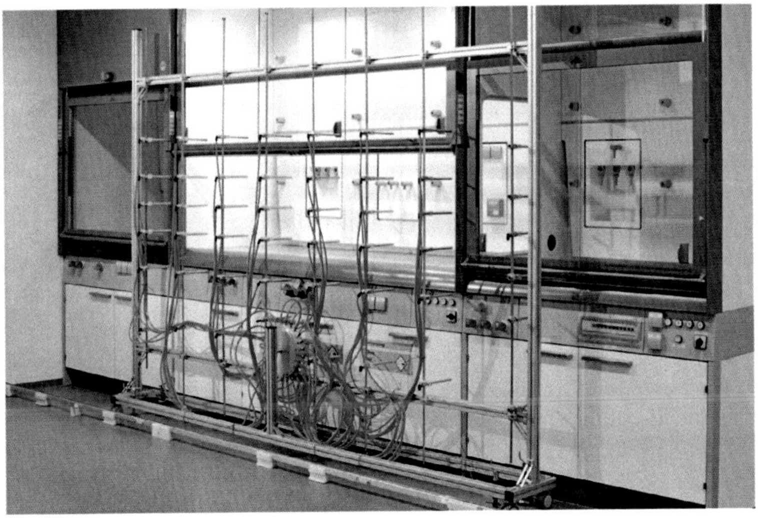

EN 14175-6, typgeprüfte Abzugsregelung nach 5.4
Messung in der äußeren Messebene

Figure 20.9 Test of robustness.

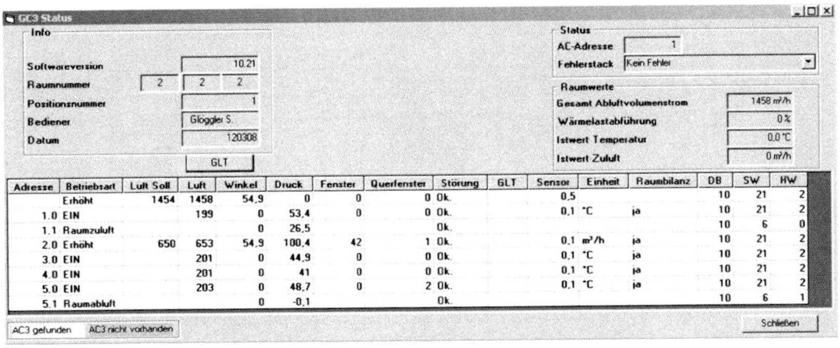

Adresse	Betriebsart	Luft Soll	Luft	Winkel	Druck	Fenster	Querfenster	Störung	GLT	Sensor	Einheit	Raumbilanz	DB	SW	HW
	Erhöht	1454	1458	54,9	0	0	0	Ok.		0,5			10	21	2
1.0	EIN		199	0	53,4	0	0	Ok.		0,1	°C	ja	10	21	2
1.1	Raumzuluft			0	26,5			Ok.					10	6	0
2.0	Erhöht	650	653	54,9	100,4	42	1	Ok.		0,1	m³/h	ja	10	21	2
3.0	EIN		201	0	44,9	0	0	Ok.		0,1	°C	ja	10	21	2
4.0	EIN		201	0	41	0	0	Ok.		0,1	°C	ja	10	21	2
5.0	EIN		203	0	48,7	0	2	Ok.		0,1	°C	ja	10	21	2
5.1	Raumabluft			0	-0,1			Ok.					10	6	1

Figure 20.10 Test chart.

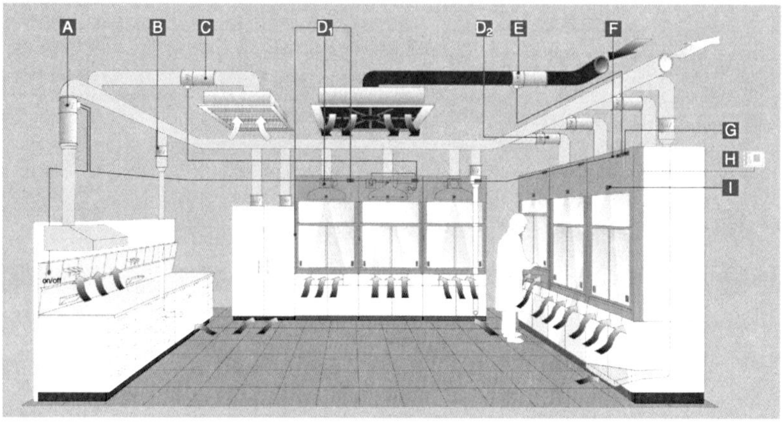

Figure 20.11 Extract monitoring with extract-air-controller.

21
Laboratory Furniture Made from Stainless Steel – for Clean-Rooms, Labs, Medical-, and Industry Applications

Eberhard Dürr

The following article presents the arguments in favor of stainless steel as the perfect material for the construction of clean rooms and similar facilities.

21.1
Areas for Stainless Steel Equipment

Industry production in:

- Semi-conductor technology
- Automotive industry
- Circuit-board manufacturing
- Nanotechnology
- Wafer manufacturing for polycrystalline
- Silicium
- Electrical engineering
- Optical industry
- Photovoltaics
- Meat processing
- Food production
- Dairy processing
- Automotive industry
- Production of medical articles
- Heavy engineering
- Microsystems technology
- Precision engineering
- Measurement facilities according to VDI/VDE 2627
- Use of satellites
- Space technology.

Medical technology:

- Operating rooms
- Implants

The Sustainable Laboratory Handbook: Design, Equipment, Operation, First Edition.
Edited by Egbert Dittrich.

- Isolation wards
- Medical dressings
- CSSD/central sterilization supply department
- Tropical laboratories
- Botanical laboratories
- Production of medical articles
- Blood banks
- Manufacturing of products that are subject to the Medical Products Act.

Pharmacology:

- Pharmacies
- Animal husbandry
- Oral medication
- Pathology, forensic labs
- Parenteralia
- Biological/gene technology
- BSL-3 laboratories
- BSL-4 laboratories
- Packaging and manufacturing of pharmaceuticals
- Research labs.

There are currently more than 6.7 billion people on earth, and an end to the trend of growing populations is not yet in sight. It is estimated that at least 9 million human beings will inhabit our planet by the middle of this century, mainly due to the population growth in developing countries while the number of people in European countries looks set to take a further fall.

An ever broader range of products and production technologies requires different types of clean-room conditions that must be validated and qualified. Even small levels of contamination during the process may already cause significant disruptions, losses, and inefficiencies, which is why the demand for clean-room-compatible facilities, equipment, materials, and production resources is continuously increasing.

In the wake of these developments, the global market is increasingly demanding a practicable, quantitative description, and specification of clean-room parameters and of all different requirements of clean-room equipment and components, regarding them as key technological performance criteria.

The certainty of clean components, clean processes, and the reproducibility of analytical results is very important in laboratories, primarily in those with GMP-requirements.

21.2 Hygienic Requirements of Surfaces

Hygiene is a cause that concerns us all!

The risks of cross-contaminations and infections in the public health sector are increasing. It therefore seems no surprise that the cleaning of contaminated surfaces is subjected to ever stricter international hygiene rules.

The cleaning of surfaces in sensitive areas such as clean rooms, manufacturing facilities for pharmaceuticals, laboratories, and hospitals is governed by more and more demanding regulations than ever before. Surfaces in these areas must not only be "visibly clean" but also free from bacteria, germs, particles, prions, and other forms of contamination, and it must be made sure that they pose no risk whatsoever for humans.

21.3 How to Clean and Disinfect Stainless Steel Surfaces

Here are some recommendations of how to clean and disinfect surfaces:

1) The first thing you need to ensure if you want to obtain surfaces which are clean and uncontaminated is – according to a study of the University of Cincinnati – that you are using the right materials in your institute and laboratory facilities.
   This study, published in the "Journal of Clinical Microbiology," demonstrates that germs can survive for up to 90 days on plastic surfaces. This is why many studies recommend the use of stainless steel products in critical areas.
2) It is important to take into account what level of cleanliness and disinfection you intend to achieve. Different microorganisms have different levels of resistance against different disinfectants, and your decontamination strategy needs to reflect that.
   Resistance for our purposes describes the ability of a microorganism to resist a certain harmful influence, to survive, and continue to grow. There are different types of resistance.
   Recent studies have revealed an increasing number of cases where microorganisms have developed high levels of resistance against certain disinfectants. For example:
   a. Food processing companies have isolated Staphylococcus aureus strains which are resistant against quaternary ammonium compounds.
   b. Salmonella isolates that are resistant against active chlorine.
3) *Adaptation*: There is a fundamental difference between resistance and the temporary adaptation of a microorganism to harmful influences. Such adaptations can occur under certain conditions of growth and are not passed on through the genes, which is why they are called *phenotypic adaptations*.
   The development of phenotypic adaptations is accelerated by the wrong use of disinfectants (e.g., insufficiently large concentrations). Once a biofilm has been created, it is often only possible to remove it with a through mechanical cleaning and a subsequent disinfection. A timely disinfection can prevent biological material from building up and from establishing a continuous biofilm. Proper application prevents problems. We shall demonstrate later that the selection of the correct detergent and its correct use are key elements of a

successful cleaning and disinfection process. It is vital to interrupt the transmission path of commonly used equipment.

Disinfectants must always be used in microbiocide concentrations. This means that they must not be applied in concentrations below the concentrations that have been established and identified by the manufacturer in laboratory experiments and in compliance with generally recognized standards (such as EN 1276). You must also produce sufficient amounts of application solution: enough to cover the entire surface you want to disinfect so that all microbes will be killed. The application of mechanical forces (wiping, scrubbing, use of foam, or of sufficiently high-flow velocities in closed systems) serves to increase the effect of the disinfection.

In addition to the concentration, exposure time is also a key process parameter. If you are required to rinse the disinfected surface with clear drinking water – which is the rule for many applications in the food processing industry (laboratory medicine, clean-room industry) – you should only do so after you have given the disinfectant the manufacturer-specified time to take effect. If no rinsing is required – for example, because the disinfected surface does not come into direct contact with any equipment – you can, conversely, use the surface, once it has dried down, even if the defined exposure time has not yet elapsed.

The pH value is another important consideration for many active ingredients. There are agents (including certain disinfectants and detergents), which can only be mixed or used simultaneously if such use has been expressly approved by the manufacturer. Otherwise, they may lose some or all of their effect, while certain disinfectants can – when mixed with other agents – even release toxic fumes.

Temperature is another key parameter for a successful disinfection process. This is specifically relevant for the disinfection of refrigeration areas, a frequent requirement in food processing. When temperatures are low, concentrations must generally be higher. This is reflected by the instructions for use that are issued by the manufacturers.

Every small concentrations of organic contamination reduce the effect of disinfectants. In many cases, this can be avoided by a thorough pre-cleaning routine, but where this is no serious option for technical or organizational reasons, the disinfectant may need to be applied in higher concentrations. This is also reflected by the new European standards for the testing of disinfectants which mandate that disinfectants must be tested once under "clean" and once under "dirty" conditions. The type of contamination must be selected by taking into account the agent's designated area of application.

Last but not least, it must be ensured that the disinfectant's range of effectiveness is adequate for its designated area of application. Otherwise, microorganisms could survive, bearing in mind the aforementioned intrinsic levels and types of resistance – if, for example, agents are used to fight mold in a specific area. It must also be ensured that the intrinsic resistance of the test organisms – on whose basis the effectiveness of the agent is established – is indeed representative for the entire group of organisms. This is also a requirement of the new European standards (EN 1276, EN 1650, etc.). If used correctly, disinfectants do not have to be changed or "rotated" regularly. If disinfectants are correctly used, they eliminate undesired microorganisms. We must not allow the development of high levels of resistance.

4) The effectiveness of any disinfection process is also determined by the roughness of the surface to which it is applied. This is why the use of stainless steel surfaces (and others) requires the establishment of roughness parameters in compliance with DIN 4287.
5) *Cleaning with alcohol*: The large-scale application of alcohol-based disinfectant sprays can create a risk of explosion and fire.
6) The comprehensive list of disinfectants issued by the German Society for Hygiene and Microbiology (DGHM) provides a detailed index of application-specific parameters.

21.4 Cleanliness Classes for Sterile Areas

The guidelines determine maximum levels of contamination in respect of particulates and CFUs depending on volume, surface size, and time.

For further information about classifications of sterile areas in good manufacturing practice (GMP) classes A to F, please see the EU Guidelines on GMP ECV 2000.

Since it is not possible to fully shut down all particle sources even in clean-room facilities (with the exception of stainless steel clean rooms), airflow engineering (turbulent mixed ventilation, laminar flows) and a combination of high-pressure and low-pressure areas are applied to prevent contaminations and cross-contaminations.

The following table shows the level of particulate emission typically produced by human beings (average per minute) (Table 21.1).

Particles include living particles, that is, bacteria and viruses (CFUs). On 1 cm^2 of human skin, such germs can be found in concentrations of up to 1000 (hands), 100 000 (forehead) and 1 000 000 (armpit). Every human being emits on average more than 10×10^6 bacteria every single day, which multiply further and can reach a number in excess of 10×10^8 within 24 h.

There are two levels of clean-room technology for different areas of application and different specifications of the designated products.

Depending on the requirements, clean rooms with different cleanliness classes following ISO or c'GMP can be provided (Table 21.2).

EU Leonardo da Vinci – Microorganisms and their effects.

Table 21.1 Particle emission.

Activities	Numbers of particles > 0.3 μm min^{-1}
Sitting and standing in motion	100 000
Sitting with a slight movement	500 000
Sitting with medium movement of body and arms	1 000 000
Alternately sitting and standing	2 500 000
Walking (3.5 km h^{-1})	5 000 000
Walking (6.0 km h^{-1})	7 500 000
Go fast (about 8–9 km h^{-1})	10 000 000
Calisthenics	10×10^6 up to 30×10^6

Table 21.2 Classification of air cleanliness.

ISO 14640-1 Classification number	Maximum concentration limits (particles per cubic meter of air) for particles equal to and larger than the considered sizes shown below (concentration limits are calculated in accordance with calculation (1) in 3.2)					
	0.1 μm	0.2 μm	0.3 μm	0.5 μm	1 μm	5 μm
ISO class 1	10	2	—	—	—	—
ISO class 2	100	24	10	4	—	—
ISO class 3	1 000	237	102	35	8	—
ISO class 4	10 000	2 370	1 020	352	83	—
ISO class 5	100 000	23 700	10 200	3 520	832	29
ISO class 6	1 000 000	237 000	102 000	35 200	8 320	293
ISO class 7	—	—	—	352 000	83 200	2 930
ISO class 8	—	—	—	3 520 000	832 000	29 300
ISO class 9	—	—	—	35 200 000	8 320 000	293 000

Note: Uncertainties related to the measurement process require that concentration data with no more than three significant figures be used in determining the classification level.

21.5 Microorganisms

The word "microorganism" derives from the Greek word for small, "micro." Microorganisms are indeed so small that you can only observe them under a microscope. Microorganisms are generally divided into four main groups: bacteria, fungi, viruses, and protozoans. Microorganisms are very common in nature and can be found in all aquatic ecosystems such as fresh water, rain water, and the sea.

21.5.1 Microbial Decontamination

Conditions:

- *Sterilization*: This is an absolute requirement, since there are no grades of sterility. Conditions: moist heat, 121 °C for 15 min and 2 bar or dry heat, 170 °C for 120 min.
- *Disinfection*: Chemical or physical destruction/removal of infectious agents. Disinfectants are (commonly chemical) agents, which kill the living forms of pathogenic germs (generally sparing spores). Disinfectants reduce or inhibit growth (breaking the chain of infection), but they do not sterilize.
- *Decontamination*: Cleaning in order to remove contamination and to eliminate the possibility of contamination.
- *Pasteurization*: A type of disinfection for materials that would be deformed or destroyed by excessive heat (e.g., milk). The one-off or repeated application of lower levels of heat over a defined period of time serves to reduce the number of viable microorganisms.
- *Germicide*: The application of chemical agents which kill microbes but not spores. Bactericides kill bacteria, virucides kill viruses, fungicides kill fungi, and sporicides kill bacteria spores.
- *Sanitation*: Reduction of pathogenic germs in order to protect public health. Procedure: mechanical cleaning or application of chemicals.

Bacteria are the world's most common type of organism. They live nearly everywhere: inside dirt, the earth, water, plants, and animals.

Many bacteria are highly useful for humans, while only a few are harmful. Some bacteria, for example, are vital players in the decomposition process, engineering the circulation of nutrients within the ecological cycle, while others are of great use for the manufacturing and pharmaceutical industries.

Pathogenic bacteria (microorganisms), on the other hand, cause illnesses. Bacteria are subdivided into different groups according to their shape (coccoids, bacilli, etc.).

21.5.2 Bacterial Spores

If exposed to adverse or life-threatening conditions, bacteria can enter a permanent vegetative state and become so-called spores. This allows them to survive when nutrients are not available in sufficient quantity and under conditions that would otherwise cause them to dry up. Spores are resistant even against common antimicrobial agents and strategies. They surround themselves with a hard protective sheath which allows the dormant bacterium to survive for periods that can last a few weeks or up to several years. If the conditions improve, the spores can convert into cells.

21.5.3
Coccoid Bacteria (Round Shape)

Examples include:

- *Escherichia coli*: coliform enterobacteriaceae, causes food spoilage
- *Staphylococcus aureus*: pyogenic (pus producing) germ, transmitted by smear infection (MRSA)
- *Enterococcus faecium (couple)*: intestinal germ, resistant against a broad range of antibiotics (VRE)
- *Streptococcus pneumonia*: causes pneumonia, meningitis, otitis media (middle ear inflammation), and other illnesses
- *Micrococcus* spp. *(tetrads)*: relatively harmless skin germ.

21.5.4
Bacilli (Rod-Shaped Bacteria)

Examples include:

- *Pseudomonas aeruginosa*: "waterborne germ," resistant against antibiotics, causes nosocomial infections
- *Mycobacterium tuberculosis*: causes TB
- *Bacillus cereus*: causes dangerous neonatal infections
- *Lactobacillus acidophilus*: known as a *probiotic* bacterium
- *Legionella pneumophila*: causes "Legionnaire's Disease," transmitted chiefly through air-conditioning systems and in showers.

21.5.5
Other Forms of Bacteria

Examples include:

- *Vibrio cholera (vibrio)*: human pathogen, causes cholera (severe diarrhea)
- *Helicobacter pylori (vibrio)*: causes ulcers
- *Leptospira interrogans (spirochetes)*: an infection that attacks the liver and the kidneys
- *Borrelia burgdorferi (spirillum)*: causes borrelliosis.

21.5.6
Fungi

Fungi are sub-divided into yeasts and molds. They are used for the industrial production of numerous useful products. They otherwise raise spoilage of fruit and vegetables

1) Yeasts (typically unicellular fungi), including:
    - *Saccharomyces cerevisiae* (baking, brewing)
    - *Candida albicans*: widespread pathogen, causes vaginal infections

2) Molds ("filamentary"):
   - *Penicillium roqueforti*: used in cheese-making
   - *Penicillium notatum*: the first antibiotic.

21.5.7 Viruses

Viruses cannot be seen under the microscope. They can infect humans, animals, plants, and other microorganisms. Viruses live and multiply in host cells. They feature either DNA or RNA, but never both. They appear in three main forms: helical, icosahedral (with 20 even, triangular faces), and complex. Examples for viruses include: rotavirus, herpes simplex, hepatitis A, HIV.

21.5.8 Protozoans

Protozoans are unicellular organisms and ubiquitous in aquatic environments or in dirt. They are classified into four groups according to the ways in which they move: flagellates, amoeba, sporozoans, and ciliates. Protozoans are of key importance for the food chain, since they eat plants and decompose organic residue. They serve to "control" the biomass and bacterial populations. Some protozoans can produce cysts under certain adverse conditions.

Examples include: malaria parasites, trypanosomes, Giardia lamblia, and cryptosporidium.

21.5.9 Waterborne Pathogenic Germs

Waterborne human-pathogenic germs are generally the results of fecal contamination. *Escherichia coli* is a typical indicator for such a contermination. Other indicators of a water contermination include nonspecific coliforms and Pseudomonas aeruginosa. Data from the early 1990s (Wikipedia) provide the following description of infections that have been caused by waterborne germs:

- *caused by protozoans*: amoebic dysentery, cryptosporidiosis, and giardiasis
- *caused by bacteria*: cholera, botulism, typhus, diarrhea, Legionnaire's disease, and leptospirosis
- *caused by viruses*: hepatitis A, polio, rotavirus, small round and structured viruses, and enteroviruses.

21.5.10 Individual Cleaning Concepts and Hygiene Regulations Plus Decontamination Measures

The following is an excerpt from the long and diverse list of surface requirements. Customized double-door systems, shelf-equipped locks, material airlocks, and personnel transfer systems.

The stainless steel facility must provide a controlled manufacturing environment and comply with the strict cleanliness requirements of the clean room where customized air-conditioning and ventilation systems, material/shelf-equipped airlocks and double-door personnel transfer systems create conditions that meet all the requirements.

21.5.11
Surface Configuration

Prevent contamination by ensuring that members of staff behave correctly.

Humans are the primary source of contamination in clean room environments. Freshly trained members of staff tend to behave correctly, but as the memory of their training fades, clean-room discipline generally deteriorates. In extreme cases, bad behavior such as quick walking – which creates turbulences in the air and upsets the laminar air flow – can even jeopardize the clean-room classification. Other rules which are frequently breached in everyday practice concern the conscientious donning and doffing of protective clean-room equipment, hygiene, and the handling of personal objects.

Hygiene → Documenting infections as an element of quality management

The removal of potential biological cakings and pharmacological substance compounds or residual disinfectants which can be safely cleaned.

Aluminum can be corroded by hardened water.

Nosocomial infections – whose numbers are rising – are indisputably relevant for the clinical stays of patients.

Cleaning: do not use any organic solvents.

21.5.12
Hospital Hygiene

Building projects in the public health sector – nursing homes, rehab facilities, hospitals, medical clinics, and so on – are a growing segment of the institutional construction market. At the same time, healthcare watchdog authorities and supervising bodies are becoming increasingly aware of the risks posed by Healthcare Associated Infections (HCAI), so-called hospital infections such as MRSA, vancomycin-resistant enterococci, and Clostridium difficile – which is why a great deal of research is focusing on this specific area. Although general hygiene is the first line of defense, improved supervision, the insulation of ventilation and filter systems, the provision of fixtures and fittings with the required levels of surface roughness, and the development of new disinfectants and antimicrobial agents are of similar importance.

The conduct of particle measurements in stainless steel clean rooms in compliance with new regulations is valid since March 2009.

Initial validation, revalidation, and monitoring of the clean-room facilities in compliance with Annex 1 of the EU Guideline.

Measurements in clean-room facilities need to be qualified by information about the facility's status. Validation measurements are commonly performed when the facility is "at rest." Revalidation measurements will be commonly performed once a year.

In this issue, the EU Guidelines for the first time refer to the series of clean-room technology standards.

ISO 14644 – under its general title of "Clean rooms and associated controlled environments" – now comprehends nine sections:

1) Classification of air cleanliness
2) Specifications for testing and monitoring to prove continued compliance with ISO 14644-1
3) Test methods
4) Design, construction, and start-up
5) Operation
6) Vocabulary
7) Separative devices (clean air hoods, glove boxes, isolators, and mini-environments)
8) Classification of airborne molecular contamination
9) Classification of surface particle cleanliness.

21.5.13 Biological Sciences

New vaccines, an exponentially growing number of new biopharmaceuticals, and hybrid medical technology represent major challenges for all manufacturers along the process chain in terms of a contamination control that can ensure uncompromised levels of product safety and product integrity. In combination with an intensified use of highly toxic and volatile substances, this will require improvements of the redundant safety systems for surfaces, facilities, and items of equipment. The members of staff must be capable of applying the required contamination control particles purposefully on these surfaces. The process reliability of the surfaces and facilities that are made from stainless steel also reflects the commercial pressures to reduce both development and operating costs on a broad basis, innovative projects cutting costs, and increasing efficiency.

General conditions for clean rooms:

- Many diverse requirements
- Good sound insulation properties.

21.5.14 Relevance

Improved safety features in line with the HACCP guidelines (Hazard Analysis and Critical Control Points).

Figure 21.1 Particel Emmission.

21.5.15
Why Stainless Steel?

All this explains why chromium–nickel steel – a material that poses no hygiene risks – has come to be regarded as a viable method of reducing the contamination of working surfaces as far as possible. In order to benefit from the excellent hygienic properties of stainless steel, the surfaces must have low levels of surface roughness and high-quality welding joints. Smooth surfaces in the micrometers range, however, are not alone sufficient for the efficient removal of contaminations. The surface charge, the surface structure, and the material's domain structure are of equal relevance.

It may be easy for stainless steel to form a wafer-thin, chemically stable passive layer, due to the chemical reaction between the chromium and the airborne or waterborne oxygen.

The way in which the surface has been structured is one of the key factors that determine whether or not a suitable passive layer will be formed.

This passive layer will regenerate on its own under normal conditions, providing near perfect protection from corrosion. In order to obtain this stainless steel effect, we are providing our surfaces with a special treatment in the 30 μm range and only use suitable types of stainless steel.

In order to sterilize surfaces, and to guarantee the reliability of highly sensitive analytical methods, all traces of contamination must be removed.

The conditions that stimulate the growth of bacteria and allow a biofilm to build up on surfaces must be prevented at all cost.

One of the biggest problems is prion contaminations, since common cleaning methods fail to eliminate prions. Traces of blood, too, can leave behind residual contaminations, despite the application of commercial detergents. This includes fibrin, which is generated by the blood's propensity to coagulate.

Our recommendation is based on the results of another research study: In order to reduce the levels of fibrin on stainless steel surfaces significantly, clean the stainless steel plates with a detergent that has been heated up to a temperature of over 40 °C.

21.6
Summary

- Chromium–nickel steel has stood the test of time as a useful, convenient, and durable material. In the past, stainless steel was mainly used for the exteriors of prestigious building projects, but today, it is increasingly applied in the construction of competitive functional applications such as laboratory technology and clean rooms.
- In the future, the areas mentioned on pages 2 and 3 will benefit even more from the advantages of chromium–nickel and other stainless steels.

- Many people are unaware of how bacteria, germs, prions, fibrin, and other contaminations act on stainless steel surfaces.
- In view of the results, the requirements and the many applications of cross-contermination, it must be the objective to reduce any contamination of working surfaces as much as possible.
- Our analysis has provided highly sensitive results which comply with the study from the *Journal of Clinical Microbiology*, the University of Cincinnati and the Institute for Hygiene at Leipzig.
- Bacteria die more quickly on chromium – nickel steel.
- Their mortality exceeds 95%.
- This contrasts with the mortality rate on plastic materials, which fails to reach even 50%.
- Chromium – nickel steel is the only material on which no biofilm is visible with the naked eye.
- The biofilm can be mechanically removed.
- Stainless steel enjoys the advantages of sound mechanical finishing.
- An enormously high economic efficiency of the products in the individual user requirements and tasks can be customized to the different parameters.
- Chromium – nickel steel is fully recyclable (100%).
- In contrast to plastic materials, stainless steel does not react to fire since it only starts to melt at temperatures in excess of 1500 °C. Plastics, conversely, melt at temperatures of 260 °C.
- Chromium – nickel steel allows reliable validations and certifications of product safety and integrity.
- The application of gas in a "worst case scenario" remains possible.
- Nearly perfect levels of sterility.
- Acids and alkaline solutions pose no risk to stainless steel surfaces.
- High levels of corrosion resistance.

22
Clean Benches and Microbiological Safety Cabinets

Walter Glück

22.1
Laboratory Clean Air Instrument, in General and Definition(s)

In today's laboratories, clean air instrument is used for contamination protection of goods/samples (in terms of airborne particulates).

A purified air stream intended to displace potential contaminated air is forced throughout the instrument's inner experimental chamber (ECh) for sample protection.

The instrument designed for protection target may cover

- the sample or,
- user's personal protection or,
- a combined one covering different protection targets.

Goods/samples intended to work with are handled inside the instrument. This inner space/chamber is called "Experimental Chamber." The place of instrument use will be called "Work Place," "Room," or "Laboratory."

"Clean Benches" are designed for protecting the goods/samples against airborne contaminants only. The performance parameters are not standardized – need to be defined due to task.

"Microbiological Safety Cabinets (MSC)" are known in different design. In the first instance, MSC protection is targeting the user protection and the surrounding (+environment). Other protection targets may have been combined in the cabinet design. Anyhow, MSC performance parameters delivering protection are standardized.

MSC in "Enhanced Design" also can be an additional protection tool for;

- the works carried out in purified surrounds, handling minute quantities of active substances such as substances with MCR potential (cytotoxic agent ...);
- high-grade protection (fail safe) of cabinet extract recirculated into the room;
- higher-grade purification of protecting air shower inside the cabin;
- contamination protection of the instrument inside;
- ease of waste filter control using high-efficient particle air (HEPA) filter cartridges.

The Sustainable Laboratory Handbook: Design, Equipment, Operation, First Edition.
Edited by Egbert Dittrich.

The technical ventilation principles of state-of-the-art MSC design are given in the following chapters.

22.2
Possible Joint Possession of "Clean Benches" and "Microbiological Safety Cabinets"

It is important to recall the intention of this description. In the following, "ready-to-plug-in" instrument is mentioned. Doubtless "Projects" built in place may also show parallel design principles mentioned in this description.

22.2.1
Minor Turbulent Purified Air Stream/Purified Laminar Air Stream

Displacing air stream for contamination protection is as effective as less "Mini-Turbulences" it carries. Turbulences in this air stream may cause whirls, backwash, or reflux of particulate contaminants.

A good quality air stream bearing only minor mini turbulences is called "Laminar Air Flow (LAF)."

Bernoulli rules law of stream/flow of air/fluid LAF is not given in defined parameters. Aside air viscosity, length of guarded stream/geometric design of air channel air velocity is paramount important for LAF calculation (Reynolds Number → <2000!).

In practice, footprint and dimension of laboratory instrument are limited and so it is with ECh of clean air instrument. Logically, an effective laminar (air-) flow in this chamber is intended to have nominal relative low velocity. A minimum level for a top-to-down forced air stream shall not be less than 0.25 m s^{-1} inside the ECh due to thermodynamic issues.

Clean air velocities known from clean room technology sites (~room) often are taken as a value for orientation. Those values, in most cases are not practicable in instruments and may cause mini turbulences. A "the more – the better" is displaced here and will limit product protection.

Laboratory instruments with protective LAFs in ECh are pretty sensitive in terms of air-flow pattern robustness.

Quick movements with hands inside the ECh may harm air flow patterns of laminar stream. Hot surfaces/open flame with high thermodynamic potential will essentially harm the air-flow pattern. The result may cause a heavy negative impact on flow conditions inside the ECh and may result in high contamination risk for goods and personnel.

Aside nominal air velocity of laminar air stream, an essential factor is the direction of the air flow patterns. A uniform air profile should be designed inside the ECh intended to flood the ECh like a piston pressing off potential contaminated air.

Air stream uniformity today is standardized with a variety of ±20% from nominal velocity. Surely, manufacturer of serially produced instrument needs to deliver tighter variation of air flow condition inside instruments as this is compared to clean room installations. This variety is rough enough to feed a bigger area with purified air. The need to keep variation closer is clear; the higher the variety inside the air stream higher is the potential of micro turbulences which may carry air-borne particles in other than intended direction.

Valid for especially vertical (from top to down) air streams inside the experimental/instrument chamber is:

- The nominal air velocity of LAF should not be less than $0.25\,m\,s^{-1}$. Velocities (higher than) $>0.4\,m\,s^{-1}$ should be recognized as carrying "Micro Turbulences" (the higher the more). Nominal air velocities in the rank of $0.3\,m\,s^{-1} \pm 10\%$ are good compromise in instruments.

22.2.2
Purified Air Quality inside Experimental Chamber

Process-air in most cases is pre-filtered by mesh or rough fiber filters (ref. EN779) that come as filter mesh, mat, or cassette filters.

Especially, supply-air for LAF inside the ECh today is filtered by (minimum) a HEPA filter in quality H 14 (Filter Standard: EN 1822 … Production and Quality Control …). As a final in row filter, it is able to deliver purified/sterile air. Some instrument design may have HEPA filters in exhaust air also.

The Example HEPA (filter) in Quality H 14 is built for trapping airborne particulate contaminants in very small size. Typically, the worst-to-adsorb size of particles is in the rank of 0.15 – 0.25 μm (MPPS → Most Penetration Particle Size). Here, the H 14 filter should have an adsorption degree of 99.995%!

If sterile work conditions are a mandatory need for processing, for example, in life-sciences labs, the particle size of 0.3 μm is interesting. It is the "Safety Particle Size" (one-third of smallest known Bacteria Size …). In this size, the H 14 HEPA has an adsorption degree of (greater than) $\geq$99.999%, which is accepted as a biological safety parameter for determination of sterile air condition. It says that these smallest known phages today are kept inside the filter three times secure by size restraint in a probability of 99.999% @ minimum.

This theoretically also means one out of 100 000 particles in the size of 0.3 μm can pass the filter (penetration – degree: 0.001%). In practice, in the last 30 years, no accident is known to have happened or can be traced due to an insufficient HEPA filtration – even in a single fault condition, an adsorption of minimum 99.97% is accepted and found to be safe enough to stop the controlled hazardous work in biological areas.

The use of H 14 HEPA filters for safety equipment is the state-of-the-art standard for biological application. Other applications may need to have other or additional filter design, for example, OEB agent processing. But even for such dangerous "Dust" processing, in the first instance, HEPA-filtered instruments bear the most benefits.

Laboratory clean air instruments with horizontal or vertical air stream (LAF) need to be designed and tested based on IEC 1010-1/EN 61010-1 (Safety of Laboratory Equipment …) and under respect of valid radio interference standard(s).

Risk analysis for those accordingly designed laboratory instruments, in most cases defines them as products with risk of electric shock in terms of the European Product Liability Regulation; following the EEC-Low Voltage Regulation.

Performance and quality criteria for internal air flow can also be traced to ISO/EN 14644-1 as a guideline.

If readings are taken accordingly inside the ECh; which is, for example, fed by a H 14 – HEPA supply air filter; ISO class 3 conditions can be validated (ref.: DIN EN ISO 14644-1). Purified air conditions better class 100 (acc. to U.S. Fed. Std. 209e) or GMP – conditions class A (acc. to Annex 1, 03/94/EEC; GMP-Regulation …).

Aside the general IEC 1010-1//EN 61010-1 standard for biological safety cabinets, additional "Performance Criteria related to Safety" are given in EN 12469:2000.

Additionally, for CMR agent/cytotoxic agents in some countries, additional national requirements are valid; for example, Germany: additional criteria given in the National Standard DIN 12980 (equals accepted in Austria and Switzerland by worker safety authorities …).

22.3
Laboratory Clean Air Instruments Intended to Protect the Samples – "Clean Benches"

Often those are designated as "Clean Room Benches" or "Hoods" even though they have no space to build a "room" or you can climb in, or are not intended to extract something from the work space. The only parallel they may have is a (HEPA~) filter.

The designation "Clean Bench" often used in America is in most cases the most matching. There are no design standards out today to build a clean bench. No regulation or limits are given for such instruments for the protecting work carried out.

This may be the reason why most clean benches known today are more or less designed related to parameter limits given in safety cabinet standards. If so, equals are incorporated and sterile work condition can be achieved in the inner instrument work area/work chamber (e.g., class "A" conditions according to GMP – guidelines …).

In case of close-to-production application (Quality Assurance …), such instrument might have additional air purification controls (~Online Particle Monitoring), anemometer probes, humidity sensors, and so on, equipped in order to rebuild production scale in clean room conditions in smaller laboratory scale.

Any mixing-up of these "Clean Benches" with, for example, "microbiological safety cabinetry" might result in serious dangerous work conditions!

It might be a fact that some design is pretty comparable but it should be clear: safety cabinets have special additional performance criteria (limits), which are responsible to support personal and environmental safety within a monitored parameter set. A Clean Bench cannot deliver equals.

Clean benches are intended for best protection of the product handled. They are intended to be used in laboratory environment designed with horizontal or vertical air flow, in addition possibly to LAF. The nature of the laboratory is a demanding design: for example, a chemical lab may require pretty special work tabletops or special high-grade stainless steel, in an opto-electronic lab where tiny optical parts are built, special light or background color is a need....

Field of application is not limited. Paramount important is: Clean Benches are not to be used for dangerous work or to handle dangerous material/agents!

The air flow extracted from such instrument is "blown" back into the room. Releases from the experiment may be part of this air flow.

If under circumstances in conjunction with Clean Benches minor amounts/minute quantities of hazardous materials need to be handled, suitable technical lab ventilation and additional personal protection devices should be part of processing.

22.3.1 Functional Principles of "Clean Benches"

The basic functional principle of Clean Benches is simple (Figure 22.1).

Via a pre-filter element, potential contaminated air (dust ...) is drawn into the system by a blower forcing the air into a plenum in front of the main HEPA – filter.

The air passes the HEPA filter resulting in purified process air. The air flow condition and air velocity is designed to flood the whole work chamber with purified process air.

Eventual contaminants/air residues are blown out of work space.

Using the core zone of purified air stream (LAF) for handling sensitive products/goods enables the formation of a sterile work condition. For example, sterile packaging, handling, pipetting, construction/mounts, concentrations, and purified storage are possible applications.

Both principles: horizontal or vertical LAF Clean Bench design deliver the same: *Product Handling Protection* (Figure 22.2).

22.3.2 Clean Benches: Upsides and Downsides of Design Principles

Horizontal instruments are always good for handling products with good overview in instrument work zone aside purified work condition. An example for atypical application may be the use in a pharmacy laboratory for preparing artificial nutrification preparates out of different sources into one feeding system for the patient under clean conditions or complicated puzzling tiny parts of an extended optical instrument under cleanest conditions.

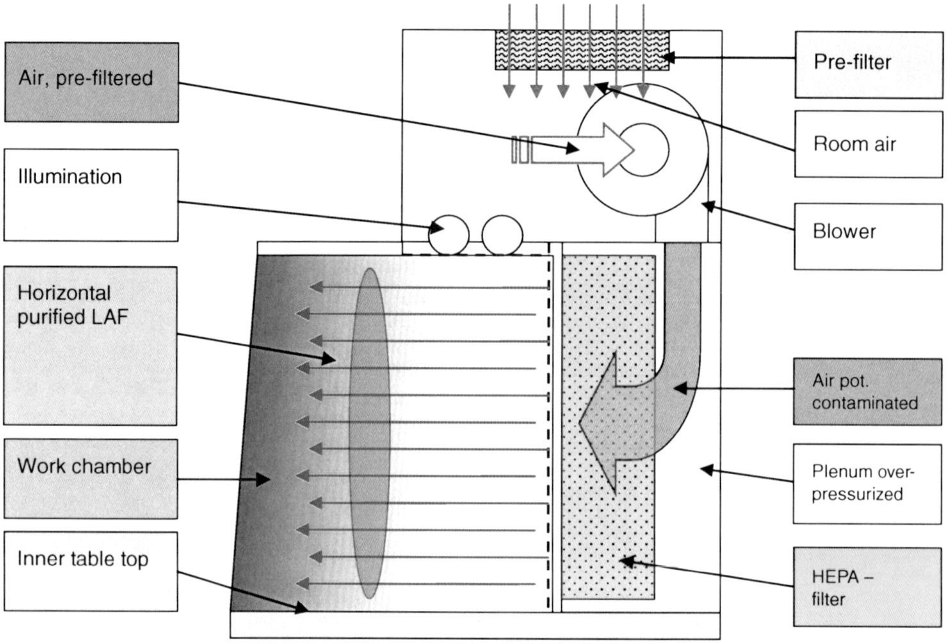

Figure 22.1 Desktop instrument with horizontal airflow, side view cut.

Vertical Clean Benches are, in terms of accessibility, not as good as horizontals. But, they offer great benefit, if lab space is an issue and smaller products need to be handled (or stored). The uniformity of vertical flows naturally is much better as an open-fronted horizontal can be. Additionally, the price difference between a modern product protection device Clean Bench and a Safety Cabinet Class 2 is not very big.

Therefore, it might make sense to think about purchasing a Class 2 device for a little more money while offering additional benefits.

What is the need to look for prior to purchase of a Clean Bench?

- They are only intended to deliver product protection.
- They are not foreseen to handle hazardous agent or material.
- They might be used as "Small Scale Clean Rooms."
- The nominal air velocity of LAF shall be in between 0.3 and 0.4 m s^{-1}.
- LAF profile patterns shall not differ more than 20% from average/nominal.
- Shall have ultimate end-standing H 14 HEPA filter(s) for supply air.
- Are able to rebuild small-scale close-to-production GMP conditions (ref. annex 1; 03/94/EEC)
- Shall be backed with CE marking/documentation by manufacturer
- If built serially – shall be available in different work lengths and possibly in different heights of work chamber
- Shall be intended to be set up in a row without sides in order to build modular purified work zones side-by-side

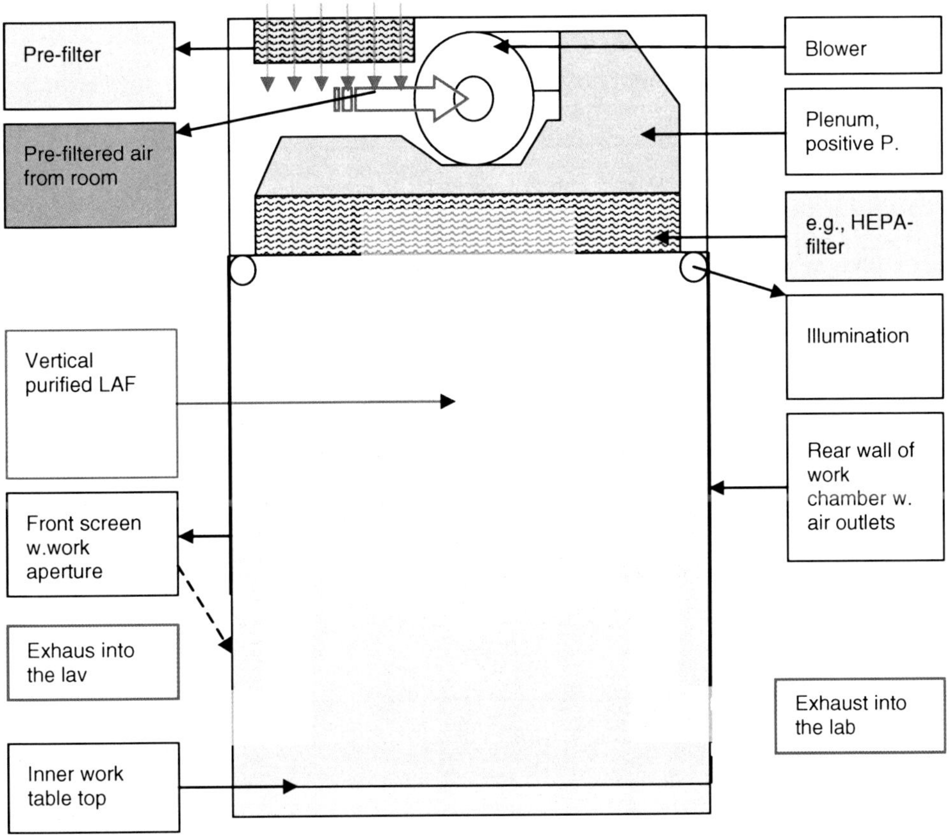

Figure 22.2 Desktop instrument with horizontal airflow, side view cut.

- Shall be capable for different special-colored illumination devices
- Shall have an ultimate ease of cleaning edge and can disinfect easily; all corners shall be rounded surfaces capable for disinfection and cleaning agents can be used according to processing parameters
- Optical and acoustical alarms shall occur if flow conditions/air velocity may vary out of limits, indicating a clogged filter element ...
- Accessories shall be available, like floor stands with adjustable height....

22.4 Microbiological Safety Cabinets

Instrument designed for use in Life Science Laboratory: Environment supporting protection of personnel and environment (laboratory) aside goods handled are designated as MSC.

Additionally, those instrument designs cannot deliver good-/product protection in the first instance, keeping potential airborne particles out of the ECh and protecting ECh against cross contamination of agents handled.

According to their first intended use, MSCs are classified in three classes:

- *Class 1*: Design for personal protection only, Nno product protection parameters at all
- *Class 2*: Design for personal and environmental protection +product– and cross contamination protection
- *Class 3*: Hermetic encapsulation design for personal- and environmental protection + product protection, no cross contamination protection defined; self controlled by negative pressure processing, fail safe design.

22.4.1 Definition of Protective-Functions

At the end of the seventies in the United States, contamination prevention knowledge was widely set on status as it is roughly still today in terms of protection against aerosols and airborne particles. With development of suitable test equipment, microbiological safety in terms of controlled work chambers/cabinets could be defined.

In parallel, the European standardization followed this with little time delay. The United Kingdom, France, and Germany are establishing standardization of MSCs while picking cherries of work carried out in the United States.

In the meantime, internationally, the standardization for MSCs is except some national special requirements and cultural assets roughly in conformity.

22.4.2 Personal Protection

In EN 12469 ("Performance Criteria … related to Safety"), criteria for personal and environmental – protection on an open-front Class 2 safety are determined by "adsorption degree" of front aperture and cabinet exhaust filter.

The cabinet shall withhold at least 10^5 particles of 0.3 μm (particle size related to safety – refer to previous description).

The first guidelines to build MSCs were given in NSF 49 (1978 final version). Today, this leading paper is reviewed a couple of times and has the status of a USA National Standard (refer to ANSi NSF 49 …).

It becomes clear cabinet design with combined protection of personal, environmental, product, cross-contamination – Protection (class 2 cabinet) is more of interest as a device having proper singular protection only.

Product protection defines cabinet's ability to withhold potentially contaminated air intake from products handled inside the ECh.

Cross-contamination protection is defined lastly by quality of LAF inside the ECh. It is a parameter defining how good the air flow inside protects the goods handled against airborne contaminants inside the ECh.

Normative definition of "Product Protection" and "Cross-Contamination Protection" in common understanding mostly is combined to "Product Protection."

22.4.3 Protective Function of Different Cabinet Classification

Doubtless today Class 2 cabinets combining the major protective functions are the "beloved" instruments for laboratory use.

Other classes may provide stronger/better singular protective functions, but they are not intended for "universal use."

MSCs are different from "Clean Benches" described in Chapter 1, especially due to following reasons:

- MSCs can provide additional protection to organizational and individual protective arms handling hazardous materials inside the ECh. They are "Personal Protection Devices," which are reflected in European Law for additional protection for handling hazardous agents
- The manufacturer needs to determine the performance parameters for his construction related to traceable testing carried out in accordance to state-of-the-Art (latest Status of Standard ...)
- All type-testing need to be performed with organism/spores to qualify instrument safety parameters
- Product and cross-contamination protection are on secondary stage importance versus personal – and environmental protection

According to the risk of the different types of biological agents in European Law for handling hazardous materials, four safety classes are defined. All life science laboratories need to be classified by the owner in one of these classes.

The safety/risk classification is orientated to risk resulting for citizens starting with low-to-none risk up to high deadly risk (risk group/class 4) in case of a contamination.

Due to the remaining risk, European worker safety law and law for handling hazardous agents require safety cabinets as an additional protection device (Table 22.1).

22.5 Microbiological Safety Cabinet Class 1

Class 1 Biological Safety Cabinets are intended to use only for personal protection. A powerful air flow inward the front aperture is acting as a barrier against aerosols and particular contaminants released inside the work chamber.

Table 22.1 What class of cabinet for what risk group ... ?

Laboratory risk level// biological safety cabinet class	Risk group level 1	Risk group level 2	Risk group level 3	Risk group level 4
Class 1	X	X	—	—
Class 2	X	X	X (1)	X (2)
Class 3	—	—	—	X
Other/additional criteria	Safety cabinet for Personal Protection is normally not required, often Class 2 is used	—	(1) Often a reinforced Technique is required such as additional exhaust connection or triple filter system	(2) Due to open fronted Cabinet additional total personal protection clothing is required

X = suitable.

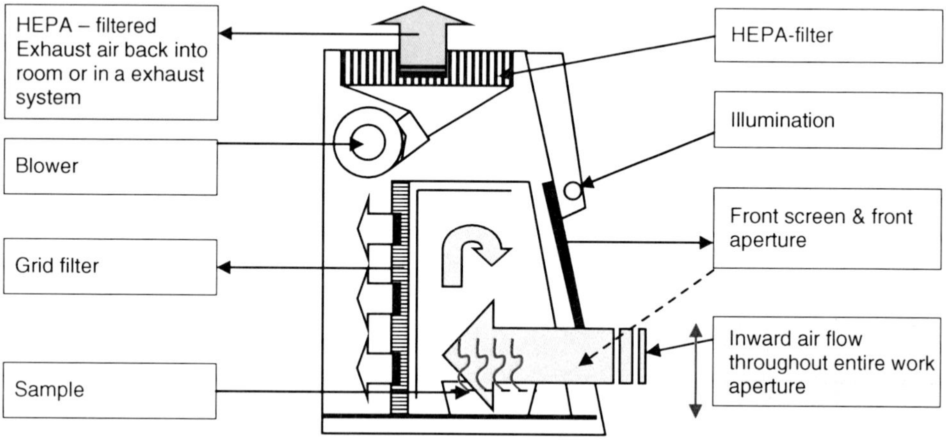

Figure 22.3 Side view, table top unit insight view.

The inward air flow velocity is typically in between 0.7 and $1\,\mathrm{m\,s^{-1}}$. A HEPA filter is responsible to collect all contamination released.

Laboratory room air – potentially contaminated is forced over the entire work area inside. Class 1 does not have any precaution for product or cross-contamination protection (Figure 22.3).

What's the need to know about buying a Class 1 – Cabinet:

Class 1 Cabinets

- Do have personal protection only (and environmental if extract is forced to the outside).

- Do have a relatively high inflow velocity (turbulent toward the inside) and herewith high extract air volume rate.
- Do have high personal protection degree on the front aperture.
- Do NOT have any product protection.

22.6 Microbiological Safety Cabinets Class 2

Cabinets of class 2 design comprise different air flows in order to achieve different protection.

Personal and environmental protection hinders a prohibited amount of aerosols and a particular contamination is released inside the work chamber during the handling of samples (e.g., pipetting). The inward air flow throughout the work aperture and integrity of extract HEPA filter are way by which this can be done.

Product protection is delivered by a downward slow best of the art LAF (supply air). It covers the product handled like a shower and delivers released aerosols and particulate contaminants to the dedicated HEPA filter(s). With the help of this air shower, potentially incoming contamination through front aperture are prevented from getting into the "safe" work area inside the work chamber.

Cross-contamination protection is achieved with a high quality LAF inside. The air shower is responsible to prevent samples handled against airborne contamination to cross over inside the "safe" work area. Harmony and equity of supply airflow patterns are responsible for good cross-contamination protection (Figure 22.4).

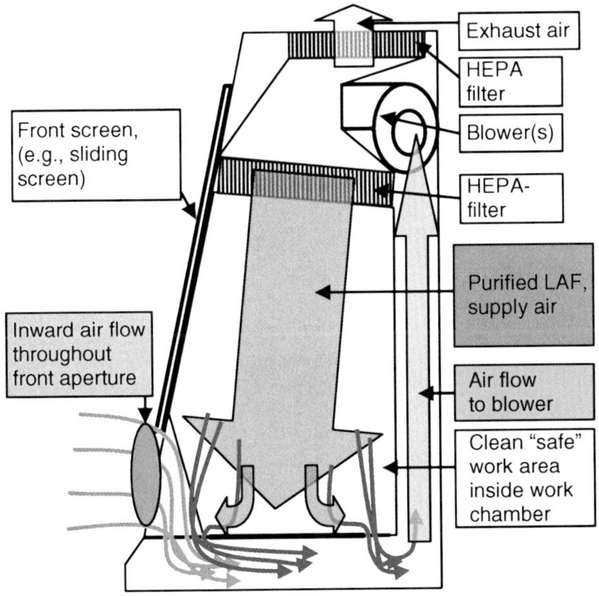

Figure 22.4 Side view, table top unit insight.

Airflow velocity into work aperture is mostly in the rank of minimum 0.4 m s^{-1}. Higher risk may need increased inflow velocity ... 0.55 m s^{-1} for a higher degree of personal protection.

Downward airflow inside needs to be balanced to the inward airflow. A higher inflow pattern is needed to increase the velocity of the supply air. Supply air patterns may become more micro turbulent with increased air velocity. Supply velocity shall be between 0.3 and 0.4 m s^{-1}. The slower the velocity, the better it is. Limits are varying and depend on the air velocity of inflow due to vertical vectors acting.

Airflow patterns naturally influence operational parameters in wide range and in ergonomics. Noise and energy consumption up to room air conditions may be influenced.

Calibration of class 2 cabinet airflow patterns is of paramount importance for personal and product and cross-contamination protection. Therefore, only trained and authorized service personnel shall calibrate/recertify filter integrity and air velocity patterns.

If nominal parameters are adjusted newly, a recertification of safety is a must have. Table 5 of European Standard En 12469 gives minimum requirements for recertification of a class 2 cabinet. Personal protection as a minimum shall become re-certified. In addition, safety may be recertified with KI-Discus method *in situ* (refer to EN 12469, test methods and annexes).

Integrity of built-in HEPA filters is also the cornerstone of safety. Therefore, loading upward airflow to blower(s) with test aerosol is a need for reliable particle counting/filter test. A laser particle monitor with a test volume of 1 cft min^{-1} (28.3 l) is a device that can take readings of 0.3 μm particle size. Clean filter side need to be scanned with iso-kinetic test probe.

It is strictly recommended to file all test results in an instrument life book for reference especially if cabinets of class 2 are used in higher risk area. Failures may be traced this way in case of an accident.

22.7 Enhanced Microbiological Safety Cabinets Class 2

22.7.1 Enhanced Safety by Means of Class 2 Cabinet "Extract Connection" on the Building Extract System

Basically systems shown in 3.8 are suitable in risk group 1–3 areas.

In laboratories there is a great need to ensure safety due to the following reasons:

- handling additional combustible solvents
- unwanted smells released by products handled
- the kind of potentially released aerosols/agent and risk of inhalation

- normally sufficient routine service intervals for re-certification of safety, but frequently upcoming higher risks must be handled and an unknown failure may have a serious and a fatal consequence.

An additional connection cabinet can lower the risk.

Releases out of cabinet extract air can be drawn to the outside without any risk (although an additional treatment may be a need).

A direct tight connection of a cabinet to the outside may be suitable for a singular cabinet having not more than 2–3 m of exhaust hose. Interaction of extract air stream and inflow/personal protection is the reason why it is not the best way to operate a cabinet in enhanced mode. It should be clearly understood that the person who installs the exhaust connection is finally responsible for the safe operation of the cabinet.

Therefore, the method of choice is to connect a safety cabinet class via a thimble duct connector to an in-house exhaust system.

The presence of a suction cabinet close to the exhaust outlet of the cabinet guarantees the extract of airflow in addition to extracting a portion of the room air, which has no risk to the outside world. In case the in-house exhaust system fails to work, the cabinet can be operated in a safe manner. In normal operation, a remaining air slit allows room air to come in. In case of failure, the cabinet can release air via this slit to the laboratory.

Thimble duct connector design shall allow 10–20% of cabinet extract airflow to get in as "Leakage Room Air." Therefore, an in-house extract system shall be able to deliver 10–20% more airflow stream on thimble duct extract connection (flange).

Modern biosafety cabinets normally are equipped with control interface, allowing control building site flaps or vents in order to adjust extracted total volume air stream due to operational need.

The building exhaust system shall never be interlocked with the cabinet. The best way is to operate the cabinet ventilation with no respect to the operation of the extraction system. Even if the cabinet airflow patterns may *in situ* be compromised, operational safety is better than a cabinet, which is shut off while handling hazardous substances inside the work chamber (Figure 22.5).

In case of a need to couple more than one cabinet to an exhaust system, for separation and individual control, every cabinet shall have its own tight closing flap (Figure 22.6).

Due to this basic design rule, the total air volume rate can be controlled sufficiently.

The extracted air volume stream shall be monitored. An acoustical and optical alarm shall occur if extraction air stream is insufficient or fails.

Today, cabinet suppliers mostly have Thimble Duct Connectors fitting on their cabinet as an accessory available in order to ease installation. Doubtless, good cooperation of effective extraction system on-site and cabinet supplier of choice is a need. Laboratory planners shall get in touch, early enough, with both in order to deliver a good job. Treatment of extracted laboratory and cabinet air may be

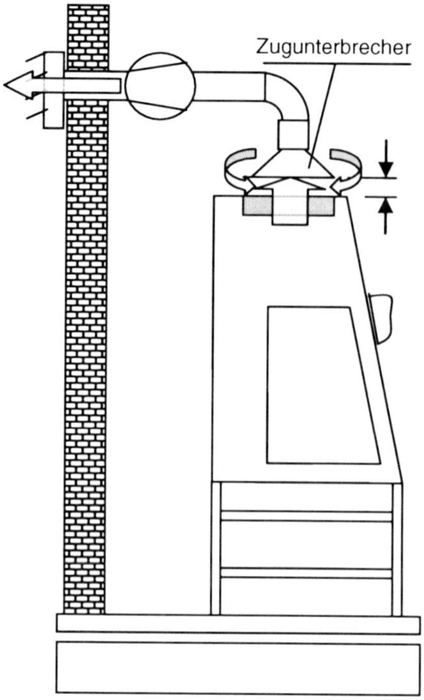

Figure 22.5 Operational principle of a cabinet ducted via thimble duct connector.

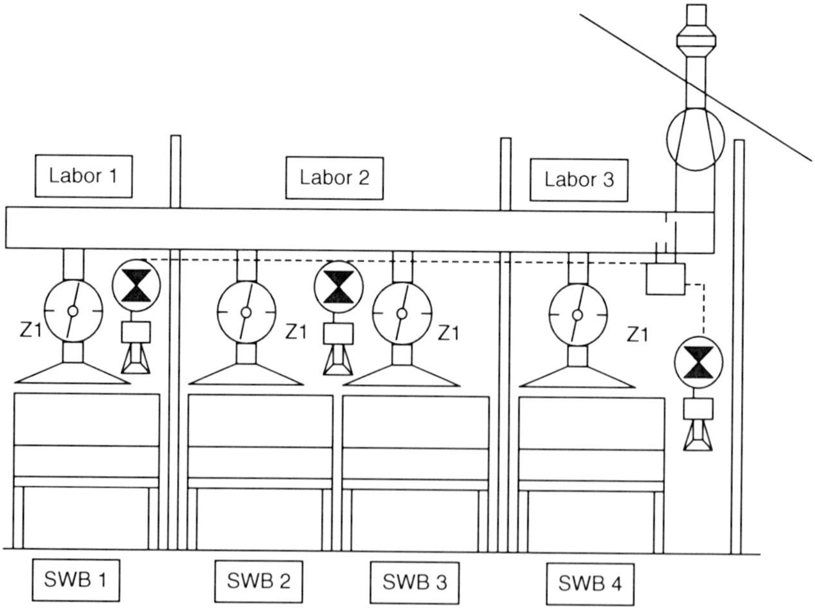

Figure 22.6 Thimble duct connector and a flap for every cabinet.

an additional need. Additional common filtration or burning may be needed to extract the air riskless into the environment.

22.8 Enhanced Safety of Safety Cabinet Class 2 by Means of Redundant HEPA Filter(s)

22.8.1 Redundant (Second) HEPA Exhaust Filter

This design is intended to avoid undetected extract HEPA filter failure during operation. A second filter after origin exhaust filter neutralizes filter failure and guarantees environmental protection and if cabinet is recirculated back into the room personal protection.

The design requires independent integrity testing of both filters in a row This design is predominantly used in the United Kingdom.

In other European countries, other designs are preferred. A summarizing HEPA filter is pre-filtering both supply air and extract air, also as a redundant filter.

An elegant design poses this filter close to the work tray in order to prevent the rest of cabinet from getting contaminated (Figure 22.7).

Direct under work surface of inner work chamber a segmented HEPA main filter-cartridge filters 99.999% of particles 0.3 μm size preventing most of cabinet

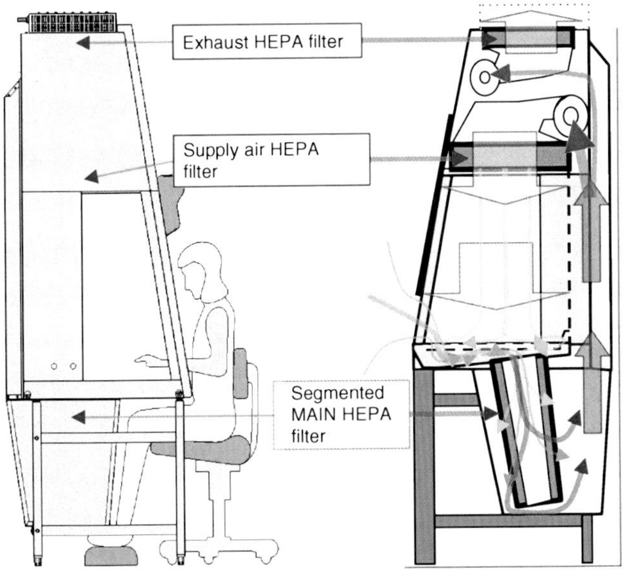

Figure 22.7 Segmented main-HEPA-filter with redundant supply- and extract – HEPA filter, functional principle.

unreachable ducts and housing from contamination. Both supply air and extract air are filtered twice in a row by equal-quality HEPA filters.

This design prevents every single fault of any HEPA in charge. No contamination can get out or get in touch with valuable goods handled.

The design today is used in handling MSR or high active substances, for example, in cytotoxic drug formulation. This is why this design often is called cytotoxic cabinet or Triple Filter Cabinet.

Even they are reinforced/of enhanced design such models often in addition are connected to the outside also due to national requirement or due to reason listed earlier.

What you shall look for buying a Biological Safety Cabinet class 2 (acc. to EN 12469):

Example:

Sole Source Description for a 1.2 m aperture class 2 – Cabinet

- Microbiological Safety Cabinet Class 2, rated electrical supply: 230 V AC, 1/N/PE; 50/60 Hz
- Length of Front Opening/Aperture: 1.2 m
- Independent tested by third-party test house in accordance to current European Standard EN 12469
- Grants additional recognition! Intended use: medical and biological Laboratories Risk Group 1–3
- Noise Level: <60 dB(A) (working mode, readings taken in 1 m distance to mid of aperture in a height of 1.5 m above floor level)
- Energy consumption: $\ll$0.4 kWh h^{-1}
- Heat rejection into environment: <0.15 kWh h^{-1} (working mode, 25 °C ambient Temp.)
- Design: Sloped front; case made of cincin steel, additionally epoxy powder coated backed for closed surface gray-white color,
- manual or electrical sliding front screen; fortunately aerosol-tight closing; with an nominal threshold if its closed/stand–by mode of 1/100 000 on 0.3 μm particle size.
- Front screen shall be designed to be opened fully for disinfection and cleaning purposes; fortunately with an additional hinged function;
- Front opening working mode: ≥200 mm
- Illumination for scientific work: ≥800 lx
- Two lead through each side, for example, 23 mm diameter with stoppers to run hoses or cables from inside to outside, for laboratory tap installations;
- Two duplex outlets each side in work chamber, maximum 5 A td.
- Work table tops segmented, modular system, fitting in an autoclave, alternatively full size work table tops shall be available by request as an accessory;
- Arm rests easing pipetting, non-blocking front grille, shall be available as an accessory;
- Optical and acoustical alarms for malfunction(s) according to standard, acoustical alarms should be mutable;

- Operator board with displays and buttons (touch pad) for microprocessor control in ergonomic position outside work chamber;
- Display for actual status and service intervals shall be incorporated.

22.9 Microbiological Safety Cabinet Class 3

BSCs class 3 are of minor importance today. In most cases, those are fixed installation due to high-risk level application. Today, a fewer offering is on the market only as a ready-to-plug-in instrument.

Often these are called *Glove Boxes* due to the manipulators' need for the fully tight enclosure the cabinet is offering. Inside they are protected by a certain negative pressure making sure no aerosols can be released to the surroundings. Manipulators and (sterile) transfer ports offer no comfortable work conditions for scientific approach.

Because the activities of day-to-day laboratories are not so important, we want to let you refer to intended literature in case of request – you also may get in touch with the author for further details.

22.10 Inactivation of Cabinet and Filters

This concerns many responsible in laboratories. How shall they run in future cabinets in their lab? Wondering about operational costs of BSCs we need to put a spot on it.

Depending on its use, the inactivation of a safety cabinet is needed prior to dispose the whole instrument or prior to filter exchange.

When ducts and rooms are hidden, it is not easy to get access. The only method is to convert to the gaseous state or vaporize the whole instrument prior to gaining access to potentially contaminated surfaces or filters.

For years, the vaporization with formaldehyde-water or -vapor followed by neutralization with ammonium vapor was used to sterilize the cabinet and its filters.

Formaldehyde vaporization is a sufficient sterilization method if it is carried out with temperature $> 65\,°C$; rH $> 80-98\%$ and a time left for ≥ 9 h.

With this the physical validation of sterilization process is easy and reliable.

Because of the carcinogenic potential of Formaldehyde, often a Hydrogen-Peroxide Vaporization is done to make the room remain clean.

Even Hydrogen-Peroxide has comparable carcinogenic potential and the same efforts are put in so that at the end of the day the disinfectant disappears only by ventilation. Application might be little more tricky and validation is not that easy, but savings in cleaning/wiping down the instrument after vaporization is much less compared to Formaldehyde processing with white reaction powder (harmless Hexamethylenteramine) after neutralization is done on the surfaces inside.

Processing both formaldehyde and H_2O_2 vaporization cannot be done for free and need's a specialist touch – no doubt. Since it is frequently operated, it is a remarkable part of operational costs.

Prior to buying or getting into a contract service with personnel, make sure they have good training and special knowledge about BSC vaporization/sterilization. The laboratory manager is responsible for "traffic safety" of cabinet and filter waste. If this is not carried out carefully, there is great risk for safety because injuries can likely happen.

After sterilization by vaporization, filters are supposed to be inactivated – in some areas, authorities request to take an additional effort in an autoclave – and then can be wasted or burned.

23
Safety Cabinets

Christian Völk

The scope of the sustainability evaluation starts on a large scale with the building and ends on a small scale, for example, with safety cabinets (safety storage cabinets). But, what part of a safety cabinet is sustainable or should be sustainable? In order to answer this question, we must first take a closer look at the topic of safety cabinets.

23.1
History – the Development of the Safety Cabinet

In 1971, the first safety cabinet for the storage of flammable liquids, according to European and national guidelines was presented. The development of this safety cabinet was considered to be a pioneering contribution to preventive fire safety. One game-changing consequence of the development was the adaptation of the national regulations for the storage of flammable liquids by the German Committee for Flammable Liquids ("Deutscher Ausschuss für brennbare Flüssigkeiten – DABF)" (1) in the form still continued today, of valid technical rules for hazardous substances ("Technische Regel für Gefahrstoffe" – TRGS) (2) 510. Furthermore, in 1984, the first DIN standard for safety cabinets for flammable liquids, DIN 12925-1 "Sicherheitsschränke für brennbare Flüssigkeiten," was published by the standards committee for laboratory equipment. Based on this standard, the first European standard, EN 14470-1 Safety storage cabinets for flammable liquids was finally born.

1) Deutscher Ausschuss für brennbare Flüssigkeiten
2) Technische Regel für Gefahrstoffe.

The Sustainable Laboratory Handbook: Design, Equipment, Operation, First Edition.
Edited by Egbert Dittrich.

23.2
Safety Cabinets for Flammable Liquids

23.2.1
Definition – Safety Cabinets for Flammable Liquids

The European standard EN 14470-1 sets out the design and test criteria for safety cabinets, which are used in laboratories to store flammable liquids in closed containers at normal room temperature. The maximum allowable size of a safety cabinet is limited in the standard to 1 m^3 internal volume (Figure 23.1).

The definition in the standard is supplemented with three focal protection objectives for the storage of flammable liquids:

1) Fire risk and fire case
   The fire risk related to the stored flammable substances should be minimized and the protection of the cabinet contents ensured in case of fire via a tested minimum duration.

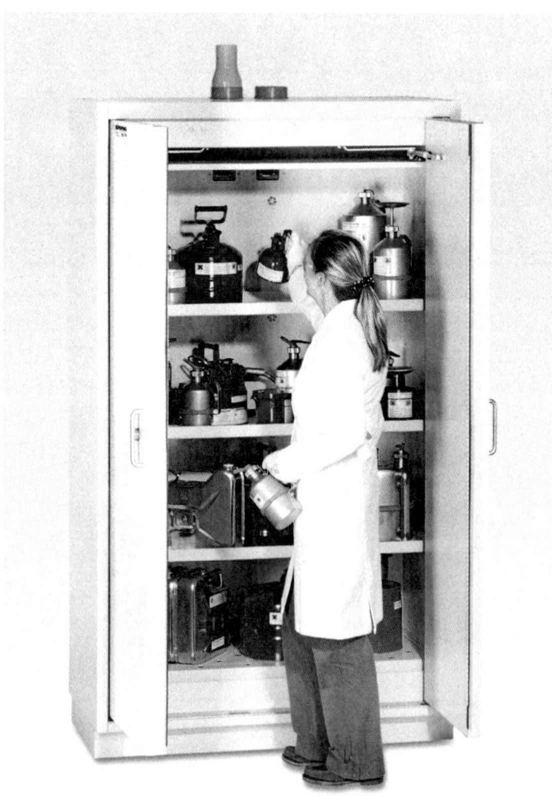

Figure 23.1 Safety cabinet type 90 for flammable liquids to EN 14470-1.

2) Fumes
   The fumes (vapor) released into the working environment should be minimized.
3) Leaks
   Possible leaks (spillages) from containers should be retained or collected inside the cabinet.

The owner (operating company) of a safety cabinet is responsible for compliance with the safety objectives. However, implementation in practice requires design measures on the manufacturer's side. The standard, therefore, describes minimum requirements for safety-relevant functions.

23.2.2
Fire Protection, Fire Resistance

The fire protection levels are divided into type classes (categories) according to their fire resistance (FR) rating. Safety cabinets with a minimum FR of 90 min are classified as rating type 90. In the standard, there are four categories in total, based on the usual fire protection ratings defined in the building regulations: Type 15, Type 30, Type 60, and Type 90. The classes (categories) are determined in a fire chamber test (fire test) as part of a type test. The following is an explanation in a simplified form:

A prototype of a new safety cabinet is exposed to flames in a furnace. The firing or heating takes place according to a standardized temperature profile. The values from this profile are based on a realistic fire.

In general, the test starts at an ambient temperature of approximately 20 °C in the fire chamber (furnace) and by analogy in the safety cabinet. The test is ended as soon as the temperature increase inside the cabinet exceeds 180 K (in this example the temperature would then be 200 °C absolute). The safety cabinet can be assigned to one of the type classes (categories) according to the time that has expired. This naturally presupposes that the increase was not reached before at least 15 min had passed. To obtain classification Type 90, the maximum temperature increase of 180 K must not be exceeded for a minimum period of 90 min. In a test for 90 min, the temperature in the furnace increases to approximately 1000 °C.

The minimum fire protection requirement specified in the EN 14470-1 standard is 15 min. This minimum level is intended to ensure in all of Europe that in case of fire:

1) the personnel have sufficient time to leave the room *and*
2) the fire service has sufficient time to put out the fire, before an extinguishable fire becomes an uncontrolled fire.

Therefore, when choosing a cabinet, the respective circumstances of the installation and use must be taken into account, for example, sprinkler system, quantities and properties of substances, and the respective relevant regulations.

Use of type 15 cabinets is not allowed in Germany and many other European member states due to the national regulations. In recent years, the highest rating, type 90, has increasingly established itself as the standard and today is the state-of-the-art standard in Germany and Western Europe.

Especially when compared to other safety cabinets, the models designed, built, and tested to European standards are therefore particularly long lasting. Safety cabinets tested to the US standards, which are used widely on other continents such as America or Asia, usually have a FR of 10 min. In practice, this has far-reaching consequences in comparison to type 90 models:

1) The investment required for safety cabinets to the EN standard is higher, *but*
2) in return provides nine times higher safety (type 90 models) with regard to the time window for the personnel and emergency services in case of fire *and*
3) the property damage caused by fires is generally minimized.

23.2.3 Pipe Penetration

The standard does not refer to pipe and cable penetrations. However, an opening through the cabinet top or rear wall is supposedly a weak point in preventive fire protection. Therefore, if it is intended to equip safety cabinets with pipe and/or cable penetrations, only tested combinations are to be used. Only if the safety cabinet with integrated pipe and/or cable penetration has successfully passed a fire test can the relevant type class (category) be maintained (Figure 23.2).

23.2.4 Door Technology

The primary objective of fire protection must not be impaired by individual functional characteristics. For this reason, in the standard it is required that the doors

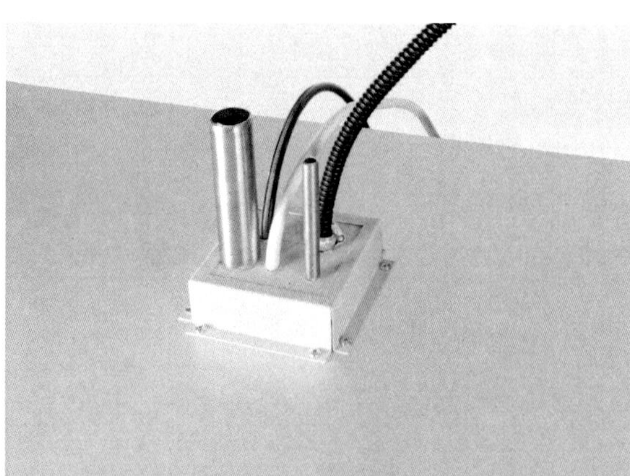

Figure 23.2 Tested pipe and cable penetration for safety cabinets to EN 14470-1.

of safety cabinets must generally be designed to be self-closing. The closing speed must remain below 20 s and the closing force on the closing edges must not exceed 100 N.

Solutions with stay or hold-open systems with temperature release elements are exempted from the continuous self-closing requirement. These must initiate the door closing automatically at an ambient temperature of 50 °C (tolerance +0/−10 °C) at the front of the cabinet. Whether the doors are side-hinged or swing-out/wing doors or space-saving folding doors, this aspect does not play any role for the technical requirement. On the other hand, it is imperative that with automatic closing in case of fire, all open door types also lock when closed. This point is also specified in the standard and must be taken into account by the manufacturer. However, when procuring or choosing a cabinet, users should ensure that these locks are positioned at an ergonomically favorable height (Figure 23.3).

In recent years, many types of cabinets have also been developed for which the term *doors* cannot be used, but which must transfer the self-closing function. It is therefore the case that cabinets with a drawer front can often be found in

Figure 23.3 Hold-open device with temperature release element for automatic door closing in case of fire.

laboratories which, in the case of a fire, must also initiate self-closing at 50 °C. In engineering terms, more is required if a combination of side-hinged or folding doors and pull-out trays (similar to a drawer) is used. In this case, it is necessary to ensure that if a fire occurs, the pull-out tray is drawn into the cabinet first and the doors are then closed safely and reliably and according to the normative requirements.

There are differences in the design, quality, and life of the door technology and therefore, also in the sustainability of the different brands.

There are manufacturers who, since the publication of the European laboratory furniture standard EN 14727 in 2006, have applied the high quality requirements to safety cabinets. The standard describes, among other things, dynamic load tests for moving components such as doors, drawers, and pull-out trays. In these test methods, for example, doors are opened and closed with 60 000 cycles and must then be fully functional. Especially in an international comparison, these safety cabinets ensure a higher degree of safety and a longer life. By comparison, within the scope of UL tests in the United States, 6000 cycles are tested, which means conversely that a 10 times longer product life is confirmed for cabinets tested to EN 14727.

23.2.5
Interior Fittings

In addition to the doors, the various interior fitting variants are also subjected to load tests to EN 14727. These test methods are defined supplementary to the general requirements of EN 14470-1 suitable for the subsequent use.

The three most popular variants must pass sucessfully the following tests among other things.

1) Shelves
   "Fixed" shelves are usually height adjustable if tools are used, but are permanently installed for normal use. Accordingly, these shelves are subjected to static tests. This includes testing their deflection (sag), distortion (torsion), stability (tilt resistance) in interaction with the maximum load given by the manufacturer (Figures 23.4 and 23.5).
2) Pull-out trays
   Compared to shelves, pull-out trays are the more convenient solution to use. In line with their functions, they are not only subjected to static tests but also to dynamic tests. The number of cycles is specified with 60 000, analogous to the dynamic door test. After the cycle series has expired, the pull-out shelves must still be fully functional. Furthermore, pull-out shelves must be fitted with double pull-out locks.

The general EN 14470-1 requirements for interior fitting variants include, among other things:

1) Shelves and their mountings must be made of non-absorbent material. However, placing absorber cloths on or in the shelf is allowed.

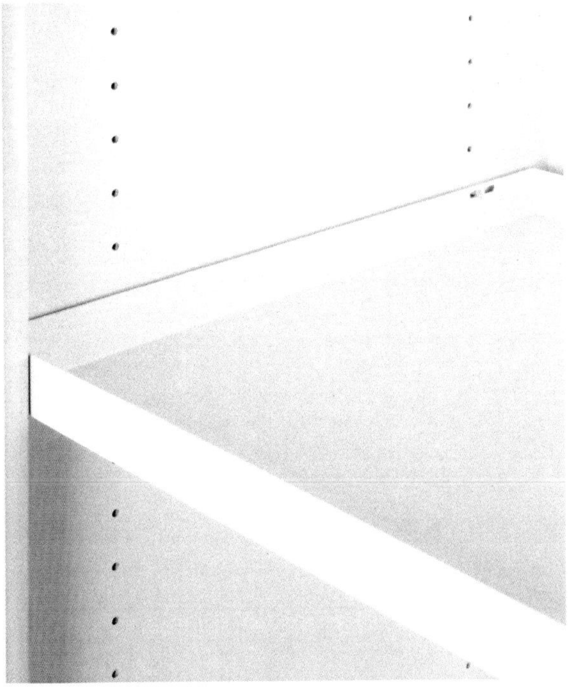

Figure 23.4 Shelf.

Figure 23.5 Pull-out trays.

Figure 23.6 Double pull-out lock, pull-out trays.

2) During the fire test, the shelves must be able to carry the weight given by the manufacturer. No damaging deformations may occur during the test.
3) The self-closing in case of fire must not be impeded by the shelves, for example, pulled out pull-out trays must not block the automatic door closing.
4) The highest shelf must not be more than 1.75 m above the floor.

In an international comparison, safety cabinets built and tested successfully to European standards fulfill the highest safety aspects and quality requirements (Figure 23.6).

23.2.6 Bottom Tray

Another important component of safety cabinets is the bottom tray (also called a *spill containment sump*) for catching spilt liquids. In the case of cabinets with shelves, they usually do not have any collection volume, and the bottom tray must be installed below the bottom standing level. The collection volume available is directly proportional to the maximum storage quantity and the largest possible container. The bottom tray must catch 10% of the volume of all containers stored in the cabinet or at least 110% of the volume of the largest container. The condition with the largest volume must be applied.

Further, the collection volume of the bottom tray must not be minimized by using the tray as a standing area. An older but ideal solution is to use perforated sheet inserts. The perforated standing areas are simply placed in the bottom tray.

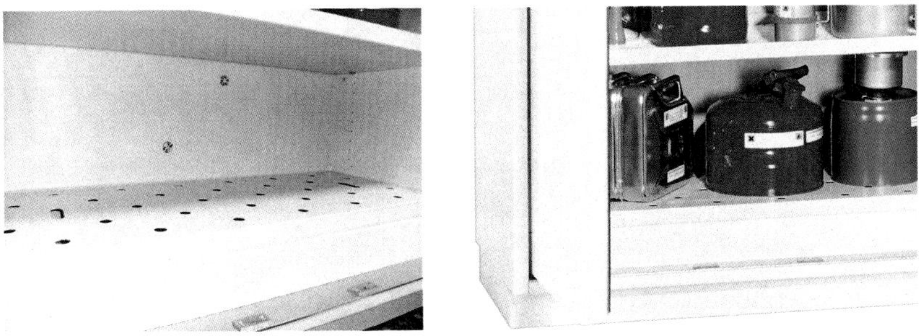

(a) (b)

Figure 23.7 (a) Perforated sheet insert in bottom tray. (b) Perforated sheet insert as standing area.

This provides additional standing area on the tray without reducing the collection volume at all (Figure 23.7).

The maximum allowable container size and the collection volume must be given according to the standard, either on the cabinet rating plate or in the operating instructions. However, this is not anchored in the international or the national law. It is therefore possible for countries to set further environmental conditions for the collection volume.

Approval under the building regulations (e.g., national technical approval), is not usually required for bottom trays in safety cabinets. Unlike trays approved under the building regulations, bottom trays are tested for leaks after the FR test (fire test), that is, they must retain their function even after a fire.

Safety cabinets with pull-out trays are a special case. In these cabinet models, the traditional bottom tray can be omitted, provided that each individual pull-out tray satisfies the aforementioned requirements.

23.2.7 Ventilation

Safety (storage) cabinets to EN 14470-1 must be equipped with ventilation connections in their design or rather when manufactured. This should enable each user, depending on the respective national regulations, to ventilate the safety cabinet. Attention must be paid to three important points:

1) The air exchange rate must at least equal 10 times the internal cabinet volume per hour. A higher rate may be necessary to reduce odors emitted by some stored substances.
2) The ventilation must be effective directly above the bottom tray of the cabinet.
3) The ventilation openings must close automatically at 70 °C (±10 °C).

Modern, forward-looking concepts, which provide sustainable solutions are already available, especially for points (2) and (3) mentioned earlier.

According to the standard, the ventilation must be effective directly above the bottom tray of the cabinet. The reason for this is simple: solvent fumes are generally heavier than air and sink to the bottom. Many cabinet models are therefore equipped with an exhaust air opening above the bottom tray and an inlet air opening in the upper part of the cabinet. This solution is sufficient to fulfill the standard. But what happens to the fumes that sink into pull-out trays or shelves higher up in the cabinet? An optimized and forward-looking solution is available for this case: Ventilation openings for inlet and exhaust air in each cabinet level or near the individual standing areas. This solution enables the fumes to be extracted directly in the place in which they arise.

Based on the standard, point (3) mentioned earlier can also be solved in all kinds of different ways. Some solutions are equipped to the standard and others are equipped to the standard and are also sustainable:

The ventilation flaps (dampers) must be inspected visually at regular intervals to ensure that they are fully functional. This can be very time-consuming for safety cabinets in which the ventilation shut-off dampers are located in exhaust air connections. The exhaust air pipe has to be dismantled for each inspection and then re-installed, air-tight, after the check has been completed. This takes time and is therefore cost-intensive. A better solution is to use safety cabinets with ventilation shut-off valves, which can be checked from inside the cabinet. There are solutions that use the colored signal light or "traffic light" principle:

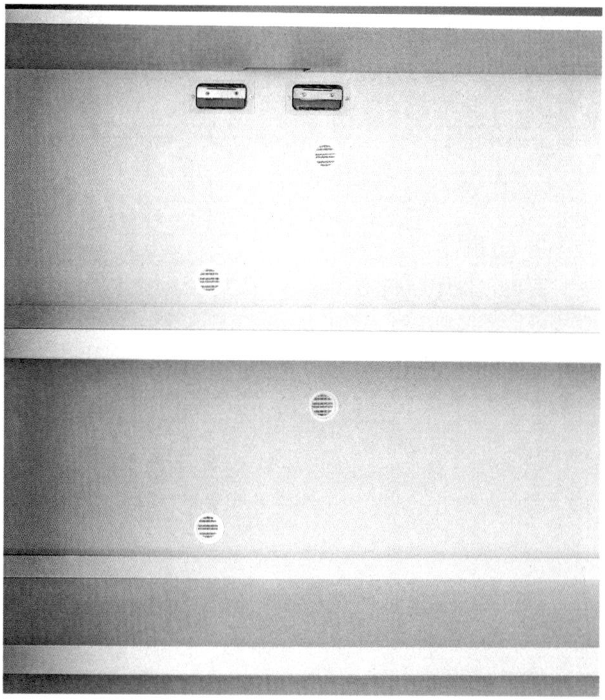

Figure 23.8 Ventilation and extraction at every level in the cabinet.

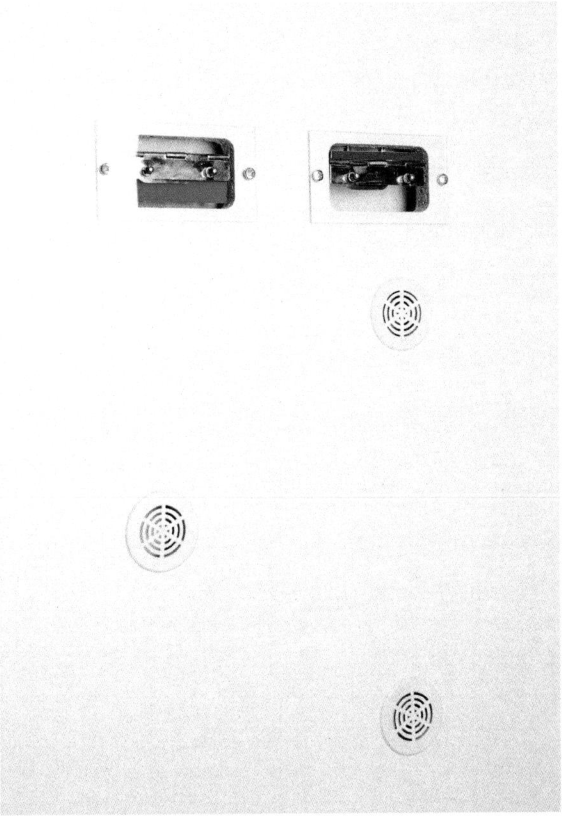

Figure 23.9 Ventilation shut-off dampers with colored signal light principle.

- if the damper indicator light is green, everything is ok
- if the damper indicator light is red, a fault exists. If a red light occurs, the ventilation shut-off damper can be replaced easily from the inside of the cabinet. A fast and reduced cost solution (Figures 23.8 and 23.9).

23.2.8 Earthing, Equipotential Bonding

The EN 14470-1 standard does not mention the topic of earthing (grounding) and it is left to the manufacturers to provide earthing to the equipotential bonding. This is reflected in the diverse range of cabinet makes available on the market:

1) without earthing option,
2) with cabinet carcass with earthing lug,
3) with standard continuous earthing concept from conductive interior fittings, connected to the outer carcass (Figure 23.10).

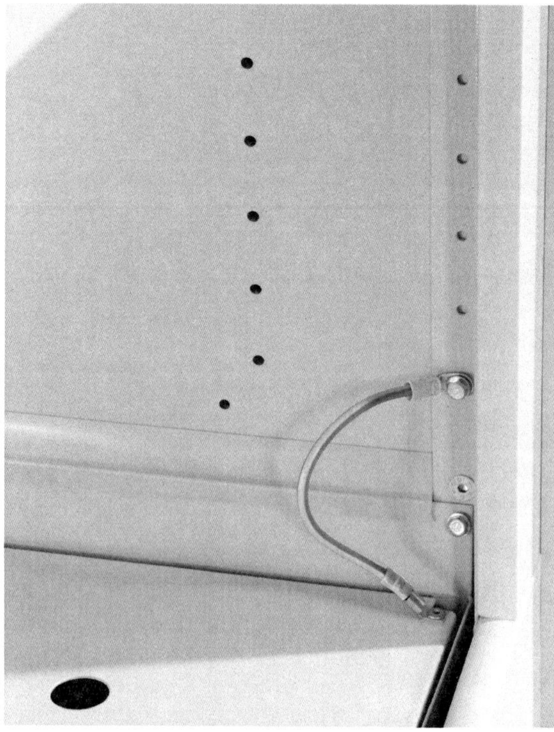

Figure 23.10 Continuous earthing concept: conductive interior fittings connected to the outer carcass.

At the national level in Germany, there are legal requirements that apply and are specified in the guidelines of the bodies responsible for industrial safety ("berufsgenossenschaftlichen Richtlinien" – BGR) and the industrial safety technical rules ("Technischen Regeln Betriebssicherheitsverordnung" – TRBS). However, ultimately the user must decide whether and how the safety cabinet is earthed as part of their hazard or risk analysis.

There is often only a small cost difference between the different makes with and without earthing. However, subsequent modification – if at all possible – is always cost-intensive. It is therefore advisable to opt for safety cabinets with a continuous earthing concept when planning and procuring your requirements.

23.2.9 Marking and Operating Instructions

EN 14470-1 does not allow much leeway regarding the marking of a safety cabinet and the requirements are described clearly. However, only the minimum requirement is defined regarding the operating instructions or user manual. Here, too, different versions are available on the market. Yet, anyone who has a modern, forward-looking hazardous substance and risk management system should insist on a minimum level of information included with the product. Contemporary operating instructions not only contain technical information but also notes on safe use, maintenance, and ideally the necessary safety documents too, for example, TÜV GS certificate.

It is advisable to request these documents, at least in electronic form, for example, as a pdf file, before purchasing a new cabinet.

23.3 Safety Cabinets for Pressurized Gas Cylinders

In addition to flammable liquids in the laboratory, in the event of a fire, pressurized gas cylinders constitute a significant hazard. When designing a new laboratory, suitable safety (storage) cabinets appropriate for the intended use must be considered.

23.3.1 Definition-Safety Cabinet for Pressurized Gas Cylinders

In 2006, the EN 14470-2 standard for safety cabinets for pressurized gas cylinders was published as the successor to the German standard DIN 12925-2 issued in 1988 and the draft version of 1998. The European standard EN 14470-2 sets out the design and test criteria for safety cabinets, which are used in laboratories to store pressurized gas cylinders at normal room temperature.

The definition in the standard is supplemented with two focal protection objectives for the storage of pressurized gas cylinders:

1) 1. Fire risk and fire case
   In case of fire, it is necessary to ensure that the cabinet contents do not cause any additional risks or contribute to the spread of the fire for a period of 15 min.
2) 2. Gases and fumes
   The cabinet design and construction must ensure that small outflows of gas are extracted inside the safety cabinet (Figure 23.11).

Analogous to the safety storage cabinets for flammable liquids, the owner or operating company is responsible for compliance with the safety objectives. Based on the standardized minimum requirements, manufacturers are obliged to offer appropriate constructive solutions, which can be used and implemented in practice.

23.3.2 Fire Protection, Fire Resistance

The fire protection of safety cabinets for pressurized gas cylinders must, for a defined period of time, ensure that the stored pressurized gas cylinders do not

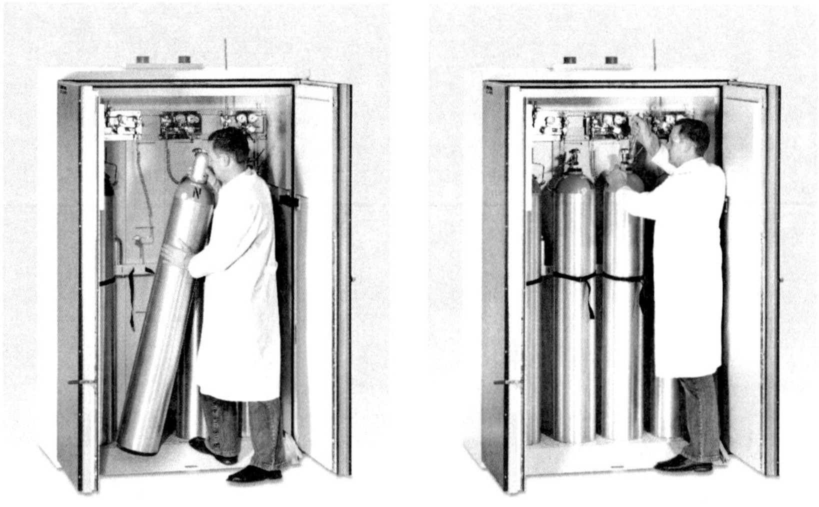

(a) (b)

Figure 23.11 (a) Safety cabinet type 90 for pressurized gas cylinders to EN 14470-2. (b) Safety cabinet equipped with holding devices for pressurized gas cylinders.

become unstable. Unlike safety cabinets for flammable liquids, the FR is not directly related to the temperature increase inside the cabinet. The fire protection class (protection level) is determined by the maximum allowable temperature increase measured at the neck of the gas cylinder.

A prototype of a new safety cabinet is therefore exposed to flames in a furnace. The firing or heating takes place according to a standardized temperature profile. The values from this profile are based on a realistic fire.

In general, the test starts at an ambient temperature of approximately 20 °C in the furnace or at the neck of the pressurized gas cylinder. For the purpose of the test, a media line leading out of the cabinet is also connected to the valve of the pressurized gas cylinder.

This makes the test more difficult, but provides an appropriately realistic test result.

The test is ended as soon as the temperature increase at the neck of the pressurized gas cylinder exceeds 50 K (in this example that would be 70 °C absolute). The safety cabinet can be assigned to a type class (category) according to the time that has expired. To obtain classification Type G90, the maximum temperature increase of 50 K must not be exceeded for a minimum period of 90 min. By comparison: in a test lasting 90 min, the temperature in the furnace increases to approximately 1006 °C.

With the publication of the EN 14470-2 standard, many laboratory operators in Germany and Western Europe opted for safety cabinets – Type G90. For the first time ever, the new highest category enables a continuous 90 min fire protection concept for the storage of flammable liquids and pressurized gas cylinders.

The 90 min concept in laboratory construction enables European laboratory equipment manufacturers and suppliers to set positively themselves apart from the rest of the world with regard to state-of-the-art standards. Especially on other continents, such as America or Asia, there are no known equivalent cabinet solutions. Yet fire protection is an especially important part of achieving sustainability.

With the high FR, the risk of fire harming people and damaging buildings is reduced drastically.

23.3.3 Ventilation

In the standard, health protection is dealt with on the same level as fire safety. The introduction to the standard, points out that the cabinet construction must enable even small gas discharges to be extracted. Further, with a standardized safety cabinet, it must be possible for the owner or operating company to store pressurized gas cylinders

1) with flammable or oxidizing gases in cabinets in which 10-fold air exchange can be set
2) with toxic gases in cabinets in which 120-fold air exchange can be set.

Due to the different densities of gases, the ventilation must take place in the upper and lower part of the cabinet interior. In order to ensure sustained fire safety, in the event of a fire, the inlet and exhaust air valves of the ventilation connections close automatically. This function is also tested as part of the fire test.

23.3.4 Insertion and Restraint of Pressurized Gas Cylinders

The standard also pays particular attention to health and safety when handling pressurized gas cylinders at the safety cabinet. In particular, it must be possible to place the pressurized gas cylinders in the cabinet as safely as possible and with the least possible effort. To ensure this, safety cabinets are usually offered with lifting devices or rolling ramps. Modern cabinet designs have standard roll-in flaps or ramps, which can be easily operated by hand. According to the new EN 14470-2 standard, in general roll-in ramps may not be made of aluminum. The interaction of aluminum flaps (ramps) and rusty gas cylinders, when the cylinders are rolled into the cabinet increases the sparking risk to an unacceptable degree (Figure 23.12).

23.3.5 Installing Pipes and Electrical Cables

For the normal use of pressurized gas cylinders in a safety cabinet, they are connected by a fitting and a media pipe fed through the top of the cabinet. For this reason, the fire protection requirements with regard to the diameter, material, and

Figure 23.12 Actuation of the rolling ramp via ergonomically positioned handle.

number of media pipes and possible electrical cables are described in the standard and are tested as part of the type test. Each individual penetration can have a negative effect on the FR of the cabinet. Accordingly, the requirements specified in the standard are very detailed:

1) The number of pipes must be limited to the necessary minimum number.
2) Each gas cylinder may only have three pipes each with a maximum diameter of 10 mm.
3) It is recommended that the pipes used be made of stainless steel or material with similar thermal conductivity.
4) In the case of electrical cables, only two cables each with maximum diameter 20 mm may be used for each gas cylinder.
5) Open pipes that are no longer used must be closed off again according to the manufacturer's instructions.

Use that deviates from the specifically defined conditions results in the safety cabinet no longer conforming to the standard and any test markings become invalid. Furthermore, fire protection can be impaired and the Type G category no longer corresponds to the real FR.

23.3.6 Marking and Operating Instructions

The EN 14470-2 standard contains information on the operating instructions and necessary markings on a safety cabinet used to store pressurized gas cylinders. Just like cabinets used to store solvents, there are substantial differences in the quality of the documentation provided by the manufacturers of pressurized gas cylinder cabinets. In the interests of sound hazardous substance and risk management, it is advantageous if the documentation provided not only contains technical information but also the necessary safety documents, for example, TÜV GS certificates. Before purchasing a safety cabinet, the documentation should be requested, at least in electronic form, for example, as a pdf file.

23.4 Safety Cabinets for Acids and Lyes

Acids and lyes are used in everyday laboratory work and due to their harmful effect on water and the environment, they must also be stored safely. In practice, safety cabinets for acids and lyes are used.

Although there is no direct standard and no official definition for these cabinets, the general quality requirements of the laboratory furniture standard EN 14727 is valid or rather applicable. Based on this and the legal requirements for the handling of substances harmful to the environment, storage cabinets for acids and lyes have been launched on the market.

23.4.1 Definition – Safety Cabinet for Acids and Lyes

This type of cabinet is similar to the safety storage cabinets for flammable liquids, that is, spilt acids and lyes are safely collected in collection trays (spillage trays) and harmful fumes are extracted. Used properly and as intended, only non-flammable acids and lyes may be stored in these cabinets, because these cabinets do not fulfill any fire safety functions (Figure 23.13).

23.4.2 Collection Trays

Acid and lye cabinets are equipped with a bottom tray (spill containment sump) for collecting any spilt or leaking liquids. The collection volume available is directly proportional to the maximum storage quantity and the largest possible container. The collection tray must catch 10% of the volume of all containers stored in the cabinet or at least 100% of the volume of the largest container. The condition with the largest volume must be applied. Depending on the manufacturer, the maximum allowable container sizes and the collection volume are given on the cabinet

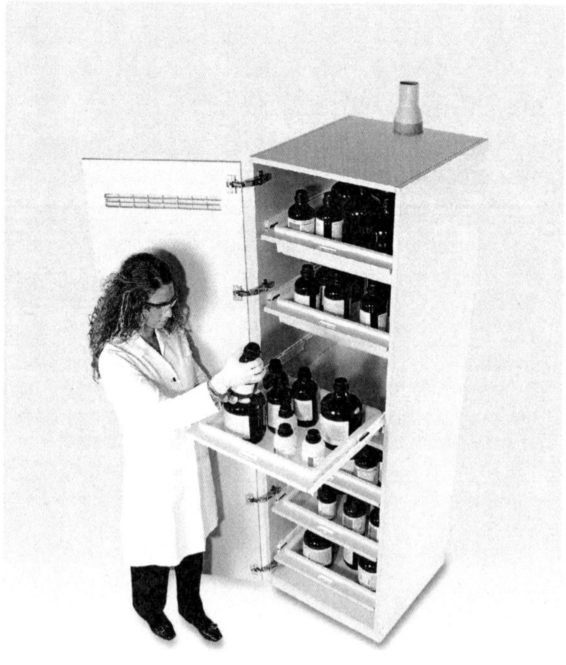

Figure 23.13 Safety cabinet for acids and lyes to EN 14727.

rating plate or in the operating instructions. However, national regulations may stipulate different requirements with regard to the allowable storage quantities and must be taken into account accordingly.

Acids and lyes are often used and stored in laboratories in small containers. This is why pull-out trays have established themselves as the favored solution for these substances. They enable clear and manageable storage of small containers and make them easily accessible.

23.4.3 Ventilation

An industrial venting system (forced ventilation) is indispensable for acid and lye cabinets for two reasons:

1) Acid and lye fumes constitute a substantial potential hazard for the respiratory tracts and health of the personnel.
2) The mostly corrosive fumes attack many materials and can cause damage to the cabinet and nearby equipment.

In order to reduce both risk factors to a minimum, these cabinet types are prepared for connection to industrial (forced ventilation) exhaust systems.

The documents provided, for example, the operating instructions contain air exchange rates recommended by the manufacturers. Nonetheless, sustained health protection can require increased air exchange rates.

23.4.4
Marking and Operating Instructions

A high-quality acid and lye cabinet requires appropriately high-quality markings and documentation. The markings should be adapted to the cabinet size, and the warning and hazard symbols should also be legible from medium distances. Furthermore, the necessary pictograms, for example, that do not extinguish with water in the event of a fire, should be attached to the cabinet as a standard feature. The marking is rounded off by the information on the possible container size with regard to weight and capacity.

This data must also be listed in the operating instructions. Ideally, for proper documentation within the scope of the hazard or risk analysis, copies of test certificates, for example, the GS certificate issued by TÜV Süd, should be printed in the operating instructions. This simplifies collation of the necessary documents and forms a good initial basis for hazardous substance management. The following tip applies here too: Before you purchase a cabinet, request the documents from the respective manufacturers, at least in electronic form, for example, as a PDF file.

23.5
Test Markings for Safety Cabinets

In modern laboratories, alongside functionality and flexibility, safety is also high on the list of priorities. Accordingly, renowned manufacturers have their products certified by independent test bodies, for example, TÜV Süd. All test markings are commissioned by the manufacturers voluntarily. In general, for safety cabinets in Germany, this is in the form of the GS marking – the national symbol for tested product safety ("geprüfte Sicherheit" – GS).

A GS tested product provides the highest security for investors, planners and designers, and users. A mistaken belief still exists that for safety cabinets designed and built to EN 14470, parts 1 and 2, that is, fire safety storage cabinets, the fire test is the most important document.

This is not so – because without a valid, successful fire test, there is no GS testing or certification. Or stated the other way around: a GS marking for a safety cabinet may only be issued if a valid fire test exists and all other normative requirements in the EN 14470-1 and -2 standards and the laboratory furniture standard EN 14727 are fulfilled.

TÜV Süd has developed the "High-Quality Seal," voluntary for manufacturers, for particularly high-quality and sustainable safety cabinets. The marking can be issued by TÜV Süd for all safety cabinets which not only fulfill the requirements specified in the relevant standards and the German equipment and product safety law but also provide additional quality and convenience. In the assessment for the "High Quality" seal, the categories: product design, user friendliness, durability, and production quality are rated and tested.

23.6 Special Solutions for the Storage of Flammable Liquids

Safety cabinets designed and built to EN 14470-1 are by definition intended for use in laboratories to store flammable liquids in closed containers at normal room temperature. This means: safety cabinets are only approved for passive storage *and* the storage of chilled hazardous substances is not covered by EN 14470-1.

But, there are now modern and innovative solutions available for both applications which are based on EN 14470-1 and whose safety is documented with the GS marking.

23.6.1 Active Storage

In active storage, vessels containing flammable liquids are used as removal or collection containers, or are opened for other purposes. Conventional safety storage cabinets to EN 14470-1 are not approved for this use. For some time now, however, certified system solutions have been available, especially for this application. With these systems, flammable liquids, for example, from canisters or drums, can be removed or collected safely and with approval. These solutions are based on fire safety storage cabinets to EN 14470-1, but which thanks to further design developments not only fulfill the requirements of the German national guidelines TRGS 510 but also other regulations. For example, guidelines such as TRbF 30 (technical rules for flammable liquids) for filling and emptying points specify requirements for earthing, ventilation, and monitoring. These are taken into account in certified system solutions and are tested as part of the issue of the GS marking. The GS certificate provides an important element for the statutory hazard and risk analysis (Figure 23.14).

23.6.2 Cooled Storage

The EN 14470-1 standard only refers to the storage of flammable liquids at normal room temperature. The topic of "chilled storage of hazardous substances" on the other hand is only dealt with in the laboratory guidelines or guidelines of the associations responsible for industrial safety (German "Berufsgenossenschaften") and does not fully reflect state-of-the-art standards. For example, based on the regulations, it is possible to store extremely flammable, highly flammable, and

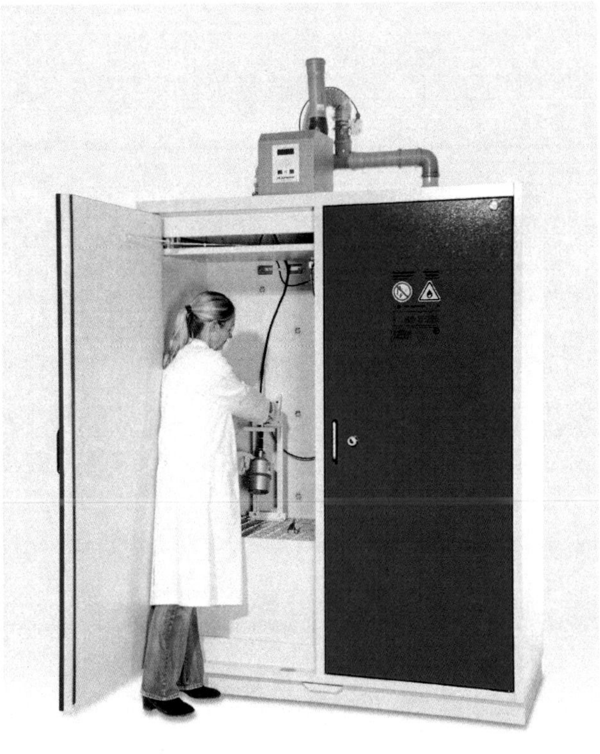

Figure 23.14 Certified system solution for active storage according to EN 14470-1 and TRbF 30.

flammable hazardous substances in laboratory refrigerators. The main criteria for these laboratory refrigerators are that the interior must be free from ignition sources and silicone. However, issues arise regarding safety aspects:

Why are fire safety storage cabinets specified for even small quantities of flammable liquids, but refrigerators/chilled cabinets without fire safety properties are approved for highly sensitive media that require chilled storage?

What use is a laboratory full of safety cabinets if a laboratory refrigerator acts as a "fire accelerant" and makes a mockery of safety?

This gave the manufacturers of safety cabinets cause to develop and present suitable solutions. The "cabinet-in-cabinet" solution has proven to be one of the most safe and most ecological and economical solutions. In this variant, a laboratory refrigerator or chilled cabinet approved for the storage of flammable liquids is surrounded by a fire-resistant safety cabinet. In this way, the elementary components: fire protection and cooling are fulfilled. The safety of these products is confirmed by test bodies, for example, TÜV Süd, who issue the GS marking and today represents the state-of-the-art standard (Figure 23.15).

Figure 23.15 Safety cabinet for chilled storage of flammable liquids to EN 14470-1, "cabinet-in-cabinet" solution.

23.6.3
Clean Room Cabinets

The construction of a safety cabinet requires materials, some of which are dust or particle producers and are unsuitable for clean rooms. The formation of dust in conventional working spaces is negligible and cannot be seen with the naked eye. In clean room technology however, these particles, which are often smaller than twenty thousandths of a millimeter, can become a negative influencing factor. In the production of semiconductors, these particles have the same effect, for example, as rocks on a motorway.

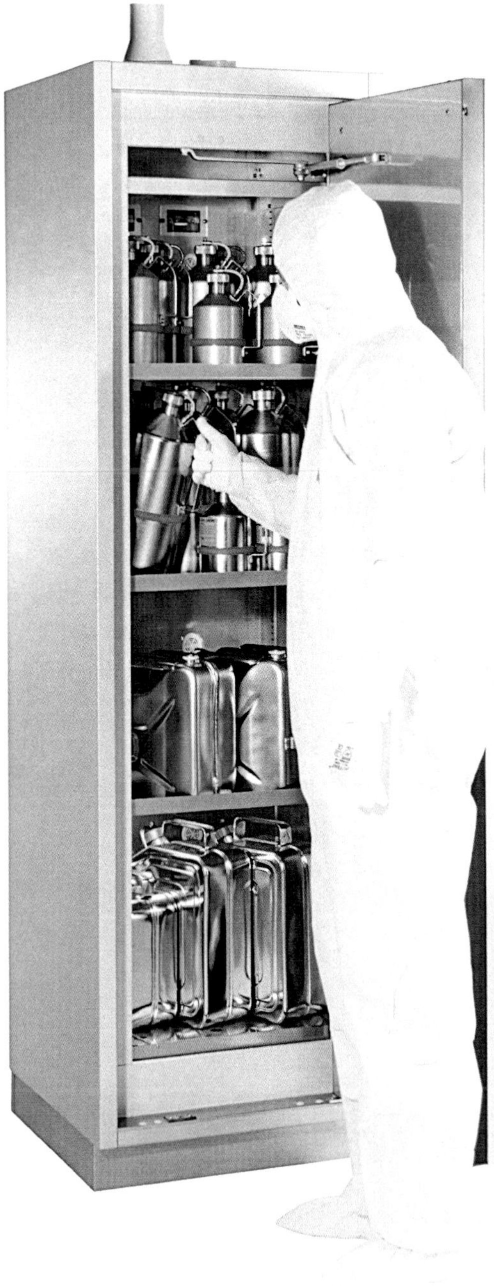

Figure 23.16 Safety cabinet for installation in clean rooms according to EN 14470-1 and EN ISO 14644-1.

Similarly, sensitive processes are also found in the clean rooms of other production sectors such as aerospace, medical engineering, pharmaceutical and food industry and set corresponding requirements for the equipment and resources brought into the production facilities. Therefore, each clean room, not to be confused with a sterile room, is classified according to the maximum allowable particle size and quantity and is subject to strictly standardized cleanliness requirements.

The solution much practiced to date was therefore either to store the media unprotected in the clean room or to bring them into the room individually each time they are required, almost dust free by passing them through an air lock in which they are subjected to an "air shower." With especially developed safety cabinets and the use of new materials, an alternative solution is now available. The safety cabinets for clean rooms can be installed as equipment for class 5 clean rooms according to ISO 14644-1. It should be ensured that the clean room classification is confirmed by an independent test body, for example, TÜV Nord (Figure 23.16).

Abbreviations

DABF	Deutscher Ausschuss für brennbare Flüssigkeiten
TRGS	Technische Regel für Gefahrstoffe (Technical rules for hazardous substances)
TRbF	Technische Regel brennbare Flüssigkeiten (Technical rules for flammable liquids)
FWF	Feuerwiderstandsfähigkeit (fire resistance)
BGR	Berufsgenossenschaftliche Richtlinie (Guidelines of the German "Berufsgenossenschaften" – the bodies responsible for industrial safety)
TRBS	Technischen Regeln Betriebssicherheitsverordnung (Technical rules for industry safety)

24
Laboratory Service Fittings for Water, Fuel Gases, and Technical Gases

Thomas Gasdorf

Laboratory fittings are precision instruments for the distribution and dosing of liquid and gaseous media. In order for them to work properly, it is important to choose the adequate fitting for the respective task. The selection of laboratory fittings is determined by a number of criteria.

24.1
Medium

The materials and techniques applied depend, to a great extent, on the medium the fittings are used for. Therefore, with regard to material selection, corrosion protection, as well as sealing and dosing technology, the properties of the media must be taken into consideration first and foremost.

To that effect, the DIN 12918 series of standards for laboratory fittings is generally subdivided into four parts: Part 1 refers to water, part 2 to fuel gases, part 3 to technical gases, and part 4 to high-purity gases. However, the classification according to one of these four parts of the standard is by far not enough for the selection of a media-compatible fitting. Regarding the supply points subsumed in part 1 of the standard, poor conductivity or a limited TOC value of the water, for example, requires special solutions.

24.2
Temperature

The selection of basic material, sealing, and shut-off assemblies is also strongly influenced by the temperature of the medium. The maximum limit for plastic fittings is approximately 60 °C. Brass is used at about 160 °C and stainless steel up to 230 °C. These are special designs with particularly temperature-resistant components. The temperature limits for standard fittings have been aligned to the

The Sustainable Laboratory Handbook: Design, Equipment, Operation, First Edition.
Edited by Egbert Dittrich.

medium and are mostly much lower: for water and gases, they are 65 and 60 °C, respectively and for vapor it is 120 °C.

24.3 Dosing Task

It must be assumed that the sealing and dosing mechanism is perfectly adjusted to the properties of the medium. Therefore, the purpose the medium is used for is of utmost importance. Often, the connection of certain devices or the rinsing of large pistons only requires the opening and closing function. In comparison to this, many analyzers or even the exact filling of small containers place higher demands on the dosing capacity.

24.4 Safety

It is obvious that safety is a major concern when fittings for flammable or toxic media are concerned. However, even a simple water fitting with an abruptly spurting jet can have dangerous consequences, for example, when rinsing a vessel with toxic, aggressive, or biologically active content.

24.5 Place of Installation

The selection of the suitable design depends on the place of installation of the fitting. Typical designs are wall- or bench-mounted installation as well as suspended installation. Moreover, in practice, it is important for the dimensions of turrets extensions of swivel arms to be variable over a broad range in order for the fitting to suit the respective situation of installation.

24.6 Ease of Installation

A decisive cost factor for the purchase of fittings is the installation. What is the use of a favorable initial price if the installation requires costly adjustments because the connection pipe is too short or the thread deviates from the local standards of the installation trade? This requirement can be met only by a modular system containing both female and male threads in various installation-compatible designs for connection and assembly. Suitable hose lines already prefabricated by the manufacturer solve the interface problem and considerably reduce onsite installation times and thus, even costs.

24.7 Materials

The corrosive properties of the medium have an essential influence on the selection of materials. The standard generally allows numerous materials for parts of the fitting that get into contact with media, provided they are corrosion resistant or protected against corrosion.

The purpose of use and the medium itself determine the fields of application of the basic types to be distinguished by material:

24.7.1 Brass

In the ideal case, the body of the fitting is a pressed part made of highly tempered brass, machined and equipped with an accurately fitting headwork. Very good surface protection is achieved by powder coating. Due to blow holes and poor surface quality, cast parts are no longer state-of-the-art. A smooth surface was achieved by a nickel-plated valve seat that has been screwed in previously. However, even with the additional processing step such a work piece does not reach the quality of a pressed part. Furthermore, the nickel-plating process has a negative effect on sustainability.

24.7.2 Stainless Steel

If a correspondingly high quality (1.4571/AISI 316) is used, the high-temperature resistant material proves to be extremely stable and particularly corrosion resistant. Fittings made of stainless steel are also suitable for vapor and offer the possibility of thermal disinfection. Exclusive design can be achieved by surface polishing or brushing.

24.7.3 Plastics

The decisive advantage of plastics such as PP or PVDF is their high resistance toward chemicals. Another feature is the ease of deformation. The use of such fittings is in particular limited by their thermal compatibility. Fittings suitable for the tough workaday life of a laboratory also require more robust dimensioning due to the material's lower modulus of elasticity. In order to achieve the same stability as with metal fittings, screw threads, for example, require larger dimensions.

24.7.4 Brass Plus Plastics

The combination of an outer brass shell and an inner plastic structure unites the requirements for strong external mechanical stress with high corrosion resistance toward the medium.

24.8 Headwork

The structural separation of fitting housing and headwork provide several advantages. The movable wear parts of a fitting are arranged in the headwork. This enables the cost-efficient exchange of damaged parts in case of a defect. The manufacturing of valve bodies results in cost advantages, because, if used for several media, they can be produced in larger series. The respective headwork can be installed according to the desired control behavior and the specific medium. The headworks need to be secured against unintentional unscrewing, for example, through assembly with a clamping torque of at least 20 Nm.

24.9 Seals

They are indispensable in the manufacturing of fittings. The quality of a fitting can be well determined by the workmanship of the seals and of their seats and grooves. Common materials for seals are different kinds of rubber, PTFE, and copper.

24.10 According to Standard

An important indicator for whether the essential criteria for a high-quality laboratory fitting are met can be found in the technical specifications of the manufacturer. There, the design according to the applicable standards and regulations should be acknowledged. In this context, the standard's requirements for laboratory fittings can be considered the minimum limit to be observed. In case safety, durability, and function are of particular importance, it may be reasonable for individual features to exceed these values.

In addition to the selection criteria mentioned, special care should be taken that the laboratory fittings meet the legal requirements and the regulations of the insurers; otherwise, the safety of the employees could be at risk (Figures 24.1–24.3).

24.11 Water

The suitable solution for each kind of water and for the required dosing task must be selected. Here, all three basic materials – brass, stainless steel, and plastic – come into question.

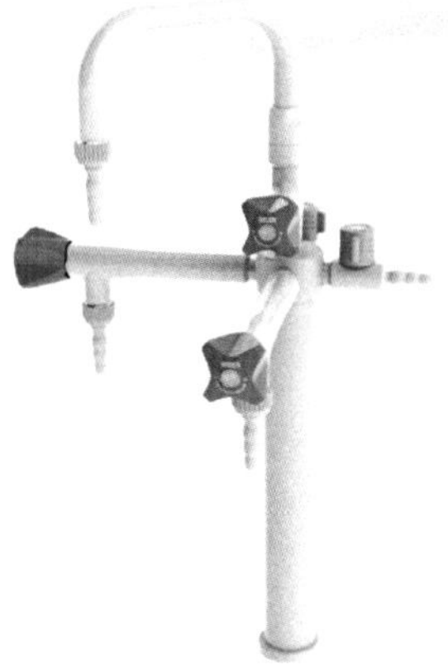

Figure 24.1 Multiple-media bench-mounted fitting. Headwork and nozzles are designed according to the medium.

Figure 24.2 Cut-away of a brass housing prior to the protection coating against external corrosion. The housing with uniform valve seat allows the installation of various headworks.

24.11.1 Brass

Due to the optimum price – performance ratio, the classical brass fitting is standard in laboratories. It is particularly suitable for the basic supply of cold and hot water. With the dimensioning of bench-mounted fixtures and swivel arms exceeding the standard, such a fitting can also cope with extreme stresses. At, for example, 95 °C, the maximum temperature can also be far above the limit value of 65°C set in DIN 12918 part 1.

Figure 24.3 Depending on medium and dosing task, various headworks with sealing and dosing techniques of different materials can be installed.

24.11.2 Stainless Steel

Fittings of stainless steel are highly temperature resistant, particularly stable and corrosion resistant. Their application is recommended for aggressive waters and extreme temperatures. The user who relies on stainless steel in areas of thermal disinfection is on the safe side. Suitably designed fittings can tolerate temperatures, for example, up to 230 °C.

24.11.3 Plastics

Regarding temperature resistance and strength, plastic is subject to far more limitations than metal. The temperature and pressure limits of standard fittings range at 48 °C and PN 4, respectively. A series designed for higher loads tolerates temperatures of up to 60 °C and pressures of up to PN 10. A special recirculating fitting tailored to ultrapure water is designed in a way that the medium can be constantly kept moving in order to avoid the formation of a biofilm. Among the huge number of plastics available on the market today, mainly PP and PVDF are used for the demanding manufacturing of fittings. The field of applications predominantly includes treated water such as distilled, fully demineralized, or ultra-pure water.

24.11.4 Shut-Off and Dosing

In practice, four technologies are particularly suitable for utilization in laboratory fittings for water.

24.11.5
Lubricated Headwork

The rotary motion of the component with rising spindle is converted into a stroke. A plate with a rubber washer is pressed against the valve seat with the inlet port during shut-off. While opening, the plate is returned and the flow is released. Advantages of this solution for water fittings are high tightness, fairly good dosing over about 1 1/2 turns of the handle, high resistance toward pollution, and calcification as well as easy replacement of defect or hardened rubber washers.

Excessive pressure on the elastic rubber may entail several disadvantages: the material is overstressed and quickly becomes brittle. After opening, the material expands and shuts off the port (self-throttling effect). It is also possible that, due to the adhesive effect caused by the elastic rubber, the outlet will be released quite abruptly when opened.

24.11.6
Ceramic Disc Cartridge

Ceramic disc cartridges have two discs with identical outlets, which lie on top of each other, free of play. One of them is fixed, the other is rotatable. The valve is opened by congruently adjusting the outlets. Since the ceramic discs are very hard, abrasion is extremely low but exists. In case of polluted media and media containing solid particles, defects are possible.

It is an advantage that, when dosing, the outlet is opened very evenly. Disadvantageous is the fact that the handle is only slightly turned between the two final positions. Therefore, the disc design with only one outlet should be selected, since here, with a half-turn of the handle (180°), it must be turned twice as far as in the case of handles with two outlets (90°).

24.11.7
Diaphragm Headwork

Diaphragm headworks are shut off by a completely tightly fitting diaphragm covering the intake. As soon as the diaphragm is lifted, flow is released. This headwork is particularly suitable for use with ultra-clean water, since it is free of dead space and the medium only gets into contact with the diaphragm. The elastomer EPDM is very well suited to being used as diaphragm material.

24.11.8
Plastic Headwork with Overwind Protection

These headworks with integrated slipping clutch, which are mostly used for demineralized waters, do no longer correspond to the current quality standards due to the exposed grease. Meanwhile, PVDF cartridges without clutch and stable enough with ceramic disk cartridges are available.

24.11.9
Ceramic Disc Cartridge

Ceramic disc cartridges comfortably and precisely mix cold and warm water. When operated by means of a wrist action handle, temperature limitation is recommended. In training and educational laboratories, even water quantity limitation is recommendable.

24.11.10
Potable Water

DIN 1988 and EN 1717 are applicable to installations for potable water. Attention should be paid to the fact that all plastics getting into contact with potable water are really suitable for that and correspond to the KTW recommendations of the German Federal Health Authority. This is only the case when they are tested and approved.

In the Technical Rules for Potable Water Installations of DIN 1988 Part 4/EN 1717 on the protection of potable water, substances are subdivided into five classes according to the hazard potential. A lot of the substances handled in laboratories are toxic, carcinogenic, or radioactive and belong to hazard class 4. Hazard class 5 includes agents of communicable diseases, such as hepatitis viruses or salmonella. In areas where substances of class 4 and 5 are handled, special measures for the protection of potable water must be provided.

Such measures are the use of collection locking devices in the supply line which shut off an entire laboratory network or individual areas from the (public) potable water network. For a small number of supply points, individual locking devices on the fittings are recommended. In both cases, care should be taken that the systems are installed in a way that the annually required functional testing according to DIN 1988 Part 8 and to the accident prevention regulations is practically feasible.

Individual locking devices for fittings are inter alia.

24.11.11
Free Draining

This means, the water can drain off freely without the possibility to absorb substances in the pipe direction. The standard requires the water jet regulator to be installed at least twice the inner diameter above the supply, in any case at least 20 mm above the highest possible water level. Also, aeration and ventilation must be secured without the possibility for hoses to be connected.

24.11.12
Pipe Interrupter

They are designed in a way to reliably separate the drinking water net from the connected system. There are two types of pipe interrupters: the semi-open type

A2, approved for hazard class 4, and the open Type A1 which can also be used in the range of class 5. The fitting outlets and connections should be designed to ensure that back-flow of the water does not occur, since otherwise the water could spurt out sideways at the pipe interrupter.

24.11.13 Backflow Preventer

Their capability to prevent the intrusion of hazardous substances is limited and thus, they are only approved for areas that do not represent a source of danger (classes 1 and 2). In areas of class 3, which includes slightly toxic substances, they may be used only for a short time and in classes 4 and 5 not at all. Therefore, they do not come into question for use in laboratory fittings for the protection of potable water.

24.12 Conclusion

Users who pay attention to whether their fittings comply with the relevant standards, such as DIN 12919, DIN 1988, and EN 1717 can ensure that the work safety requirements of the competent institutions are fulfilled.

24.12.1 Cooling Water

Here again, the fulfillment of the requirements set out in DIN 1988 is the prerequisite for safe operation. Material selection and design must correspond to the composition of the cooling water or the coolant and to particular minimum and maximum temperatures probably required.

24.12.2 Circuits

In order to prevent faulty operation in a closed cooling circuit, it is recommendable to use a fitting that opens and closes flow and backflow simultaneously. The design to be selected should be able to avoid that heating up during standstill, that will lead to impermissible pressure rise and thus to damages. The smaller expansions usually occurring are compensated by the plastic hoses that are normally used to connect the cooling device.

The half-open circuit is widely used. Here, only the flow is adjustable or capable of being shut off. The backflow will be equipped with a non-return valve (backflow preventer) which opens automatically at a certain differential pressure.

24.12.3 Temperatures and Volume Flow

Temperatures of flow and backflow along with the volume flow are determined by the amount of heat to be absorbed and by the desired final temperature.

Cryostats mostly require a temperature far below the freezing point of water. Fittings and connecting pipes must be resistant to the special additives in the cooling water. In order to avoid the formation of condensate, all component parts able to adopt a temperature below the dew point must be adequately insulated.

This, however, causes problems in case of metal diffusers in the laboratory fume cupboard, since here, externally applied insulation makes little sense.

24.12.4 Pressure

Centralized cooling water treatment has proven to be efficient and practical for laboratory buildings. The line pressure for the supply of several floors, which results from building height and required volume flow, can damage sensible cooling devices. Thus, it is often reasonable to adjust each individual point of use by means of a pressure reducer.

24.12.5 Quick Connects

Quick connects considerably increase ease of use. In order for the cooler and its connecting hoses not to lose coolant when disconnecting from the circuit, at least the connecting parts mounted on the hoses should be provided with an automatic shut-off.

24.12.6 Treated Waters

The demands on the quality of treated waters rise with the increasing optimization of analysis technology. Therefore, the lab fittings used do not only have the task to dose the medium, but also to preserve it. The medium must not be changed by the lab fitting, neither regarding conductivity, nor with regard to TOC value or other important parameters like pyrogens. Most suitable are designs of plastic, where the medium does not get into contact with grease. Chemically, nickel-plated brass fittings do not fulfill these requirements. Moreover, nickel that enters into the solution causes an incalculable environmental pollution.

24.12.7 Vapor

Vapor, in the past mostly used for heating, has almost completely disappeared from laboratories today. The "waste" product from production processes is only

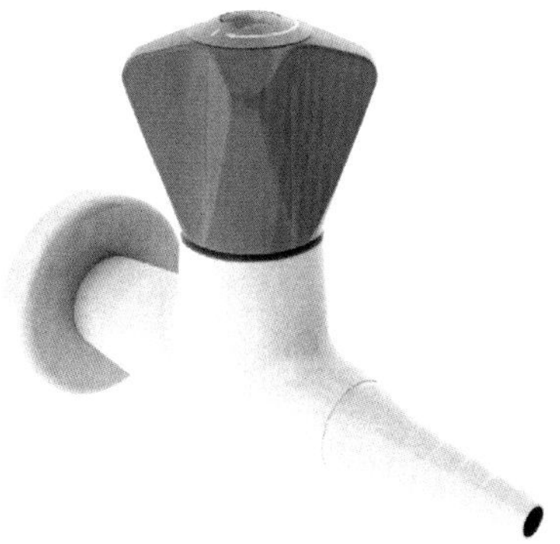

Figure 24.4 Wall-mounted fitting for water. The powder coating develops a homogeneous surface, which is UV stable and provides high-quality corrosion protection. The removable grommet allows the connection of lab devices.

occasionally used in research. Due to this development, water vapor often contains corrosive components and varies in temperature and pressure.

Common fittings tolerate temperatures of up to 120 °C because of the plastic coatings applied. Special designs even allow 230 °C, however, they are strongly limited in their geometry and thus, in their installation options (Figures 24.4–24.7).

24.13 Burning Gas

Due to the explosive and toxic properties of gases, safety is of utmost importance for supply facilities, pipes, and fittings.

The competent German authority for the compliance with accepted rules for the handling of fuel gases required by legislation is the Deutsche Vereinigung des Gas- und Wasserfaches e.V. (DVGW) (German association for gas and water). Based on the rules and regulations of the DVGW and related standards and guidelines, the DVGW tests and certifies products, companies and individuals. In practice, this means that gas installations may only be carried out by authorized specialist companies in connection with precisely determined installation techniques and certified products.

The material in use for burning gases may differ from country to country and depends on the content of sulfur.

What certified lab fittings, according to the DVGW worksheet G 260 (fuel gases), should look like is set out in DIN 12918 Part 2. The limits for pressure and temperature are 0.2 bar and 60 °C. With a fitting body made of pressed brass with subsequent powder coating, the requirements regarding strength and corrosion protection are completely fulfilled.

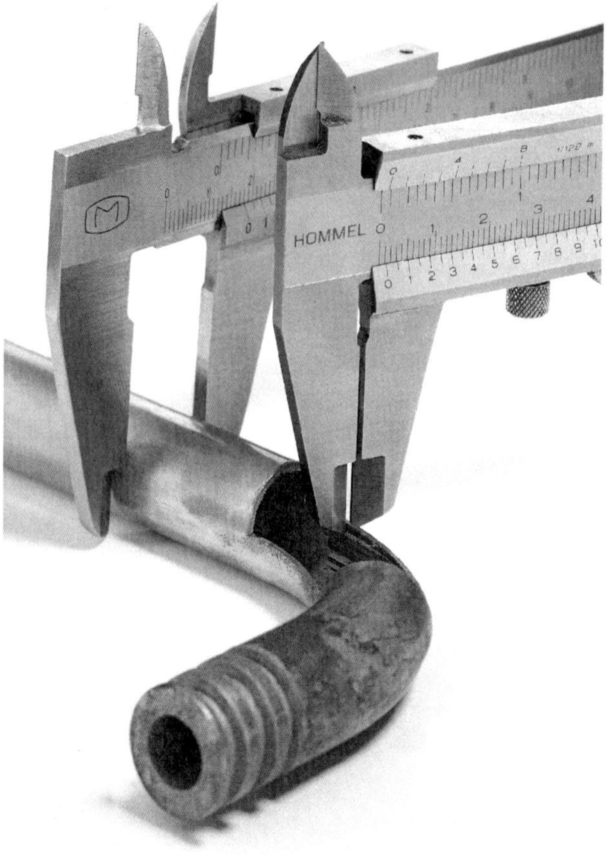

Figure 24.5 Users of fittings with swiveling outlets, wall thicknesses of over 1 mm and outer diameters, which, with 18 mm, exceed by far the values predetermined by the DIN standard, are on the save side. The stable brass tube resists even strong mechanical stress. Even after incorporation of a thread, the specified minimum wall thickness of 0.75 will be securely met.

Under normal circumstances, the fittings are used to feed gas burners. Since dosing occurs at these devices, mostly only the open and shut function of the fuel gas fittings is used. This requirement is very well fulfilled by a ceramic disc cartridge with two diffusers. A press-down protection protects against undesired opening. Only when pressed down, the fitting can be opened by turning it anti-clockwise by 90°. The valve position must be clearly visible on the handle. The handle position can be marked, for example, by arrows on the label or by two wings on the handle base body. When the valve is closed, the marking is diagonal to the flow direction, and when the valve is open, the marking is in flow direction.

Figure 24.6 View on three different headworks of lab water fittings. The left hand side shows a lubricated headwork in the middle, while ceramic disc cartridges are shown on the right-hand side. The headworks on the left are made of brass, those on the right are of plastics. Due to the two outlets in the headwork on the right, it is hardly suitable for dosing, as the handle can only be turned by 90°.

Figure 24.7 If according to DIN 1988 Part 4/EN 1717 individual backflow prevention for the protection of drinking water is required instead of separate supply of lab water, this can only be adequately achieved by means of pipe interrupters. The figure shows pipe interrupters of type A1, which meet protection class 5, and those of type A2 with protection class 4.

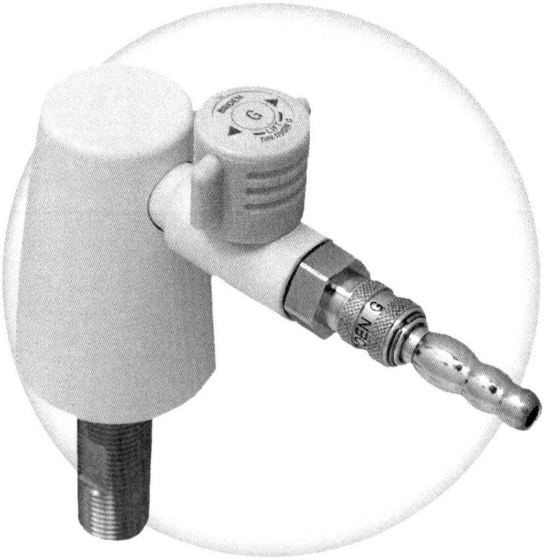

Figure 24.8 Fitting for fuel gas suitable for educational purposes with quick coupling and nozzle hose connection.

Another mechanics with very high safety comfort is provided by the so-called Lift-Turn type. Here, the handle must be lifted before the fitting can be opened by turning it anticlockwise by 90°.

In particular, when used in scientific classrooms, it is important for the handle to be not only attached but also secured by a screw against removing.

24.13.1 Quick Couplings

Quick couplings used in the educational applications must have a shut-off unit. Adequate corrosion protection for such stressed components is nickel plating. Uncoated brass couplings are not suitable for robust operation in schools and teaching.

The shut-off units in quick couplings, however, are generally unsuitable for permanent shut-off. For this purpose, the tap of the fitting must be used (Figures 24.8–24.10).

24.14 Technical Gases up to 4.5 Purity Grade

In labs for education and research, a lot of technical gases are used. Their demand-driven provision depends on the respective intended purpose and the physical properties of the individual gases. In this case, relatively high demands are placed

Figure 24.9 A press-down protection protects against undesired opening of the fuel gas fitting. Only when pressed down, the fitting can be opened by turning it anticlockwise by 90°. The red pins clearly indicate the valve position. When the valve is closed, the marking is diagonal to the flow direction, when the valve is open, the marking is in flow direction. The signal effect is additionally enhanced as the pins protrude when the fitting is opened.

on the dosing capacity of lab fittings in contrast to fuel gas fittings. Even a rotational movement of 360° is far from sufficient. The adjustment range should be several rotations of the handle to enable the exact adjustment of the flow rate to the connected devices.

Usually, the fittings are made for technical quality 2.0. Special cleaning of all surfaces in contact with the medium allows utilization up to a 4.5 purity grade (Figure 24.11).

Figure 24.10 View on the shut-off unit of a fuel gas fitting with quick coupling. If the nozzle hose connection is not attached, the shut-off unit prevents the gas from escaping despite the opened tap.

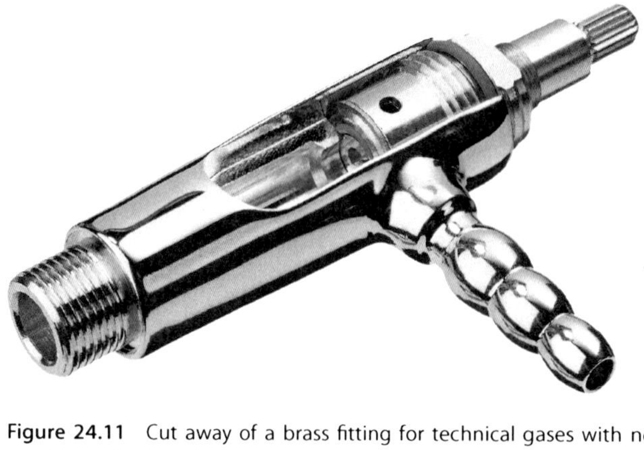

Figure 24.11 Cut away of a brass fitting for technical gases with needle headworks for precision adjustment.

24.14.1
Several Types

This requirement can be well fulfilled with needle headworks. The user can choose between various types to find a solution for most different tasks: headworks with high-precision adjustment are suitable for the control of smallest amounts. In most cases, fine adjustment is selected as standard. The control of larger flows, as for instance in the case of vacuum, can be carried out with "High-Flow" headworks.

The amounts flowing through the respective valve openings are shown in the flow diagrams of the manufacturers.

24.14.2
Normative Framework

For fittings according to Part 3 of DIN 12918, pressure limits of 10 bar and temperature limits of 60 °C apply. In opened and closed position, the handles shall resist an actuating torque of at least 6 Nm (Figure 24.12).

24.15
Vacuum

For vacuum, plastics such as PVDF and PP, but also brass are suitable. In most applications, coarse dosing is absolutely sufficient. Integrated backflow preventers prevent the mutual influencing of different vacuum chambers in a network (Table 24.1).

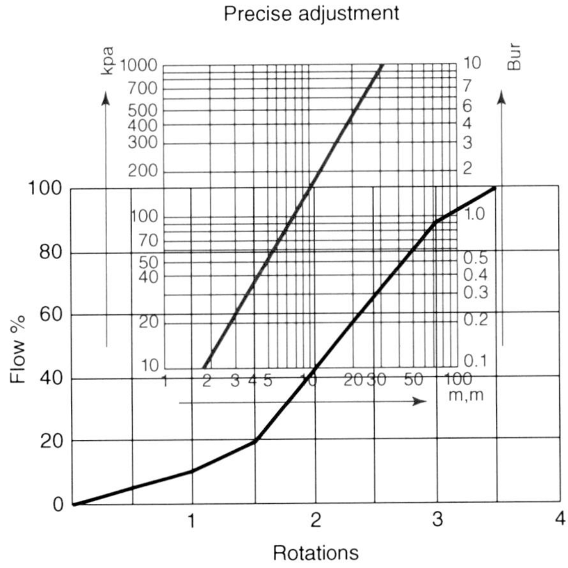

Figure 24.12 Flow and regulation of needle valve.

Table 24.1 Matrix for laboratory fittings.

Service	Media	Pressure	Temperature	Specials	Pipes	Material	Headwork/dosage	Mounting	Type	Handling
Water	WPC	PN10	Maximum 65 °C	EN1717	—	Brass	Grease chamber cartridge	Wall	Fixed spout with hose coupling	Handle
	WPH	—	—	Back Flow Preventer, collectice use	Pipe interrupter EA2	Stainless steel	Ceramic disk cartridge 180°	Suspended	Swivel spout with hose coupling	Lever handle
	WPC/WPH	—	—	Back Flow Preventer single use	Free drain	—	Ceramic cartridge	Bench	Swivel spout with aerator	—
	WNC	—	—	—	Pipe interrupter EA2	—	—	Pipe	—	—
	WNH	—	—	—	Pipe interrupter EA2	—	—	—	—	—
	WNC/WNH	—	—	—	—	—	—	—	—	—
Cooling water	WCF	PN10	To be specified	Condensate	Closed loop	Brass	Grease chamber cartridge	Wall	Fixed spout with hose coupling	Handle
	WCR	—	—	—	Half open loop	Plastic	Ceramic disk cartridge 180°	Suspended	Qich disconnect both sides shut off	RFV
	—	—	—	—	—	Stainless steel	Ceramic disk cartridge 180°	Bench	—	—
	—	—	—	—	—	—	Ceramic cartridge	Pipe	—	—
	—	—	—	—	—	—	Dosage piston	—	—	—

Demineralized water	WDC	PN10	Maximum 60 °C	Conductivity	—	Stainless steel	Niederschrauboberteil	Wall	Fixed spout with hose coupling	Handle
	—	—	—	TOC-value	—	Plastic	Diaphragm	Suspended	—	—
	—	—	—	Pyrogene	—	Brass + plastic	—	Benc	—	—
	WCC	—	—	—	Recirculation		—	Pipe	—	—
Steam	WST	—	Maximum 120 °C	—	—	Brass	Diaphragm	Wall	Fixed nozzle	Handle
	—	—	Up to 230 °C	—	—	Stainless steel	Needle headwork	Bench	—	—
—	—	—	—	—	—	—	—	Pipe	—	—
Gas	G	PN0,2	Maximum 60 °C	—	—	Brass	Ceramic disk cartridge 180° with push down	Wall	Fixed nozzle	Handle
	LPG	—	—	—	—	—	Ball valve, lift-turn	Suspended	Schnellkupplung	—
	CH4	—	—	—	—	—	—	Bench	—	—
	C3H8	—	—	—	—	—	—	Pipe	—	—
	C4H10	—	—	—	—	—	—	—	—	—
Technical gases < 2.0	CA	PN16	Maximum 60 °C	—	—	Brass	Precise needle headwork with 3 turns	Wall	Removable nozzle	Handle
Technical gases > 4.5	N2	—	—	—	—	—	Full flow headwork with 1.5 turns	Suspended	Hose nozzlecap nut	—
	—	—	—	—	—	—	Precise needle headwork with 3 turns	Banch	Quick connect	—
Vakuum	V	10E-3 mbar	—	—	—	Plastic	—	Pipe	—	—

25 Gases and Gas Supply Systems for Ultra-Pure Gases up to Purity 6.0

Franz Wermelinger

25.1 Gases and Status Types

Gases and gas mixtures of high purity are required above all for analytic applications in laboratories. High-quality fittings and systems must be used with pure gases to ensure the quality of these gases from the source to the tapping spot.

Typical properties of ultra-pure gases, which affect how they are handled are as follows:

1) Inert
2) Flammable
3) Oxidizing
4) Corrosive
5) Toxic
6) Cryogenic (combustion is also possible in this state).

Gas purity of ultra-pure gases in the laboratory area is normally higher than 4.5. Lower qualities are designated technical gases.

All available gases, ultra-pure gases, and the associated mixtures are classified in purity classes or levels. These purity levels have been developed according to their production and feasibility over decades. Consequently, there are different purity types depending on the gas type. Depending on the production process, post-cleaning and decanting procedures, many individual gas types are only available in a few purity classes. The known figures such as 4.5, 5.0, 5.5, 6.0 serve as abbreviations of the purity percentage in volume of a cubic meter of gas; the other elements are impurities. Manufacturers normally submit specifications, which also indicate the maximum permissible impurities. For example, these are the maximum quantity of steam, oxygen contents, hydrocarbons, or CO_2.

The Sustainable Laboratory Handbook: Design, Equipment, Operation, First Edition.
Edited by Egbert Dittrich.

25.1.1 System Explanation

The first number always describes the quantity of figures 9 before and after the dot in %, and the second number stands for a tenth percentage of purity (Table 25.1).

25.1.2 Examples

Oxygen 4.5 = oxygen with a purity of 99.995%
Specified remaining impurities are:

- $H_2O \leq 3$ ppm mol^{-1} and
- CnHm ≤ 0.5 ppm mol^{-1}.

Nitrogen 6.0 = nitrogen with a purity of 99.9999%.
Specified remaining impurities are:

- $H_2O \leq 0.5$ ppm mol^{-1}
- $O_2 \leq 0.1$ ppm mol^{-1}
- CnHm ≤ 0.1 ppm mol^{-1}
- $CO \leq 0.1$ ppm mol^{-1}
- $CO_2 \leq 0.1$ ppm mol^{-1}
- $H_2 \leq 0.1$ ppm mol^{-1}.

The purity threshold 4.5 for all gases has become the standard for simple classifying and selecting gas systems and fittings.

However, there are gases that are not available in high-purity classes, that is, manufacturers classify them in classes lower than ultra-pure.

25.1.3 Basic Principles

To be able to guarantee this gas quality until the point of use and reliably provide users with the required properties of gases, a few points for handling gases, bottles, fittings, and systems need to be considered and clarified in advance.

Table 25.1 Purity matrix.

Percentages	Decimal shorthand
99.5	2.5
99.995	4.5
99.998	4.8
99.999	5.0
99.9995	5.5
99.9999	6.0

Consequently, systems must be planned with gas suppliers and manufacturers of gas supply systems and fittings, as well as only being built and installed with the participation of proven experts.

25.2
Material Compatibility

Material compatibility is to be clarified precisely for building networks as well as only for use from bottles.

For example, copper is not suitable when acetylene is used and especially not with gases that have corrosive properties. Material compatibility has to be checked for every component. This starts with bottle seals and continues to the basic units of fittings, O-rings, connection parts, dial gauges and their interior parts, as well as membranes, springs, balls and pipes, or capillary tubes.

25.3
Connection Points

In the ultra-pure gas section, it is advantageous to design and plan the fewest possible sealing and connection points for supply systems and individual fittings. State-of-the-art technology employs connection points as well as transitions from fittings to pipes with metallic sealing points or avoidance of polymer sealing or even those containing hydrocarbons.

Seals are also potential leakage points where gas can escape, which can result in economic loss on one hand and is a safety problem, especially with combustible or toxic gases, on the other hand.

Another undesirable or unacceptable effect of leakages is diffusion, that is, the migration of contaminating gases and substances into the system. If diffusion takes place in a system, the problem is often that the connection technique of the gas supply systems has been designed incorrectly or produced entirely using metal.

Despite high counter-pressure, foreign molecules can infiltrate a system through a nonmetallic membrane via particle pressure.

25.4
Impurities

Additional undesirable substances in gas supply systems, especially for systems that are operated using oxygen and other reactive gases, are oils and greases, that is, hydrocarbons.

Consequently, it makes sense to design all components free of oil and grease. Special attention is to be paid to the pipes. Standard sanitary pipes always have some grease from the production process (drawing lubricant). These are

Figure 25.1 Ultra-pure gas shut-off valves including particle filter.

completely unsuitable for gas supply systems in laboratories. Pipes are to be used that are suitable for defined cleaning processing and according to DIN design with certificates. The fact that such pipes must be stored, labeled, and closed at the end correctly should be a matter of importance (Figure 25.1, Tables 25.2 and 25.3).

25.4.1 Particle

Practical experience shows that particles from the bottles and lines frequently soil or damage the systems. To counter this, various high-pressure filters have been developed for bottle connection tubes in recent years.

We also recommend retrofitting filter systems, which can be built into lines directly (online) with a shut-off valve in new piping systems or as an upgrade. Among other things, these are designed for effective space-saving, so that no bypass lines need to be built, and are equipped with new filter elements that can filter out very small particles.

For gases with higher degrees of purity, the bottles of gas suppliers normally also have better quality on the inner side. The objective is to minimize the surfaces, so that no foreign substances can stick and no particles can be carried along with the gas flow.

In the meantime, bottles are available in the market, which have an integrated filter system in the bottle.

25.5 Supply Systems: Central Building Supply/Local Supply and Laboratory Supply

The basis for designing and planning gas supply systems for laboratories requires consultation of local norms (including their interpretation or understanding),

Table 25.2 Specification for pipeline material for the stainless steel pipe to be used.

	ES1	ES2
Designation	Quality stainless steel pipe	Quality stainless steel pipe post-cleaned
Application	Technical gases (acetylene, ammonia, etc.)	Ultra-pure gases up to 5.0
Material	1.4404/1.4435 (316 l) or higher quality	1.4404/1.4435 (316 l) or higher quality
Production form	Seamlessly extruded according to DIN 17458 or ASTM A269	Seamlessly extruded according to DIN 17458 or ASTM A269
Tolerances	According to DIN 2462 D4/T3	According to DIN 2462 D4/T3, D3/T3, or ASTM 269
Hardness	Inspected < RB80	Inspected < RB80
Interior roughness	—	Labeled according to dimension Ra < 0.8 – 1 μm
Purity	—	Inside without oxide paint or tempering color, free of organic impurities and foreign particles
Packaging	Closed at the ends with caps	Closed at the ends with caps and shrink-wrapped in foil
	ES3	ES4
Designation	Quality stainless steel pipe post-cleaned, interior surfaces hardened and tempered	Quality stainless steel pipe post-cleaned, electro-polished
Application	Ultra-pure gases up to 6.0	Ultra-pure gases up to 6.0 and highly corrosive gases
Material	1.4404/1.4435 (316 l) or higher quality	1.4404/1.4435 (316 l) or higher quality
Production form	Seamlessly extruded according to DIN 17458 or ASTM A269	Seamlessly extruded according to DIN 17458 or ASTM A269
Tolerance	According to DIN 2462 D4/T3	According to DIN 2462 D4/T3
Hardness	Inspected < RB80	Inspected < RB80
Interior roughness	Labeled according to dimension Ra < 0.4 – 0.6 μm	Labeled according to dimension Ra < 0.2 – 0.35 μm
Purity	Inside without oxide paint or tempering color, free of organic impurities and foreign particles	Inside without oxide paint or tempering color, free of organic impurities and foreign particles, electro-polished
Packaging	Closed at the ends with caps and shrink-wrapped individually in foil, labeled as ultra-pure gas pipe	Closed at the ends with caps and shrink-wrapped individually in foil, labeled as ultra-pure gas pipe
	ES5	
Designation	Double-wall pipe, if required with leakage monitoring of interior and exterior pipe	

(continued overleaf)

Table 25.2 (*Continued*)

	ES1	ES2
Application	Hazardous ultra-pure gases (explosive, corrosive, and/or toxic) at difficult-to-access spots, escape routes, common rooms, and work rooms	
Material	1.4404/1.4435 (316 l) or higher quality	
Production form	Seamlessly extruded according to DIN 17458 or ASTM A269	
Tolerance	According to DIN 2462 D4/T3	
Hardness	Inspected < RB80	
Quality	According to the requirements of gas purity (cf. ES1-4)	
Packaging	Closed at the ends with caps and shrink-wrapped individually in foil, labeled as ultra-pure gas pipe	

Table 25.3 Specification for pipeline material for the copper pipe.

	CU1	CU2
Designation	Ultra-pure gas pipe	Special ultra-pure gas Cu pipe
Application	Supply with noncorrosive ultra-pure gas up to 5.0	Supply with noncorrosive ultra-pure gas up to 5.5 in systems with especially reduced grease contents, for example, gas chromatographs with ECD detector
Material	SF-CuF37 (material: 2.0090.32) according to DIN 17671 sheet 1	SF-CuF37 (material: 2.0090.32) according to DIN 17671 sheet 1
Tolerances	According to DIN 1786/EN1057	According to DIN 1786/EN1057
Purity	DIN 8905 part 2, residual grease content < 0.2 mg dm^{-2}	DIN 8905 part 2, residual grease content < 0.1 mg dm^{-2}
Packaging	Closed at the ends in box or bag	Closed at the ends with caps and shrink-wrapped in foil with N2 filling in box or bag

internal norms of companies, integration of expert consultants from the local fire protection or fire prevention authorities, construction regulations, empirical values of planners, user data, and their building managers, as well as possibly facility management.

Unfortunately, due to lack of expertise and saving in the wrong place, planning can be assigned to the wrong trade or even a supplier not involved in the sector, who then makes the overall concept a lot more expensive or designs it incorrectly with so-called safety surcharges, whether in respect to the material selection or the line dimensions.

25.6 Central Building Supply (CBS)

Design and construction of the central building supply (CBS) starts with bottle storage (gas bottle stocks as well as full and empty bottles). This location is not to be confused with the central bottle site, from where the building is supplied. It is advantageous to have these two rooms located next to one another for logistics reasons.

Fire protection and ventilation in particular should be considered here. Careful consideration is also required to determine whether a location in the basement is permitted (gases heavier than air, e.g., propane or CO_2). An optimally planned site also affects the cost of the gas pipeline distribution network.

The size and type of bottle stations result from the consumption quantities or result from user queries. In this context, it must be considered that discontinued needs (procurement peak) determine the size of a system. Whether a gas bottle station can be operated completely automatically, semi-automatically, or even manually must be clarified in each individual case. It is advantageous to design systems as completely automatic switching stations at operation of analytic systems, incubators, and so on. At the same time, the option should be provided to set the idle-state alarm on the building control system. In this context, responsibility for the gas bottle system is to be considered to the greatest extent possible. Correct system operation, especially bottle replacement, determines gas purity at the point of use in the long term.

This work can also be outsourced to gas suppliers, as the option exists within the context of a gas management system to equip systems with sensors and look "into the customer's bottle" from another site via a network and consequently monitor the fill level and gas supply.

Two important parameters are the performance requirement of the system, that is, $Nm^3\ h^{-1}$, and the pressure that is to be fed into the system.

As previously mentioned, the volume output must be calculated based on user needs with reserve for unknown and future factors.

This volume output also determines the size of a bottle station. The recommended value is that the system need not switch more than once per week, that is, switch from the operation side to the reserve side, otherwise, a system that is too small has been selected. Various system types are available in the market from a basis system with 2×1 bottle to 2×4 bottles per side and all the way to high-performance systems, which can be operated with up to 2×2 bottle bundles on each side.

This also optimizes bottle rental and logistics costs.

A few system manufacturers also specialize in mono-block or modular systems, which require very little space and can be expanded.

25.7
Pipe Networks and Zone Shut-Off Valves with Filter

Piping systems are dimensioned from data as described later in 25.9. Pipe diameter is calculated from the required minimum pressure and output volume. Various calculation models as well as tables and flow curves are available in the market. Empirical values of specialist companies are not to be underestimated. In practice, pipe networks often have dimensions which are too large, resulting in high investment and operation costs. It is advantageous when the main line is designed as a ring pipe and the outlets (lateral branches) are kept small.

Without referring to standards and maximum permitted gas speeds, the proven recommended value is to design the ring pipe in a standard laboratory building on a floor with a maximum diameter of 18 mm and the rising mains from the bottle stations to the ring pipe with maximum 22 mm. The outlets in the laboratory can be constructed with maximum 10 mm. Empirical values have shown that gas pipelines (lateral branches) can be operated when they are much smaller; 12–15 tapping spots have been operated in energy columns without pressure loss using 6 mm copper pipes.

It makes sense to divide the gas supply lines into specific sections for maintenance work; guidelines also exist for this. One section per floor and in front of a laboratory has proven to be a good choice; a zone shut-off valve with filter can also be installed at this spot. These can also be retrofitted for soiled or old systems.

The pipeline materials must be adapted to the gas properties on principle. Tables and specifications of gas companies provide information about this.

Gas supply systems are mainly constructed using copper. The reasons for this are the easy working nature of copper as well as the fact that copper pipe is available in soft quality as well as ultra-pure quality. Seen overall, there are no technical reasons to only use stainless steel for gas purity 6.0. With general models and corresponding working, copper can be used without problems for gas supply systems up to 6.0.

A prerequisite for a high-quality gas supply system is the precise definition of the pipe material. Standard sanitary copper pipe is not suitable, because it does not provide the required purity; of course, stainless steel pipe does not do this either. The pipe must precisely fulfill the required residue grease values and freedom from particles. This is to be ensured via the complete supply chain and working, and recorded in certificates and logs. The materials are one part of this, and the working, the other indispensable part. Only experienced specialist companies with skilled fitters can be assigned such tasks. The individual work steps such as storage of pipes, connection parts and fittings on the construction site, separation of pipes, intermediate storage, temporary closing of open pipe ends and fittings, and proper working of system parts is always to be done according to the manufacturer instructions.

For cost-optimized design, the connection technique is decisive in addition to the correct pipe selection and diameter. For example, pipes with dimensions which are too large cannot be laid using a customary clamping ring system, but instead have to be soldered, something that can result in soiling in principle. If a rust-free variant is specified, regardless of the reasons, you need to be aware that you can only work with suitable clamping ring screw connections until a limited pipe diameter, and that these can be very expensive. The pipe coupling systems should be certified and free of oil and grease in the same way as the pipes.

Another very good pipe connection technique involves orbital or TIG (tungsten inert gas) welding, provided that very high quality is required or corrosive or toxic gases are fed into the systems.

It should also be noted that certain norms permit pressing pipe systems according to the manufacturer instructions. However, this working type is not optimum for gas qualities >4.5, because the pipes and fittings are not available in the desired post-cleaned qualities and the connections are also produced with O rings (always greased).

The construction site must also be in a clean condition in principle to be able to carry out this work.

Unfortunately, there are often construction sites at building shells where you can still expect to find work in wet conditions, cutting and creation of dust, and even migration of hydrocarbons. Ultra-pure gas installation work is not to be performed in such cases. The building should already have been cleaned and it's best if painting has been completed. If the correct pipe brackets have already been installed for CBS in a building shell, drilling work should be superfluous. To that end, pre-fabricated components such as aluminum brackets or booms are available as well as full-coverage aluminum grids. Ideally, all attachments within the context of floor-overlapping, pre-fabricated ceiling systems (media ceilings) are installed, which provide an attractive appearance of the media supply system in addition to high-quality workmanship.

Acetylene and CO, substances that react with copper or nickel, as well as all corrosive and toxic gases are exceptions.

Because flexibly designed laboratory workstations are increasingly constructed as comprehensive energy systems with media columns from above and consequently media feed can also be designed flexibly, the installation of gas pipelines as "mobile installations" is a good choice. This means eliminated stationary, isometric, and rectangular components in favor of partially mobile pipe routing. In turn, this is only possible using high-purity copper pipe. As a result, columns or lights with tapping spots can be adapted to workstations as well as shifted 20–30 cm to the side.

25.8 Fitting Supports and Tapping Spots

Various options are available for fitting supports for laboratory gases. Comprehensive media supply systems should always be considered, that is, the design for liquid media, such as operation water, laboratory cooling water, softened water, electrical systems, and vacuums employs the same principle for all. As a result, it is possible to use system-specific, comprehensive units from one "mold" from one manufacturer. Comprehensive systems can be media columns, service wings, cells, or similar things.

The material is selected at the tapping spot analogous to material selection of the pipe system. Use of different metals and purity levels within a pipe network is to be avoided. Aluminum or brass is suitable for most fittings and tapping spots, and stainless steel only used in exceptional cases.

In addition, the details of fittings for ultra-pure gases must be specified in consultation with users. The following indispensable questionnaire clarifies the required, desired properties with respect to legal security, operation reliability, and above all, functionality of the tapping spot:

Installation site

- Gas type, gas state, and temperature
- Purity class at point of use
- Pressure-regulated or nonregulated tapping
- Pressure ranges of each tapping spot (if the pressure range is designed incorrectly, the tapping spot can be unserviceable)
- Service areas (for example, these are glove boxes in a laboratory if the gas feed including the correct tapping spot must be structured separately. If very precise doses are required in a laboratory, fittings with low outlet pressure as well as a micro-dosing valve are to be provided at the outlet).
- Outlet geometry, that is, connection system and possible basic accessory features (valve rake, volume gauge, nozzles, etc.).

Very simple, safe, and practical connection systems are available in the market for tapping spots. They make it possible to connect all kinds of tubes all the way to rust-free steel capillaries safely and appropriately without the necessity of tinkering in a laboratory to find solutions (Figure 25.2).

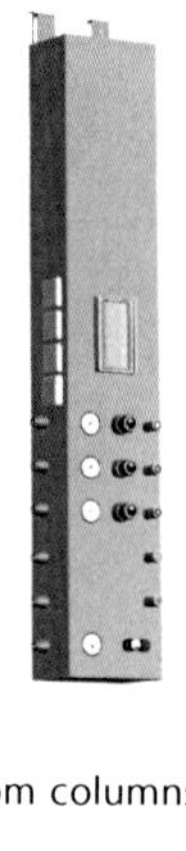

Figure 25.2 Clean room columns.

25.9
Local Laboratory Gas Supply

Local laboratory gas supply systems are especially suitable when there are different users or experiment procedures limited in time. It is imperative that the staff are able to work quickly and safely, even during short-time experiments and especially with the often used hazardous gases. Consequently, even untrained users should be able to use all connection techniques quickly and safely. Local systems are to be designed analogous to the key figures of central building supply; the smaller execution makes them considerably less expensive. Depending on the manufacturer, local systems can be reused and are less tied to one location. As a result, inappropriate and noncompliant changing of standardized connection supports of the reduction valves can be avoided.

25.10
Surfaces - Coatings

Due to the increasing trend to use highly active substances in laboratories, increased attention must be paid to the fittings. Tapping spots must comply with all requirements of a standard laboratory as well as have the appropriate properties, so that they are not damaged by users in daily, hard routines in laboratories, or clean rooms during cleaning. Particularly, the latter systems

Figure 25.3 Joint-less built-in panel.

Figure 25.4 Joint-less built-in panel, pre-wall assembly.

require increased attention and expertise. The design of fittings for these laboratories must be appropriate above all with respect to the surfaces, freedom from crevices, installation potential in separating walls or columns installed in clean rooms, resistance to corrosion, as well as the possibility of operating them when protective clothing is worn (Figures 25.3 and 25.4).

25.11 Inspections

The complete systems are to be inspected. This already starts with the selection of the correct system partner or fitter. The execution specifications, as well as the installation and inspection instructions of the manufacturer, are to be provided to the system partner or fitter before awarding a contract. Detailed inspection instructions of the overall system are to be provided in advance, so that the company can produce the system correspondingly and correctly, and includes the inspection instructions when calculating its quote.

Inspection of the overall system includes the specifications of the interfaces of the trade, rinsing specifications, inspection for gas mix-ups, tightness inspection with detailed date of the inspection medium, inspection pressure, duration, leakage rate, and data on the logs. It is advantageous to use data loggers with compensated temperature loggers.

25.12 Operation Start-Up and Instruction of the Operating Staff

To be able to celebrate successful conclusion of a project, proper, comprehensible instructions, and start-up of the overall system is imperative. Users will be thankful if they are provided with the facts of their new work systems. This also includes providing the operation explanations and specifications of fittings, as well as the information about procuring spare parts.

Figure 25.5 Installation of a clean room with column solution.

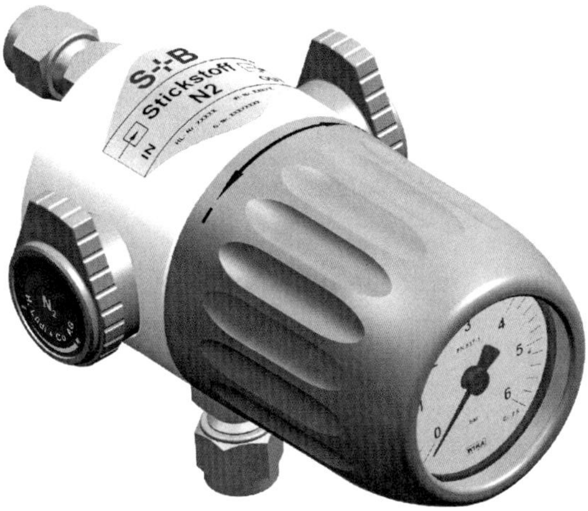

Figure 25.6 Regulated pre-wall tapping spot with shut-off and outlet valve.

Maintenance of such systems is legally prescribed. This is regulated by the information of the manufacturer or must be performed cyclically in the company by it or authorized companies. This includes gap-less documentation of maintenance work (Figures 25.5–25.11).

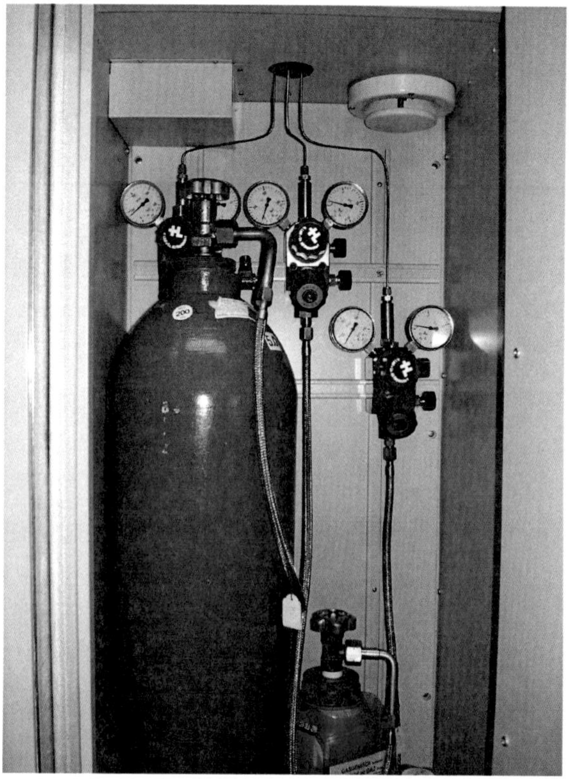

Figure 25.7 Small bottle gas station in a cabinet.

Figure 25.8 Gas bottle stations.

Figure 25.9 Extensively equipped ultra-pure gas medium columns (ETH Zurich).

Figure 25.10 Extensively equipped medium columns.

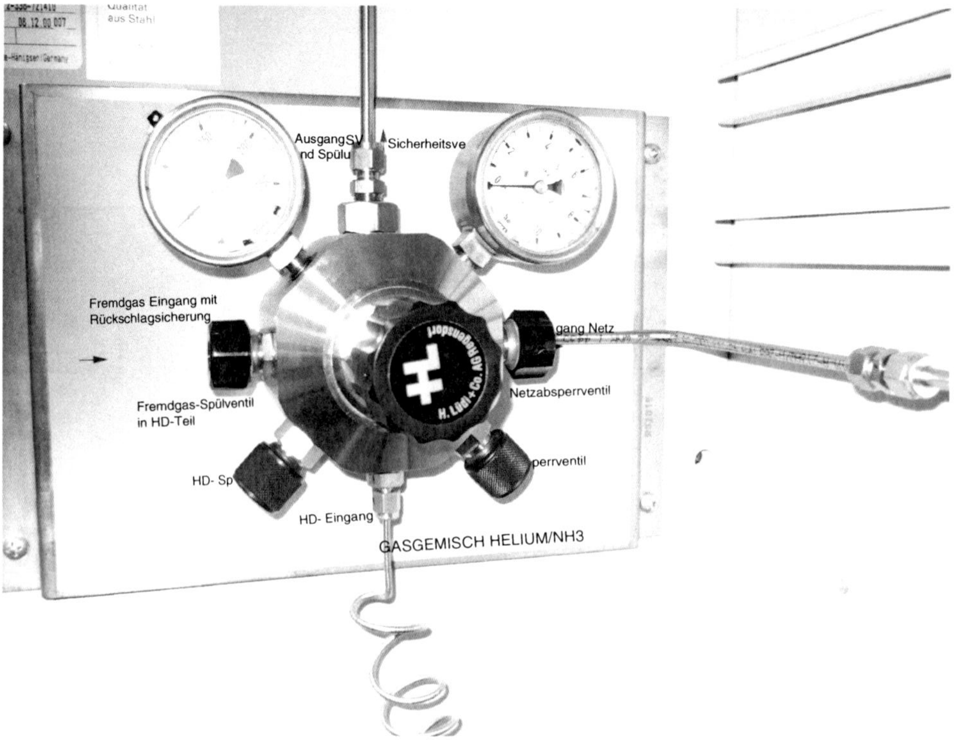

Figure 25.11 Bottle gas station for corrosive gases.

26
Emergency Devices

Thomas Gasdorf

26.1
General

Where threat of accidents like fire or contact with acids, alkalis, solvents, and other chemicals exist, rinsing with water must be provided as a first aid measure. Extensive washing of 10 to 15-min duration can often prevent worsening of the impact of the accident. It is a prerequisite for an immediate first aid that appropriate and functional safety showers are easily accessible.

26.1.1
Where and how?

The directive BGI/GUV-I 850-0 "Safe working in laboratories" (see Chapter 5) gives hints.

There are prescribed:

- A body shower at the exit
- An eyewash either in the vicinity of the shower or in the sink-mounted.

The number of body and eyewashes should be chosen so that accessibility is guaranteed within a maximum of 5 s.

The access is to be kept clear.

26.1.2
Special Fittings

Functional and performance requirements are high, thus only specially designed equipment must be considered, and fully meet the minimum requirements of EN 15154. The requirements of the technical detail for safety showers can be found in Part 1 and Part 2 for eye-washers. The scope of Part 1 of showers takes care of use in laboratories; Part 2 for eye-washers includes all fittings with connection to water supply.

The Sustainable Laboratory Handbook: Design, Equipment, Operation, First Edition.
Edited by Egbert Dittrich.

Commercially designed showers for bathrooms and toilets, just as likely as those solutions, where only individual components are replaced are not applicable. On the safe side, a user is only with systems designed with respect to construction and materials fully in accordance with standards and directives.

26.1.3
Identification

The locations of showers shall be marked with the sign of rescue first aid facilities, that is, E05, and eye-washers with the sign E06. By combining the rescue sign with the appropriate directional arrows, the escape routes are reported to the emergency safety showers.

In the preparation of the escape and rescue plans (according to all national regulations), the marks for locations of first aid facilities should be included . (See: Accident prevention regulation GUV-V A8 "Safety and Ge sundheitsschutzkennzeichnung" at work.)

26.2
Body Showers

Body showers provide a precise spray pattern as well as a minimum usage. The fitting must be simple and quick to set in motion by the injured occupants and shall not be self-closing.

The body shower is installed in such a way that the bottom of the shower head is 2200 ± 100 mm from the ground level. From the fall line of the shower head center circle, a radius of 200 mm must remain free from obstacles (free radius).

26.3
Eye-Washer

With respect to the eye-washers, it is particularly important that the spray is constant and independent of line pressure and falls over only at a height of 10 – 30 cm, in order to wash the eyes properly. It must be possible to keep the eyes open with both hands.

Ideally suited for use is the one-hand eyewash. These one- and two-head models can be used with or without the bracket. With the help of a longer hose, it is possible to treat injured persons lying on the ground.

26.4
Emergency Shower Combinations

This is a compact unit of body and eye shower.

The eyewash in this case can use the free radius of the body shower.

26.5 Hygiene

Emergency showers shall be supplied with potable water or water of similar quality. For potable water, the units are covered by the Food Act, so that only appropriate and approved components, equipment, and materials may be used. Thus, during installation, it is important to choose appropriate products in compliance with the standards and certified directives.

Enforcing these high standards also allows the possibility of providing tempered water. Depending on the duration of sprinkling, tempered water prevents hypothermia. The temperature shall be 15–37 °C.

26.6 Testing and Maintenance

Emergency showers are regularly, that is, at least once a month checked for reliable functioning and operation. In addition, the existing devices should be fully inspected and serviced to indicate proper operation of the equipment in addition to the monthly checks for a year. This has to be done by maintenance and inspection of safety showers by knowledgeable person who can also be related to operational activities.

This review should be made in consultation with the hazardous substance list, which requires to be updated once a year. It can be deduced that the first aid system shall be annually reviewed by the safety officer at least once to make sure it complies with the current requirements or whether additions are necessary. Regarding the equipment with safety showers, it needs to be verified that work and/or jobs have been added, which require emergency showers. Only if in the device, control and maintenance of safety showers meet the necessary standards, an immediate emergency assistance is guaranteed. This also is required in terms of the operator's liability.

26.7 Complementary Products

The standard series EN 15154 Part 3 opens with "Emergency showers without water" and 4 "eyewash without water" further constructions of emergency shower devices.

These products are not connected to the potable water pipe network and are loaded by containers with diverse volumes. They allow flushing of the injured person during transport (Figure 26.1–26.6).

Figure 26.1 It is particularly important for all emergency showers, that the spray pattern is designed for effective rinsing. In addition, eye-washers must be operated in such a way both hands are free to hold the eyes open.

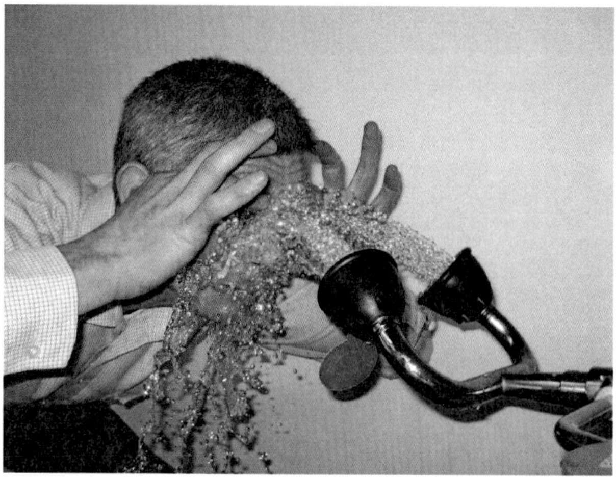

Figure 26.2 Emergency shower combination.

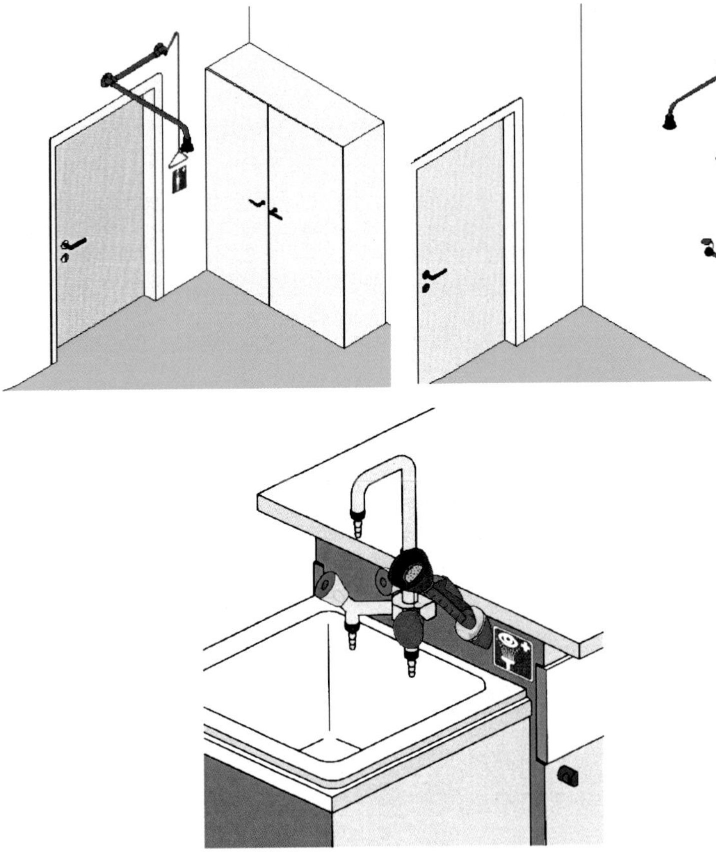

Figure 26.3 Locations.

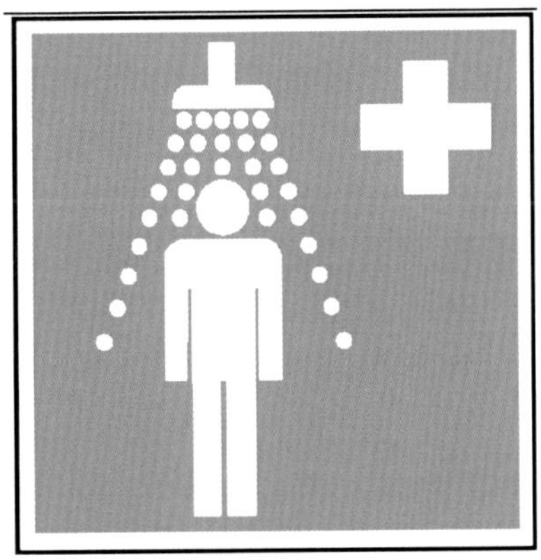

E05 Notdusche

Figure 26.4 Emergency showers.

Figure 26.5 Eyewash stations.

Figure 26.6 Funnel for the monthly check.

Part IV
Sustainability and Laboratory Operation

Sustainable planning processes are crucial. But they must be proofed and executed according standards. Certification systems give certainty and ensure comparability and quality.

At the same time the occupants in laboratories own the key to operate the building on the planned sustainability level. They are able to boycotte sustainable features or internalize and bring them to success.

The Sustainable Laboratory Handbook: Design, Equipment, Operation, First Edition.
Edited by Egbert Dittrich.

27
Sustainability Certification – Assessment Criteria and Suggestions

Egbert Dittrich

For office and laboratory buildings, certification systems are designed with the aim of ensuring sustainability to a certain degree. These systems originated from social and political demands and the resulting requirements of operators, tenants, and users of laboratory buildings, be they due to environmental or economic reasons, to take sustainable actions in the interests of their occupants. It should be noted that both reasons are not in contradiction, but together with sociocultural properties they only reflect the concept of sustainability.

Sustainability is therefore not only isolated compliance or not falling below certain metrics that characterize the ecological footprint, but acts together with economical operation and the needs of occupants working in the laboratory. Sustainability can be described as the best maxim for any kind of action in the coming decades without exclusion. The high consumption of energy of all kinds and consumables, as well as the handling of hazardous substances, forces laboratories to look for sustainable strategies that lead to particularly good results in terms of cost reduction and reduced burden on the environment. The reduction of exposure to pollutants as well as thermal and acoustic loads in the workplace leads to further increase in efficiency, improved well-being, and good team performance.

Certification and the formulation of certification criteria are independent if an audit is performed; in general, approved regulations help planners in the implementation. All major certification systems place demands on the architecture, building technology, and the resulting planning processes. Deliberately, consideration of the work and the laboratory processes carried out in the buildings is omitted. Considering the dominance of energy in laboratory processes, the effort involved in their exploitation, and the abundance of expensive consumables, a general overview of the entire types of cost alone can provide authoritative and reliable results for the construction and operation of a laboratory building. That is, successful certification of the building and planning based on the criteria is the only necessary condition to operate a building by sustainable means. Sustainable laboratory operation must be ensured continually with respect to the constantly changing new tasks of sustainable management and appropriate individual strategies.

The Sustainable Laboratory Handbook: Design, Equipment, Operation, First Edition.
Edited by Egbert Dittrich.

Sustainability in the laboratory, in the modern sense, is a complex mixture of different strategies and measures that combine interdependent linking of human factors, economic quality, and ecological quality at a high-performance system, and thus brings about improvement in the efficiency.

To clear the distinction, it should be mentioned that sustainability for a laboratory in principle was made from two equal approaches: the building with its materials and technical structures, and the operation, which also includes the appropriate use of the installed technical engineering services and lab equipment. It should not surprise anyone that the consumption of energy and materials in the laboratory is considerably higher than those demanded by the building.

27.1 Certification Systems

Among the number of currently used green building certification systems, the following are the most well known and are employed for relevant laboratory buildings. All systems seek to evaluate as many criteria as possible quantitatively to avoid subjective influence of the audit. They measure and assess physically the influences of the building on the environment. Here, we present the most used systems, at the time of writing this book:

LEED or Leadership in Energy and Environmental Design.

Developed in the United States in 1998, this system is the most widely used. It currently distinguishes between different relevant standards for laboratory building occupancies:

LEED NC – new construction and renovation
LEED CI – Office interior fittings
LEED EB – existing buildings
LEED CS – Structural.

Depending on the usage type, each is each assigned to a particular category. For example, associated with the type of use of LEED NC, following seven categories are allocated:

Sustainable sites (assessment of the exposure of people, the environment, and land through the construction process as well as the building itself);
Water efficiency (water consumption during use);
Energy and atmosphere (review of building-specific energy consumption, the sustainable use and management with the resource "electric power," the refrigerant used for the air conditioning, and the volume of renewable energy used);
Material and resources (evaluation of the building material used, their source, their recycling ratio, and operation of sustainable forest management for the protection of human health and environment);

Indoor environmental quality (review of construction materials and other items in terms of their ingredients for the protection of human health and environment; review of dealing with tobacco smoke during operation and fresh air and the protection of building materials and ventilation system components);

Innovation and design process (assessment of innovations and novelties in the project that meet the previously considered subject areas, as well as the implementation of a trained LEED specialists and LEED Accredited Professional (not compulsory but recommended with regard to scoring and implementation)).

Regional priority (promotion of local environmental aspects).

The implementation of the LEED rules, which are based on the American Standard ASHRAE, is not entirely straightforward. The criteria are basically designed for office buildings. Despite the small lab-specific categories criteria, the high popularity of LEED shows that marketing of certification is essential. BREAM (Building Research Establishment Environmental Assessment Method) has been applied since 1990, and there are many system variants. The system has been adapted with the help of HEEPI (Higher Education Environmental Performance Improvement) for university laboratories.

The standard criteria in the various types of use are as follows:

Category 1: Management (planning processes)
Category 2: Health and well-being
Category 3: Energy (during use)
Category 4: Transportation
Category 5: Water (during use)
Category 6: Materials and waste (for construction)
Category 7: Land use
Category 8: Pollution (emissions during use).

In addition to those of HEEPI, other features have been added: for example, energy measurement methods without affecting safety aspects, reduced air consumption of fume cupboards, cooling, heat recovery, integrated planning, adequate design, access on campus to special transport, and waste disposal systems.

The important difference to LEED consists of further consideration of lifecycle cost, which shows that BREEAM reflects the real situation of a laboratory building at a higher scale than LEED. The DGNB (the German Sustainable Building Council)

system was extended by the certification criteria to laboratory buildings in 2013 and represents of all the systems based on the most technical details.

It systematically considers the following aspects:

environmental quality
economic quality
sociocultural aspects
technical quality

process quality
site quality (in terms of laboratory buildings, the site quality was not changed but the criteria were recast).

Because of the dominance of energy consumption of the laboratory processes depending on the type of laboratory and considering mainly the safety of serving technical building equipment, a virtual laboratory building is implemented in the assessment to set a planning benchmark; however, laboratory process-dependent consumption data are not included in the evaluation.

A building that complies with the relevant guidelines, regulations, laws, and standards (valid at the time of review) meets the bronze standard. Thus, a method, as described in the ENEV 2009, was selected.

The DGNB system evaluates the quality of the building based on the improvements made in each profile with respect to the virtual laboratory building, respectively, to the extent possible with quantitative design criteria. In the DGNB system, lifecycle costs play an important role. The scheme considers uniquely communication areas and other comfort criteria or planning process quality. This is required because individual criteria can cancel each other.

A most important and essential entry-level condition or precondition is required. Prior to the certification process, an operating concept and a risk analysis must be submitted. It must be ensured that operators and users are not excused to consider laboratory-specific measures, such as those that influence sustainable operation and minimize through continuous assessment and monitoring of consumption data, the operating costs, despite the certification. Another difficulty may arise because the laboratory activities change after a change of the user, which may require frequent modifications to the criteria for the realization of sustainability through another mix of measures; that is, largely user-independent planning would be helpful.

The DGNB system meanwhile has been adopted by a number of European countries. The acceptance depends on a variety of criteria, and an assessment guarantees a balance between the three criteria of ecological quality, economical quality, and sociocultural aspects. In general, only buildings commonly assessed with these criteria seriously can be compared. With respect to the increasing trend in renting out laboratory buildings for tenants, it becomes more and more important for operators that the rents of laboratory space is approximately 7–8% higher for buildings certified on a high level.

On the other hand, realization of sustainability is not for economic reasons alone. If sustainability becomes a new paradigm, when acting as human beings, a holistic approach must be considered. With respect to laboratories, the integration of the work processes of the occupants focused on the laboratory apparatus and the plug-in-units in use is the other part, besides the building itself. Experts are aware of devices used in laboratories that have an enormous demand and consumption of energy and consumables as well as natural resources. Plug-ins such as laboratory washers, autoclaves, fume cupboards, freezers, and so on, generate a much higher impact during their lifecycle than the building construction itself. It is to say that the EGNATON council[1] works on this aspect. The criteria reflect the DGNB system and match with the assessment. Once the system is established, it is suitable to replace individual and incomparable rating systems of public and industrial tenders. It considers in a typical way the balance of sustainability-inherent criteria. If the specifying bodies follow the suggestions of EGNATON CERT, they are to invest in highly sustainable equipment that are ecologically uncritical and economical to use throughout their entire life cycle, work safe, are well designed, and, last but not least, have perfect performance.

27.2 Individual Strategies to Implement Sustainability

Basically, this work is intended to provide general approaches to achieving sustainability in the laboratory. Planners, operators, and especially laboratory management and the users are not absolved of their obligation to examine all processes in the laboratory for sustainability as part of the known approaches. It is always important to first meet the requirement of industrial work safety. In this respect, the party responsible should be in consensus with the experts to find solutions that are always asked again, especially when changes occur.

Sustainability with its different properties is constantly moving forward and needs to be updated. As part of the DGNB criteria, the components of the cost group 400 (HOAI) are put to test. The test for suitability and usability most likely will clearly challenge the environmental data and CO_2 footprint considering planners and auditors – also the manufacturer. In addition, there are other material properties that can be put to test thoroughly, that is, the ability to up-cycle or, better, the possibility to use the material after removal for other purposes.

In general, the enormous variety of lab buildings does not allow the provision of fixed concepts. Each lab is different and requires individual sustainability strategies. No matter what the individual task of the lab is, certification of the building and the technical equipment provides an excellent precondition to success. For sure, the behavior of the occupants, their handling of sustainability features, and their motivation are unpredictable and therefore are issues to keep an eye on.

27.2.1 Planning, Design, and Simulations

In many countries, it is not usual to design a lab. Lab planning is executed by suppliers of the equipment. Furthermore, architects carry out the planning of the interior. No question is asked about the skills of vendors and architects and their years of experience. But to design a contemporary lab with its high requirements, experienced professional lab planners are absolutely essential.

1) EGNATON – European Association for sustainable Laboratories.

By the way, certification systems such as DGNB will award extra credit points if an adequate collaboration of the planning team can be proven.

The complexity of laboratory buildings can be only partially managed without computer-based simulations.

It is recommended that the necessary planning security, especially toward clearly optimized energy consumption, is achieved by simulation calculations (CFD)[2] under the assumption of realistic parameters in respect of the planning process.

The simulation of the different planning steps shall mainly take care of the ventilation and the total energy consumption during season changes but also the lighting concept.

These very important issues require the study of different technologies, such as cooling methods. Furthermore, with respect to the pause and working time but also orientation of the building and detailed positioning of equipment, fume cupboards, facilities, and so on, quite significant optimization needs arise.

At the same time, safety factors can be identified with the help of simulation programs in advance to prevent oversizing – or the opposite.

27.2.2
Benchmarking

The banal realization of controlling, to identify positive or negative trends only with reliable and similar data acquisition, applies equally for laboratories.

In this respect, both the technique of capturing the consumption data – sensor, measuring devices, meters – and also the subsequent data processing and technical software are important.

As part of the long-term recording of the values, the accuracy of the measurement method is less important; that is, the capture device itself is important, but the continuity that results in a trend, in turn, demands action. The enormous variety of laboratories initially requires the determination of the important data to be collected. Especially, when comparing several similar buildings, the principle must be that the population of the buildings to be compared should be as large as possible. Only when at least 8–10 buildings within a data platform can be considered comparatively with reasonable certainty, one can gain from the best practice cases to draw conclusions about the causes of the performance. The construction of benchmarking platforms fails – as is unfortunately sometimes still common in university buildings – if no adequate software is available.

The continuity requirement extends not only to the measurement method but also to the reference and design parameters. Especially, comparisons of benchmarks from different cultures (e.g., USA vs Germany) to significant differences may fail. For example, the question of which areas are added to get the gross or net floor area is not part of a consensus, thus making any comparison at this point meaningless.

2) CFD – Computational Fluid Dynamics.

Basically, the statement that without benchmarking the sustainability of the operation of a laboratory building cannot be ensured appears to be valid.

27.2.3
Measuring and Control

The operation of a large laboratory building without real-time monitoring is also risky for security reasons.

Regarding the monitoring scope and scale, it always depends on the laboratory type and the individual laboratory situation. Especially, the handling of hazardous substances, for example, gases heavier than air, requires customized measurement and control concepts. The inertia of sensors is to be considered.

The more heterogeneous and the larger the group of laboratory workers is, the more difficult it is to monitor the consumption and to discipline the people. Regardless of the presence of control loops – for example, laboratory ventilation and fume cupboard control – the occupants should be familiar with the techniques and should be drawn attention to by the responsible technician to the deterioration of the overall performance.

Extremely problematic and difficult to solve is the reduction of air exchange during the night in research laboratories. The presence of one single user prevents the intended reduction. Therefore, it is recommended to restrict the operation to one part of the lab during the night or, if possible, to monitor the laboratory processes of spatially separated locations.

A laboratory building needs to be quickly adapted to different user needs. Therefore, in modern building control, individual consumers are connected within the building control system using a computer network. Measurement results transferred online via bus systems (e.g., fieldbus) again allow immediate reaction and automatic control.

Measurement, control, and regulation are prerequisites for sustainable laboratory management.

27.2.4
Ventilation and Cooling Concept

Ventilation takes at least 40% of the total energy consumed in a laboratory building. Therefore, all reduction measures affect consumption drastically. Because of the fact that laboratory ventilation is not for the comfort of the people in the first place, but rather to protect them from effluent and hazardous gaseous substances, this protection target is of primary importance with respect to energy savings. Thus, any reduction of air changes previously requires a qualified risk assessment.

Nevertheless, the interpretation of laboratory ventilation does not mean that frequent air change automatically also means high safety of the people. Flow tests show the opposite result: that is, large volumes of air with high temperature gradients increase the rate of the sinking air strongly and among the outlets constrict without mixing of the air in the inhaled range of users to improve or carry away high concentrations of pollutants, which can typically occur on laboratory benches.

To that extent, one can say that only intelligent laboratory ventilation, which can be largely adapted to the conditions, produces both safety and efficiency-enhancing comfort. With respect to the following article from Gordon Sharp, it appears to be evident that considerable types and wide areas of laboratories have clear situations and work with well-known substances. In case of alerting detected emissions, the occupants must know what to expect. If the substance in use is not hazardous, they can rely on a ventilation system upon demand. In all other cases, such technology is unpredictable. Time lags must be considered as well.

Calculations reveal that each year in laboratories a climate prevails without refrigeration for more than 1700 h, which is so uncomfortable that it hardly can be worked in. However, because cooling can take place only with very low efficiency on the laboratory ventilation, other ways should be sought to eliminate thermal loads. It should be remembered, however, that without some $\Delta +$ of supply air to the room air, the flow is suboptimal. It is primarily the responsibility of the ventilation planners to select an appropriate cooling method under the technical possibilities. In this, local factors and local energy price levels play a role.

Basically, ventilation and cooling are so inseparable and must be planned with individual concepts.

In the definition of planning, parameters should always be attempted; extreme planning significantly influences factors that require the users to be restricted to the areas. This is absolutely necessary. If a few laboratory technicians work with hazardous substances, diluted and disposed of by the ventilation in the event of an accident, it should be done in a specially designed corresponding part of the laboratory in order to drive out the air, without sacrificing the highest performance anywhere. At the same time, it is also appropriate when, in the danger zone, diffusers are used that achieve the protection objective with minimum air changes.

27.2.5
Working Conditions

The working conditions are a typical complex of sustainability, which reflects the human factors, cost, and the environment to the same extent. This includes, as shown in the previous chapters, features of a good, draft-free laboratory ventilation but also other individually occurring influences such as noise, temperature, dust, or ergonomic conditions. The realization of having to do research in interdisciplinary teams is due to give these teams the architectural preconditions, for instance, to be able to communicate in an inspiring setting. Thus, there is an extended concept of working spaces, away from traditional laboratory situations toward creative zones, that allow the intellectual interaction for groups in which electronic communication tools as well as socialization and promotional drinks and snacks are offered. The question of the sustainable design of the laboratory workplace, that is, the working conditions, is synonymous with asking the question of how future laboratory work is structured at all. A few years ago, writing areas in the lab were the exception, because little was written. Today, academic writing in research laboratories already includes almost 60% of the work; thus, of course write-up zones no longer are only a negligible makeshift. The usual spatial separation with eye contact in the laboratory will probably no longer exist in the future. And, at some point, the total separation may become possible with the aid of a laboratory monitoring system via data networks with few users dominating and controlled by robots even from another city or country or continent.

27.2.6 Consumables

No other assets affect the cost structure as strongly as the consumables. First, space for storage is required. The cabinets and storage spaces within are air-cooled areas, directly in high installed zones; therefore not only the investment but also maintenance costs per meter square are also at the highest, which makes little sense. In order to keep the path to the storage areas short, adjacent rooms without exterior walls or windows should be used.

That is to say that the organization of the storage space itself must be professional. Storage systems should be aligned with software and access in order to be aware at any time what material and how many numbers are stored at which place. Many institutional research laboratories can achieve savings if they use business-like procurement strategies. Such strategies, of course, include supply agreements, which allow only small storage rooms for consumption goods at the place of consumption, that is, customer locations. Sustainability criteria in the context of supply chain management with respect to the procurement of typical laboratory materials do not differ in any way from other industrial operations. Right here is where personal responsibility starts, coupled with the manifested and structured monitoring of suppliers and their products.

28
Reducing Laboratory Energy Use with Demand-Based Control

Gordon P. Sharp

Modern research laboratories now operate with fewer fume cupboards, rely more on computational chemistry, and have lower thermal needs due to reduced plug loads. In fact, lab energy usage is 5–10 times an office's usage [1] and the HVAC load in labs can go up to 60–80% of a lab building's energy. By reducing the entry of outside air (OA), the largest energy driver in a lab, by installing demand-based control (DBC) of a lab's dilution ventilation flow or Air Changes per Hour (ACH), it is possible to dramatically reduce lab building energy use in some cases by up to 50%. This in turn, can even help to achieve Net Zero Energy operation of a lab even in regions with harsh climates like Abu Dhabi, UAE.

Today, the minimum dilution ventilation flow or air change rate is often the dominant factor for determining supply and extract airflow volumes in laboratories. The three main drivers of airflow that impact the lab room's use of supply airflow are the hood or exhaust device flows, the thermal loads, and the minimum dilution ventilation or ACH rate as shown in Figure 28.1. To achieve reduced lab flows, energy and initial cost, all flow requirements need to be minimized.

28.1
Reducing Fume Cupboard Flows

For all but very low hood density labs (less than one fume hood per 2000 sq ft or about 200 m^2), the use of VAV or Variable Air Volume hoods is recommended to reduce fume cupboards' flows when the sash of the hood is lowered or closed. Some other helpful suggestions are the following:

- Low face velocity fume cupboards are helpful, but not required.
- It is also helpful to use an 18 in. (45 cm) design opening to limit fume cupboards' flows.

The Sustainable Laboratory Handbook: Design, Equipment, Operation, First Edition.
Edited by Egbert Dittrich.

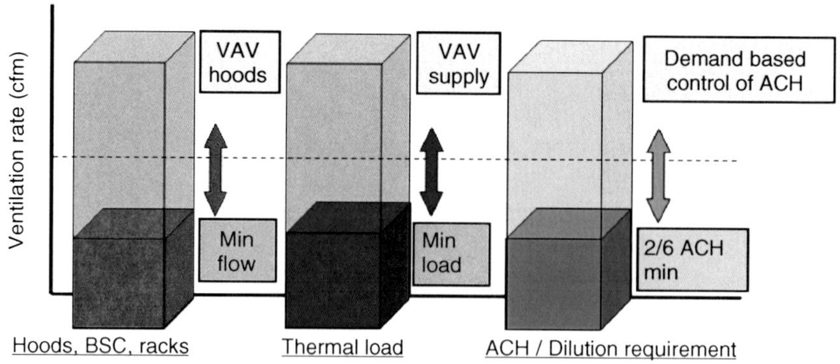

Figure 28.1 The three drivers of lab room airflow.

- Lab protocols should encourage good sash closing habits to save energy and increase lab safety.
- For users that cannot be educated around lowering the sash or do not want to be bothered with this, then use automatic sash closers or automatic face velocity setback of the hood when the user is not in front of the hood.

For labs with moderate to high hood density (greater than one fume hood per 65 m^2), the new AIHA/ANSI Z9.5-2012 Laboratory Ventilation Standard [2] provides new guidance on minimum (sash closed) flow of a VAV fume cupboard that allows it to be reduced from past recommended levels:

- The prior recommended minimum flow for a 6′ (1.8 m) fume cupboard minimum was ~250 cfm (425 $m^3 h^{-1}$).
- The new recommended range for a 6′ (1.8 m) fume hood minimum is as low as ~100 cfm (175 $m^3 h^{-1}$).
- For most labs, this means that the minimum hood flow is in the range of 2–4 ACH or less.

28.2
Reduce Thermal Load Flow Drivers

As per Labs 21 and University of California Davis study [3, 4], the average plug and lighting load are typically 2.5–3 W ft^{-2} (25–30 W m^{-2}) with perhaps only 20% or less of labs having loads greater than 4 W ft^{-2} (40 W m^{-2}). For these typical average rooms, daytime normal thermal loads are lesser than 4 ACH of conditioned supply airflow and for nighttime, the use of a temperature setback control should push required supply airflow down to 2 ACH or less.

For high thermal load labs and where even more energy-efficient operation is desired, it is desirable to decouple the cooling requirements from the airflow requirements through the use of local hydronic cooling approaches, such as chilled beams or fan coil units. Furthermore, these approaches in combination with DBC can often provide significantly lower first or capital costs by reducing the supply airflow requirements and size of the supply and exhaust fans. Although not required, using hydronic cooling such as chilled beams can be advantageous at least when it is coupled with the low ACH design produced using DBC.

28.3 Vary and Reduce Average ACH Rate Using Demand-Based Control

Due to the factors mentioned earlier, a lab's minimum air change rate or dilution ventilation airflow is often the dominant or controlling factor for determining average, and in many cases, design values for supply and extract airflow volumes in laboratories. Air change rates are often set to a single value between 6 and 12 ACH for a laboratory with no hard guidelines or standards to rely on. The truth is, there is no single correct rate of air changes for even a specific lab room. The dynamic nature of any individual lab space precludes that one "correct" value is appropriate at all times or for all conditions.

The "correct" value varies based on the specific conditions of the lab at a given point in time. For example, if a spill of a solvent or volatile chemical occurs, or chemists are doing work on a bench top that should be done in a hood, a higher room air change rate is desirable. In a spill situation, a rate above 6 ACH, such as at least 8 ACH to as much as 12–16 ACH, can provide superior dilution performance at the time of the incident and for a period of time afterwards [5]. When the situation calls for it, a higher air change rate is much more effective in reducing contaminant levels quickly [6]. However, the far majority of the time or about 98–99% of time [7], lab room air is clean, and a minimum of 2 ACH would be "correct." Diluting clean room air with clean supply air achieves no benefit and wastes significant amounts of energy.

Consequently, the best approach regarding minimum air change rates for labs is to determine the rate based on the real-time quality of the air in the laboratory. This approach allows airflow in a lab to vary based on all the situational factors affecting lab airflow, as opposed to only the status of the hoods and the thermal load. Implementing a dynamic approach to controlling minimum air change rates requires the ability to measure a unique set of indoor air parameters, such as total volatile organic compounds (TVOCs), particles, carbon dioxide, and humidity, and to integrate this information with the building management system.

Until now, such an approach known as *Demand-Based Control* (or CDCV) (Centralized Demand Control Ventilation) [8] and similar to demand control ventilation (DCV) but applied to laboratories and vivariums, has not been feasible or cost effective primarily due to the quality and quantity of sensors necessary to safely implement this approach. In addition, the associated calibration and maintenance costs rendered it impractical to populate a large number of sensors throughout a facility.

28.4 A New Sensing Approach Provides a Cost-Effective Solution

However, a proven sensing architecture now exists that can implement DBC in a practical, reliable, and cost-effective manner. This new sensing approach, sometimes called *multiplexed sensing* as shown in Figure 28.2, changes the age old paradigm of sensing, and minimizes calibration and maintenance expenses. Instead of placing multiple sensors in each sensed space or area of a building, this networked system routes packets or samples of air sequentially in a multiplexed fashion to a shared set of sensors. Every 40–50 s, a sample of air from a different area is routed on a common air sampling backbone to the same set of sensors, known as a *sensor suite*, for measurement. These sequential measurements are then "de-multiplexed" for each sampled area to create distinct sensor signals used for traditional monitoring and control. Typically, about 20 areas can be sampled, with one set of sensors, approximately every 15 min.

Calibration and maintenance expense is minimized due to the limited number of sensors and their centralized grouping in one location. The calibration process can be easily accomplished through an exchange program, whereby, a set of factory-calibrated sensors can periodically replace the onsite sensors, such as every 6 months. The system is, therefore, assured to operate at peak performance with minimal or no disruption to the facility's operation.

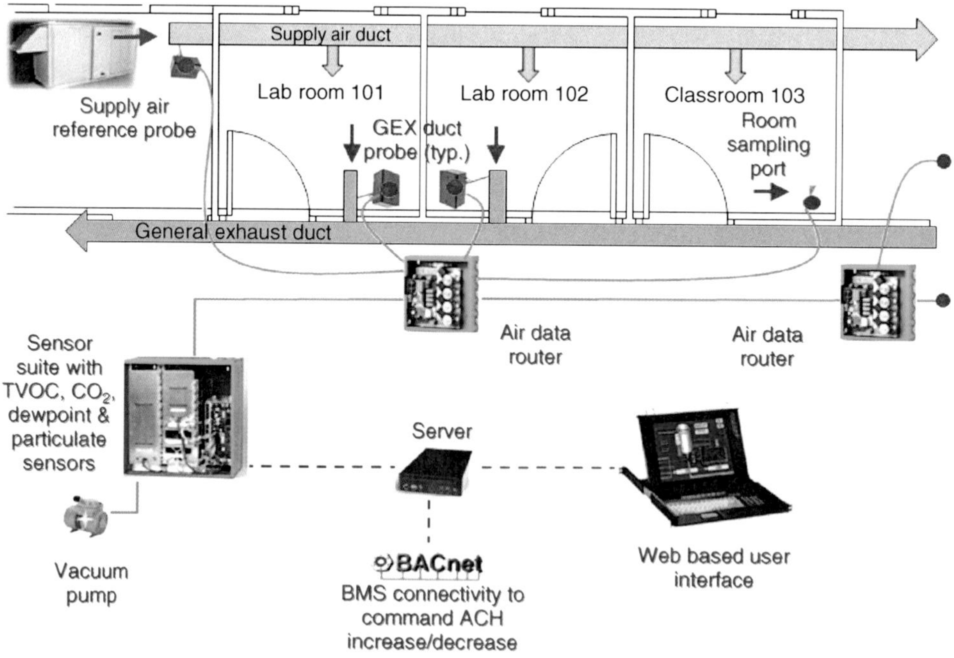

Figure 28.2 Multiplexed sensing, a new cost-effective solution.

A key innovation in this multiplexed sensing system is the development of a structured cable containing low voltage power and signal conductors, data communications, and a unique, one quarter inch hollow conduit. This sample conduit provides the transport media through which the air packets are passed one after another, similar to data packets on the communications network used in building management systems. To substantially eliminate air packets from contaminating subsequent packets, the inner walls of the air sample conduit are made of a mixture of an inert fluoropolymer and a unique nanotechnology material known as *carbon nanotubes*. These micron long chains of carbon atoms form inert, electrically conductive strands that, when mixed with the plastic fluoropolymer, form an electrically conductive matrix that prevents the buildup of static charges that might impede the flow of particles and aerosols. The resultant sample conduit can accurately transport and preserve the air sample's properties to support accurate measurement up to 500 ft away from the centrally located sensor suite.

This multiplexed sensing approach can measure almost any air parameter of interest. For laboratories, the use of a PID or photo-ionization detector type of TVOC sensor is very beneficial for accurately detecting literally hundreds of commonly used laboratory chemicals that can volatize and become a safety concern. Combining this sensor with a laser-based particle counter to identify aerosol vapors, plus other specialty sensors such as *Metal Oxide Semiconductor* (*MOS*), TVOC sensors afford detection of the far majority of airborne chemicals of concern.

In this dilution ventilation control concept, when an increase in contaminants in a space is detected, the airflow in the room is increased as appropriate up to a maximum level of between approximately 8 to as high as 16 ACH depending on the capacity and range of the room's airflow control devices. The lower value of the dynamic range when the air in the lab is "clean" can be set to either 4 ACH during the day and 2 ACH at night or even 2 ACH as a minimum flow all the time. Although a value such as 2 ACH at first glance might seem low, analyzed further, it is more than three times the typical occupied office OA ventilation level.

Finally, this DBC concept can also generate a net reduction of first cost due to the ability to downsize or effectively right-size the HVAC mechanical system due to a reduction in the peak required airflow through the reduction in the lab's minimum flow requirements. In one case, for a 75 000 sq ft lab area, a gross reduction of approximately $1 025 000 of HVAC system first cost was estimated by reducing the minimum air change levels in the labs and using a more realistic thermal load assessment.

28.5 Demand-Based Control (DBC) Improves Beam Use

When chilled beams or fan coil units are used in lab rooms having a 6 or 8 ACH minimum dilution ventilation flow, the room will typically have a large amount of overcooling and thus, a large/amount of required reheat energy use. This is because most labs, as noted earlier only need 2–4 ACH of airflow to meet the cooling load. Not only is there a large amount of reheat required but the chilled beams or fan coil units are rarely needed producing an inefficient result with a duplication of equipment for cooling. However, if DBC is used to bring the room minimum flow down to 3–4 ACH during the day and 2 ACH at night, then the amount of overcooling and required reheat energy is drastically reduced. If further cooling is required, above the 2–4 ACH, the chilled beams or fan coil units can now appropriately provide this peak cooling requirement without impacting the required outside airflow. As a result, the HVAC system can be downsized since the lab room's thermal load is decoupled from airflow, so the air system can be resized to 2–4 ACH based only on the dilution ventilation and fume hood exhaust requirements.

Occasionally, to try to prevent large amounts of overcooling, "Neutral air" or air at about 20 °C is provided to the lab so that the chilled beams or fan coil units provide all the cooling. However, reheat still needs to be provided although now it must be provided at the supply air handler. There are some approaches to use various heat recovery (HR) approaches such as dual wheel designs or "wraparound" coils, but all these systems typically require some reheat energy and significantly increase the capital costs of both the air handlers and the chilled beams or fan coil units since the room cooling approaches must now provide all the room cooling. Again, the complexity of adding HR, increasing the chilled beam's size, as well as that of the supply and exhaust fans can all be eliminated by using DBC to reduce the supply airflow to better match the amount of required cooling. As a result, this is a good example of where the combination or "whole" (DBC and chilled beams or fan coil units) has greater advantages than sum of the parts.

28.6 A Few Comments on New Lab Ventilation Standards and Guidelines

In the US and Europe, there are no prescriptive requirements for air change rates in labs other than the ASHRAE 62.1-2010 [9] fresh air requirements for university/college laboratories that corresponds to about 1.2 ACH (0.18 cfm ft^{-2} or 0.91 cfm m^{-2}). In terms of recommended levels, standards are also significantly moving away from the prescribed values of the past to a more performance-based approach based on the specific requirements of a given lab space. Chapter 16 of the new 2011 ASHRAE Applications Handbook [10] has new guidance and recommendations on minimum levels for lab air changes and control approaches for ventilation. The handbook recommends DBC of air changes and severely limits the use of an occupied/unoccupied control approach for dilution ventilation or at least, one that is not coupled with DBC or as the handbook puts it, "real time sensing of contaminants."

As per new 2011 ASHRAE Applications Handbook, Lab Chapter 16:

> Occ/Unocc Control Scope Limitation Discussion:
> There should be no entry into the laboratory during unoccupied setback times and occupied ventilation rates should be engaged possibly 1 h or more in advance of occupancy to properly dilute any contaminants.
> Active/Demand-Based Control Discussion:
> Reducing ventilation requirements in laboratories and vivariums based on real-time sensing of contaminants in the room environment offers opportunities for energy conservation. This approach can potentially safely reduce lab air change rates to as low as 2 ACH when the lab air is "clean" and the fume hood exhaust or room cooling load requirements do not require higher airflow rates. Research by Sharp [7] showed that lab rooms are on average "clean" of contaminants in excess of about 98% of the time.

28.7 Case Studies

28.7.1 Case Study 1: Arizona State University's Biodesign Institute

An installed case study example of DBC and multiplexed sensing can be seen at the new Biodesign A and B buildings at the Arizona State University in Phoenix, Arizona, that was R&D Magazine's Lab of the Year in 2006 and also a LEED (Leadership in Energy and Environmental Design) Platinum facility. The building was initially designed with a minimum ventilation rate of 12 ACH. It was decided to reduce the air change rate by a factor of 3 down to 4 ACH using DBC when the lab air is clean, and to increase the airflow to about 16 ACH when contaminants are sensed in the lab. This was successfully tested in a pilot project in 2007 and then implemented in 2009 in over 200 lab spaces and another 90 vivarium spaces. The net result of this 2009 retrofit was an energy savings of approximately $1 million annually for this 33 000 gross m^2 facility.

In addition to safely saving energy in laboratory and vivarium facilities, multiplexed sensing technology can also be used in nonlab areas such as offices, classrooms, healthcare facilities, and many other buildings where the control of OA is important to save energy. Strategies such as DCV and other OA control approaches, such as differential enthalpy control that can require significant numbers of sensors, can now be applied cost effectively and reliably for many types of facilities. At the Arizona State University, for example, this technology has been implemented in about 25 buildings including classrooms, offices, a student center, libraries, sports arena, and so on.

28.7.2
Case Study 2: Masdar Institute of Science and Technology (MIST)

The Masdar Institute of Science and Technology or MIST is located in Masdar City in Abu Dhabi, UAE. Designed to be one of the world's most sustainable facilities with a near net zero carbon footprint, the MIST 1A and 1B buildings comprise mixed use lab, office, classroom, and residential space over about 150 000 m^2. The MIST 1A facility, as shown in Figures 28.3–28.5, is now completed and the MIST 1B facility is currently under construction. The region experiences severe climate with very high temperatures and Relative Humidity (RH) similar to India but even higher RH.

MIST used DBC in its classrooms and office areas to reduce ventilation based on determining occupancy by sensing carbon dioxide. In the MIST labs, DBC is being used in conjunction with either chilled beams or fan coil units to reduce lab air change rates to 2 ACH day and night when lab air is sensed to be clean. When contaminants are sensed, purge airflow rates of as high as 14 ACH are commanded to provide greater dilution than fixed minimum ACH of 6–8 ACH for safer operation. The lab's VAV fume cupboards were initially designed to go as low as 150 $l\,s^{-1}$ when the sash was closed, but due to the new ANSI Z9.5 standard, a VAV fume hood minimum of 45 $l\,s^{-1}$ was to further increase energy savings.

DBC is providing high impact savings for this project that with the completion of MIST 1B will in combination with MIST 1A be saving approximately 9000 MWh per year.

Figure 28.3 MIST 1A views. Courtesy of Fosters and Partners.

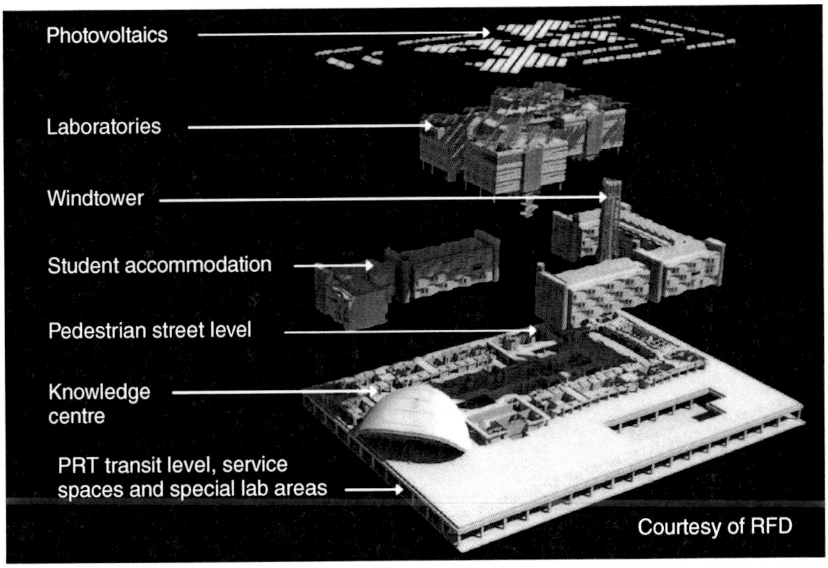

Figure 28.4 MIST 1A layout and lab spaces.

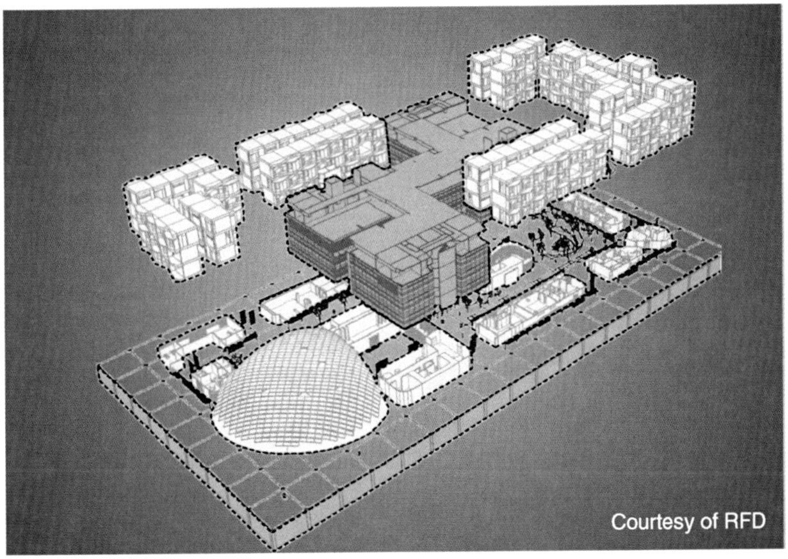

Figure 28.5 Further detail of MIST 1A lab spaces.

28.7.3 MIST (Abu Dhabi) Energy Savings Analysis Example

A new sophisticated lab energy analysis tool was recently developed that can be used to analyze the savings of both DBC plus many other energy efficient design

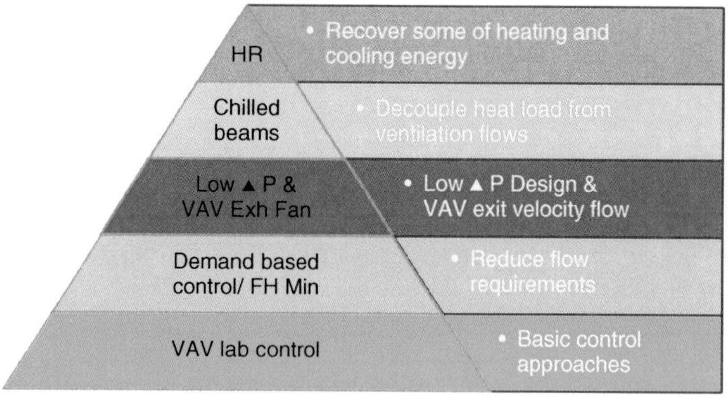

Figure 28.6 Holistic view of relative impact of multiple energy efficient approaches.

approaches for labs such as chilled beams, multiple types of HR, variable exhaust exit velocity control, and so on. This tool allows the analysis not only of these energy saving approaches alone but more importantly together to get an accurate view of the interaction of these approaches when used together.

Figure 28.6 shows a holistic view of multiple lab energy saving approaches. One of the messages illustrated by this figure is that the foundation for lab energy savings is a VAV control system that by itself may not enable significant savings but is required to enable other approaches to reduce lab flows and lower capital costs. Next most important is DBC and the related use of low VAV fume hood minimums, and if needed, automatic sash closers at least for those labs with a moderate to high density of fume hoods. The other approaches listed earlier in DBC can further increase savings such as has been mentioned for chilled beams and fan coil units.

HR has an impact that varies significantly based on climate with the greatest impact for very cold or very hot climates. For humid climates, enthalpy wheels may have a good benefit, however, they cannot be used with any fume cupboard extract mixed into the room extract so these extract streams must be separated so the enthalpy wheel can be used with just the room extract. Note, however, that even in very favorable climates to HR, reducing lab flows to at least 2 ACH at night and 4 ACH during the day will usually still save more than three or more times of what can be saved with an enthalpy wheel.

The collective impact of these energy saving approaches can be seen in the following analysis that uses the climate of Abu Dhabi. Note that the energy impact of climates such as Mumbai and Chennai are 10–20% worse in terms of cooling costs than that shown for Abu Dhabi. This energy analysis assumes a baseline model with a fixed 6 ACH minimum airflow and energy cost of \$0.12/kWh (INR 6/kWh) for electricity and \$0.03/kWh (INR 1.6/kWh) for heating. Finally, the model assumes typical cooling loads and a facility of 12 500 m^2 of which 5000 m^2 is lab space and 3000 m^2 is office space.

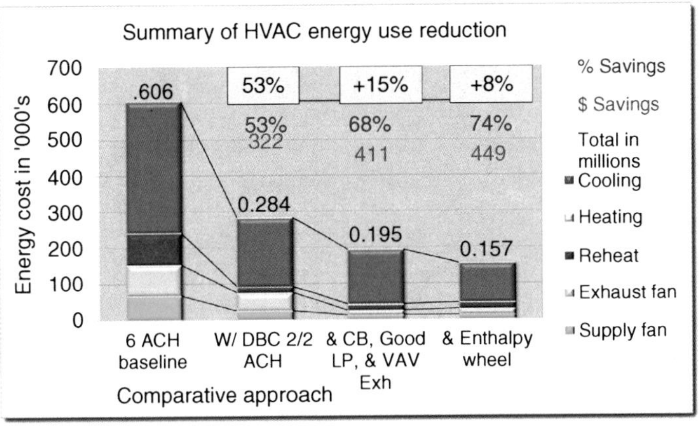

Figure 28.7 Abu Dhabi energy savings example.

Figure 28.7 shows the results of the lab energy analysis. The base case using 6 ACH shows an HVAC energy use of $606 K distributed between supply and extract fan power, reheat, and cooling as shown. Note that for climates like Abu Dhabi and much of India, this HVAC energy use represents 65–80% of the total lab building energy.

The figure shows that using DBC similar to what was done at Masdar reduces the energy use by far the most by $322 K or by 53%. An additional energy savings of about 15% or $89 K can be achieved by the combination of adding chilled beams, good low pressure drop design, and the use of a variable extract velocity extract fan control system. Finally, if an enthalpy wheel HR system is added to the HVAC system, an additional 8% of savings or $47 K can be achieved. By using the combination of these different energy saving approaches, a total energy reduction of 74% or $449 K can be achieved.

If the HVAC building energy represents a total of about 75% of the total building energy, then DBC alone will reduce the total building energy in this example by about 40%! If the other indicated approaches are included, this brings the total building energy reduction to 55%.

28.8 Capital Cost Reduction Impacts of Demand-Based Control

In addition to very significant reductions in energy use, the reduction of the lab airflow rates by DBC can also lower first or capital cost. Even though a single lab may go to higher flows, when contaminants are detected, this only happens typically about 1–2% of the time. As a result, there is considerable diversity in the system and the peak flow can in many cases be significantly reduced particularly when the cooling load is decoupled from the supply airflow. For example, a 20 000 m^2 lab project for the Cal Poly Center for Science and Mathematics in California used the combination of DBC and chilled beams to reduce the first cost of the project even including the cost of the chilled beams and DBC by $716 000.

For the MIST 1A and 1B projects that are trying to achieve a near net zero energy status, not only were there significant reductions in the HVAC capital cost but the annual energy reduction of 9000 MWh translated into a reduction of about 4 MW in Photovoltaic panels that would have been required to generate this energy. Eliminating the need for these solar PV panels saved the project almost $20 M in first costs. This tremendous first cost savings for Net Zero or Near Net Zero Projects is because the high relative cost of renewable energy puts a large premium on the energy efficiency of the project.

28.9
Conclusions on Lab Energy Efficient Control Approaches

The safe reduction of the airflow rates to as low as 2 ACH in a laboratory including an approach such as DBC is by far the single largest impact on reducing lab building energy use sometimes by as much as 50%. Decoupling thermal loads and ventilation when used with DBC is also another important step to reduce lab energy and even more significantly to reduce first or capital cost. While HR can provide good paybacks for certain climates, reducing ACH usually has much better paybacks and saving impacts. In fact, low ACH lab design using approaches such as DBC is a new paradigm in lab design and should be the foundation for the energy-efficient design of laboratories.

References

1. Sarter, D. *et al.* (2008) Laboratories for the 21st Century: An Introduction to Low-Energy Design, August 2008, *www.labs21century.gov* (accessed 28 November 2014).
2. AIHA ANSI/AIHA Standard Z9.5-12. (2012) *Laboratory Ventilation*, American Industrial Hygiene Association, Fairfax, VA.
3. Mathew, P. *et al.* (2005) Right-Sizing Laboratory HVAC Systems, Part 1. HPAC Engineering, issue 9.
4. Mathew, P. *et al.* (2005) Right-Sizing Laboratory HVAC Systems, Part 2. HPAC Engineering, issue 10.
5. Klein, R.C., King, C., and Kosior, A. (2009) Laboratory air quality and room ventilation rates. *Journal of Chemical Health and Safety*, (9/10).
6. Schuyler, G. (2009) The effect of air change rate on recovery from a spill. Presented in Seminar 26 at the ASHRAE Winter Conference.
7. Sharp, G.P. (2010) Demand-based control of lab air change rates. *ASHRAE Journal*, **52** (2), 30–41.
8. Abbamonto, C. and Bell, G. (2009) Does Centralized Demand-Controlled Ventilation (CDCV) allow ventilation rate reductions and save energy without compromising safety? Presented in Session E2 at the 2009 Labs21 Conference.
9. ASHRAE ANSI/ASHRAE Standard 62.1-2010. (2010) *Ventilation for Acceptable Indoor Air Quality*.
10. ASHRAE (2011) Chapter 16, Laboratories, in *ASHRAE Handbook – HVAC Applications*.

29
Lab Ventilation and Energy Consumption

Peter Dockx

29.1
Introduction

If we are talking about sustainable laboratories, most of the time people mean "energy-friendly" laboratory buildings. People often have a mismatch in understanding sustainability for laboratories. Everybody is talking these days about sustainability and people are often confronted with domestic solutions, sometime with solutions for office buildings. This means, energy reduction by good insulation, use of daylight, maybe controlled ventilation but also often for office buildings, natural ventilation. The possibility is for people to open there windows to do more of free cooling, free ventilation.

Laboratories have a complete other user profile. In these types of buildings, ventilation is one of the most important utilities. Why is ventilation so important in a laboratory?

- Ventilation has to protect people from hazardous goods.
- Ventilation has to protect the product from getting contaminated.
- Ventilation has to keep a stable environment for equipment.
- And final, ventilation has to give comfort to the people working in a laboratory.

All these parameters give the importance of ventilation in a laboratory building. Since we have to "ventilate," we can only use 100% fresh outside air. You can imagine that it is not safe for people working in a laboratory if we use air coming out of another laboratory to ventilate back again their lab. Also, cross-contamination is a possible problem then. With 100% fresh air, we do not only need to be able to bring in the air but we also have to condition that air. Depending on the outside air conditions, we have to heat up or cool down the air. We have to humidify or to dehumidify the air. Depending on the geographical location of the lab, the energy demand will look completely different. Next figure shows the energy demand of the same laboratory, same users, but located in different areas in Europe. You can see the difference on annual basis for heating and cooling depending on the location while the internal plug-load (room electricity) and fan energy are the same because of the usage of the lab (Figure 29.1).

The Sustainable Laboratory Handbook: Design, Equipment, Operation, First Edition.
Edited by Egbert Dittrich.

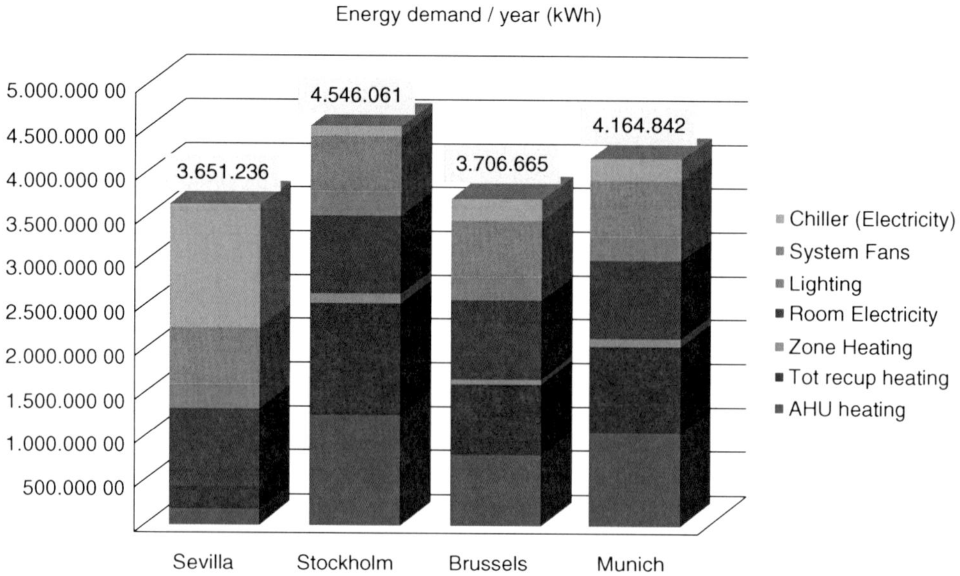

Figure 29.1 Energy demand/year for Seville, Stockholm, Brussels, and Munich.

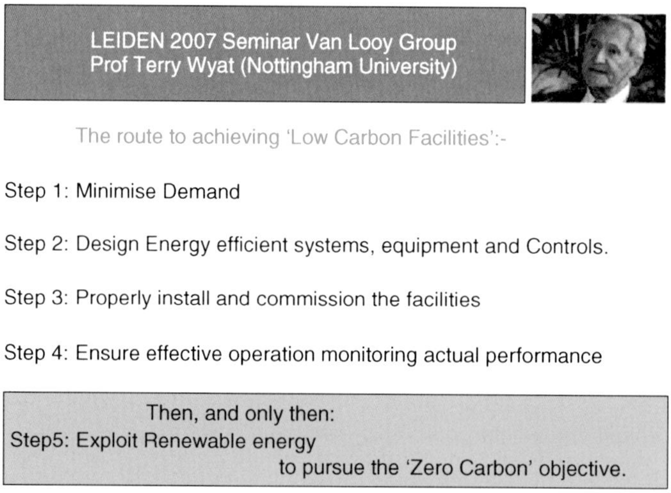
LEIDEN 2007 Seminar Van Looy Group
Prof Terry Wyat (Nottingham University)
The route to achieving 'Low Carbon Facilities':-
Step 1: Minimise Demand
Step 2: Design Energy efficient systems, equipment and Controls.
Step 3: Properly install and commission the facilities
Step 4: Ensure effective operation monitoring actual performance
Then, and only then:
Step5: Exploit Renewable energy
to pursue the 'Zero Carbon' objective.

Figure 29.2 Slide from Prof. Terry Wyat (Nottingham University).

To address the different aspects of the energy consumption of a laboratory building, we often refer to a slide from Prof. Terry Wyatt from the Nottingham University (Figure 29.2).

We take this slide to guide us through this chapter, and touching in that way, the different aspects of sustainability in the laboratories.

29.2 Step 1: Minimize Demand!

What does that mean in the case of a laboratory building? Just like we explained in our introduction, ventilation is the most important driver if it comes to energy consumption in a laboratory building. If we can reduce the amount of ventilation, we can reduce the energy consumption. Like stated in the introduction, ventilation in a laboratory is used for a lot of different aspects.

Each aspect can drive the amount of ventilated air needed and it is not always in time the same aspects that drive that amount. You can have fume hoods, ventilated cabinets, storage cabinets, snorkels, laminar airflows, and so on, that extracted air out of your laboratory to protect people from hazardous goods. All that extracted air needs to be compensated by fresh air to keep the ventilation in balance.

29.2.1 Fume Cupboard Control

A fume cupboard is built to protect people in front of the fume cupboard from hazardous goods handled inside the hood. By means of extracting air out of the fume cupboard, air will enter the fume cupboard by the glazed front. Depending on the amount of extracted air and the opening of the front (sash) of the fume cupboard, you get a certain entrance of air into the hood. Depending on the design and the amount of extracted air, the fume cupboard is able to keep hazardous fumes inside and protect people in front of the hood. How efficiently a fume cupboard can "contain" hazardous good is captured in what they call the *Containment factor*. The better a fume cupboard can hold a certain containment with less air, the better the containment performance of a fume cupboard is. Due to this, two different fume cupboards with the same amount of extracted air and the same opening in the front can have a different containment factor. Al this is regulated in the EN EN14175 (see other chapters).

Before this norm, we often used the entrance velocity in meters per second to indicate the amount of airflow needed from a fume cupboard. The entrance velocity in the opening multiplied by the square meters of the openings resulted in the airflow needed for the fume cupboard ($m\,s^{-1} \times m^2 = m^3\,s^{-1}$). The US and US related companies were used to use as an average $0.5\,m\,s^{-1}$ as a good standard. In trying to reduce the amount of airflow needed for a fume hood, more European (German) related companies were reducing entrance velocities to $0.4-0.35\,m\,s^{-1}$. With the new European norm, manufactures are redesigning their fume cupboards even trying to lower down the velocity, but keeping the same containment factor (Figure 29.3).

The air extracted by a fume cupboard is the multiplication of the entrance velocity multiplied by the opening of the sash. So to reduce the amount of air, a fume cupboard demands means to reduce both the entrance velocity and the sash height. Reducing the entrance velocity depends on how the fume cupboard is

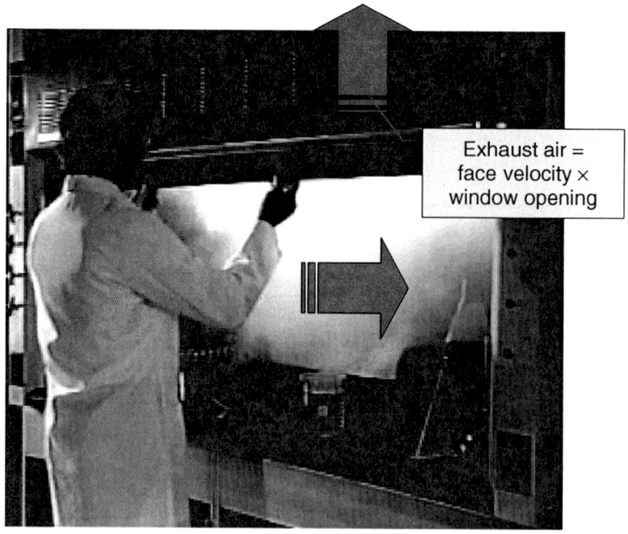

Figure 29.3 Fume hood.

built and how good the fume cupboard is in containing contaminants with a lower entrance velocity. The opening of the fume cupboard depends on the sash height.

If the amount of airflow in your lab is driven by the air extracted by the fume cupboard, it is good to install fume cupboards with a variable airflow system. There are types of fume cupboard controls that can control the amount of air extracted by the fume cupboard in relation to the opening of the sash. If the sash is nearly closed, the extracted amount of air is minimal. If the sash is opened, the amount of extracted air will be greater. If you have several fume cupboards in a laboratory, you can reduce the amount of airflow, and your installation, by applying a simultaneous factor (also called *diversity factor*). For example, if you have a lab with 10 fume cupboards and 5 people to operate them, you can take into account as airflow needed: 5 fume cupboards with maximum flow and 5 fume cupboards with minimum flow.

29.2.2 Biosafety Cabinets

Depending on the required biosafety protection needed, biosafety cabinets are recirculating 100% of air or only 70% of air. In the last case, 30% of the air is entered in the opening and extracted on top of the cabinet. The air extracted back into the room is high efficiency particulate air (HEPA) filtered before its entrances back into the room. Nevertheless, often costumers want to directly evacuate the air coming from the extract. So, depending on the fact that the biosafety cabinet is on or off, the airflow for the extract and supply of the room can change.

Also, switching on or off other types of extract in a lab has its influence on the amount of air needed for a lab.

29.2.3 Temperature Control

Also, changes in airflow can result in temperature changes in the lab. If we have a lot of fume hoods exhausting air from the room and we have to compensate this by supply of lot of air in the lab, it can be that it is getting too cold in the lab. In this case, we need to heat up the supply air. On the other hand, it can also be that there stands a lot of equipment in the laboratory and that we need to increase the supply air to control the temperature in the lab.

29.2.4 Minimum Amount of Air Changes

To keep the environment safe for people to work in, Health and Safety officers often require a minimum ventilation rate. Independent from the "air demand," there must be a minimum of supply air to the lab. The required demand can change depending on the fact that the lab is occupied or not. During nighttime, the air change rate may be lower because the fact that nobody is working in the lab. The most used number of minimum air changes is around 8 air changes per hour or 25 $m^3\,h^{-1}\,m^{-2}$. This number is more or less the same for a room height of 3 m. In trying to get laboratories more energy friendly, there are a lot of questions coming up for trying to reduce the amount of air changes. It is still the question "What is the right number then?" The answer is not so easy to say. It depends on the type of laboratory you have, how it is occupied and definitely, how efficient is your air ventilation of the room. There are also systems in the market that samples the air in the room every 10–15 min and adapt the air change rate of the room in accordance with the pollution measured. When using this type of system, the basic air change rate in the rooms is lowered to 6 or 4 air changes per hour. When a higher concentration of a pollutant is sampled, the system automatically increases the air change rate (see previous chapter). The system becomes also more popular by H&S officers because now they have real-time data about the air quality of the laboratory.

All the examples given show that "air management" in a laboratory can become complex. Since the demand of air can change from minute to minute, special control systems have been developed to control all the different airflows in a laboratory. We call this "Lab-Control" systems. Today's systems are mostly "bus" operated systems. The connections between the different valves and controllers are based on a network structure. This gives the advantages that it is easy to add controllers and valves when setups in a lab changes.

29.2.5 How to Integrate the Lab-Controller in a Smart Way in Our Ventilation System

Classic ventilation systems are based on what we call an "antenna" system lay-out. This means that we start from the air handling unit with big ducts and as we go further and further from the air handling unit, ductwork is spreading, the amount of air reduces, and the size of the ducts are getting smaller and smaller. This is a classic setup for offices or other kind of buildings where we do not expect great changes in airflows.

Like we have discussed in the previous chapter, ventilation in a laboratory is a dynamic process. Airflows change minute by minute depending on the requirements of that moment. In a classic air distribution system, ductwork must be sized to handle the worst case, also at the end of the system. Most of the time the ducts are then oversized.

Another situation is when there is a change in the setup of the lab. If for example, a lab near the end of the system gets two extra fume cupboards, and this was not calculated in the duct size at the moment of installation, often these situations give problems in installing new fume cupboards, and ductwork is too small.

In case of installing ductwork in an "antenna" system, it is better for a laboratory system to install ductwork as a "ring" system. This means that air is supplied from the air handling units to the end-users by means of a "ring." Ductwork is layed out in such a way that it supplies the end-user from two sides. In this case, there is no "end of the network." Adding extra extract is easy and like a kind of "plug and play" system.

In big installations, it also has the advantages that often size of ductwork can be reduced by multiplying the system with a certain diversity factor. This means, for example, that not all of the extract systems are in use at the same time for full load. Like we discussed in the previous chapter, the amount of airflow in the lab depends on the use of fume cupboards, are their sashes open or closed, or snorkels being on or off?. We then can predict that the full capacity of airflow is not needed. We can multiply the total amount of installed extract system by a coefficient lower than 1 to overall dimension of the air handling systems. The factor can be different on room level, floor level, or building level.

For example, we have a laboratory building with a lot of fume cupboards. In each lab, we have eight fume cupboards. We can allow to use them in one lab all at the same time for doing setups. The room diversity is then 1. But, we also say that it will not happen then, all five labs on one floor will use all their eight fume cupboards at the same time. We can say that they will use a maximum of 30 fume cupboards from the installed 40 on one level. The diversity on the floor level is then 0.75. If we have three levels of lab in that building, we can also imagine that they will not use at the same time more than 60 fume cupboards. The building diversity is than 0.5. This means that we will size the installation on room level at 100% room capacity, the ductwork on floor level will be sized for 75% of the floor capacity, and that the central air handling system will be sized for 50% of the total building capacity.

Another advantage of setting up your air handling distribution in a "ring" network is that it can easily cope with the dynamics of this system and that you also can size your ductwork for maximum 75% on floor level and 50% at the plant room. Since it is the setup of a ring system that you come with, your air from both sides mean that the biggest size of ductwork in your building is only sized for doing

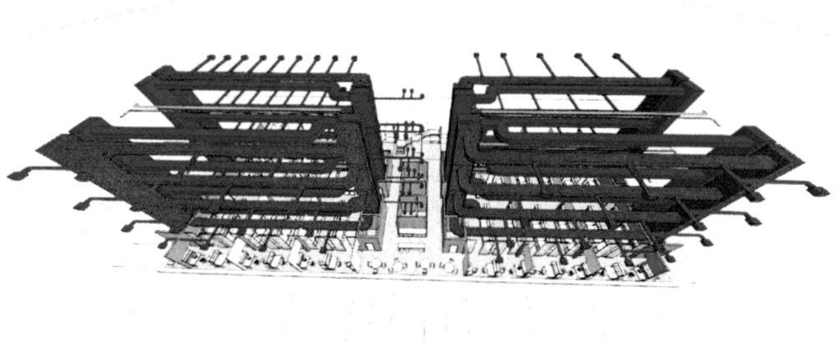

Figure 29.4 HVAC ring design in laboratory building.

75% × 0.5 = 37.5%. This reduces enormous size of the ductwork and makes coordination with other installations more easy (Figure 29.4).

29.3 Step 2: Design Energy Friendly Systems

29.3.1 Energy Recovery

Like we already addressed in our introduction, laboratory ventilation systems always use 100% of fresh air. When designing energy-friendly systems, the first thing we have to think about is to recover energy that we otherwise waste. The air extracted from a building contains still a lot of energy, certainly also a laboratory building. There are different so-called heat recovery systems in the market; one already better adapted for use in a laboratory building than another one.

29.3.2 Rotary Wheel Heat Exchangers

In this type of heat exchangers, heat is transferred from the extracted air to a heat adsorbed material coated on the fins in a wheel. Turning around this wheel to the supply side of the air handling unit, the plates will give back the heat to the cold air in the supply. Advantage of this system is that we, meanwhile, can reach high efficiencies. Big disadvantage with the system is that it not only recovers heat from the supply but depending on the coating, also chemicals that we donot want. For this reason, Germany's VDI 6022 Blatt1 doesnot allow use of these types of heat exchangers. There are meanwhile manufactures that produce special coated heat recovery wheels that would not have this disadvantage.

Another disadvantage for using this type of system is that you have to always put supply and extract units on top of each other (or next to each other). Since the airflow demands of laboratories most of the times are high, it becomes more or less impossible to install this type of units in a technical area.

29.3.3
Plate Heat Exchangers

There are also heat exchange systems using a "plate heat exchanger." The hot extracted air is passing a "plate" where, on the other side of the plate it is passing the cold supply air. The hot air is giving a part of its heat back via the plate to the cold supply air. This is a system that is quite popular in domestic installations. For laboratories, it has the same disadvantage as the former system. Supply and extract need to be incorporated on top or next to each other. Also, for laboratory applications, the units need to be adapted in what they call "pressure reverse" mode. This means that the fan in the extract system needs to be installed after the heat exchanger in airflow direction, and that the fan in the supply system needs to be installed before the heat exchanger. In the case of a leak between the supply and the extract, contamination from the supply air by extract is being avoided. Plate heat exchangers are popular when using also "adiabatic cooling" systems. We will come back later to this subject.

29.3.4
Twin Coil Heat Exchangers

In this type of system, coils are installed in the extract and the supply airflow. Water is "heated up" in the coil in the extract air, transported to the coil in the supply air where it heated up the supply air. This type of heat exchanging has the advantage that there is no risk of contamination between supply and extract. For this reason, this system is optimally suited for installing in a laboratory environment. Even the energy coming from the exhausted air in fume hoods can be recovered. Disadvantage of the system is that the efficiency of this system is most of the time lower than that of the systems mentioned earlier.

With this system, you always need still the conventional heating and cooling coil in your supply unit. You can avoid this if we add a heat exchanger in the loop of the twin coils. The extra heating or cooling energy is then added in the water loop. This saves us not only space in the supply unit but also energy, since the total pressure drop in the air handling unit will be lower (Figure 29.5).

29.3.5
Adiabatic Cooling System

With adiabatic cooling systems, we mean systems that cool down the air by evaporating water in the air stream. We can classify them in two systems, the direct adiabatic cooling system and the indirect system.

With the direct system, we introduce water in the supply air by means of a humidifier. The water introduced will be evaporating. This evaporation will extract heat from the air and cools it down. This can be very effective; the disadvantage is that it brings also more moisture in the supply air which is not always a good thing.

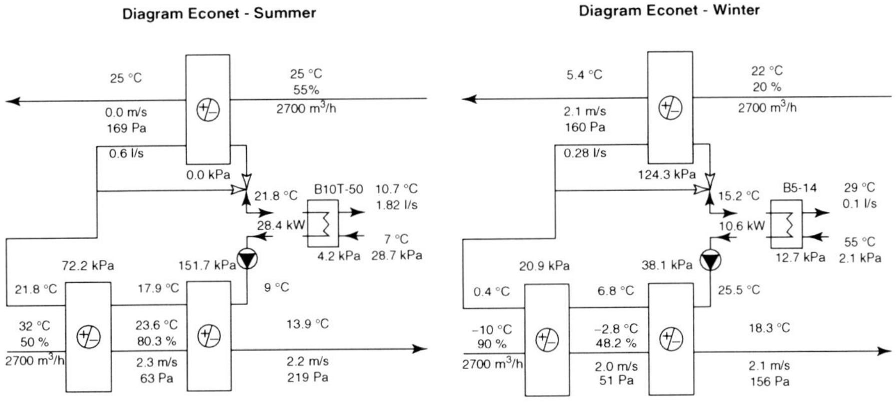

Figure 29.5 Schematics of heat exchanger in twin coil loop.

In the indirect system, we do the same but, instead of humidifying the supply air, we do that we the extracted air. The extracted air will also then be cooled down. Taking this air through a heat exchanging system, like described earlier, can cool down the air in the supply (Figure 29.6).

For example, we have air extracted out of a laboratory building at 22 °C (55% Relative Humidity (RH)). We humidify that air so it cools down until about 17 °C. We bring that air in a heat exchanger to cool down outside air from 26 °C. Depending on the efficiency of the heat exchanger, the air in the supply will then be cooled down too, by example, 19 °C. We can supply this air to the lab back again or if necessary cool it further down. We have cooled down the supply air with 7 °C without using expensive cooling energy. You see that the efficiency of the indirect systems depends on the efficiency of the heat exchanger. There are special units in the market that are built around this system to have a high efficiency. These units can also have an additional advantage during wintertime if you need to humidify your air and you also want to do this with an adiabatic humidification system instead of a steam humidifier. Due to the high efficiency of the heat exchanger, you don't need much heating energy to get the supply air on the right enthalpy line for adiabatic humidification. On a year basis, the energy is much less than the energy you needed in case of steam humidification.

29.3.6 Extract Systems

Also, the extract systems themselves can have a big impact on the energy consumption of a laboratory building. The actual purpose of the extract system is to dilute the contaminated air from the laboratory that it doesnot harm any other air intake from its own or other building. The best way to study this influence is with wind tunnel analysis. This can be done by Computational Fluid Dynamics (CFD) or in a real wind tunnel with a scaled model of the building and its surrounding (Figure 29.7).

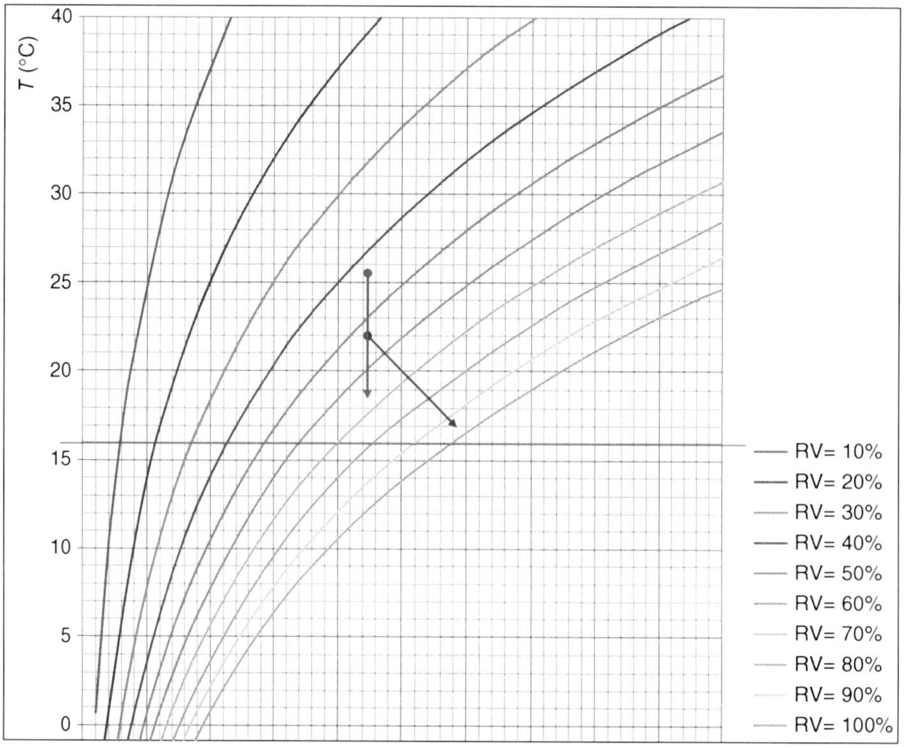

Figure 29.6 Mollier diagram with indirect adiabatic cooling.

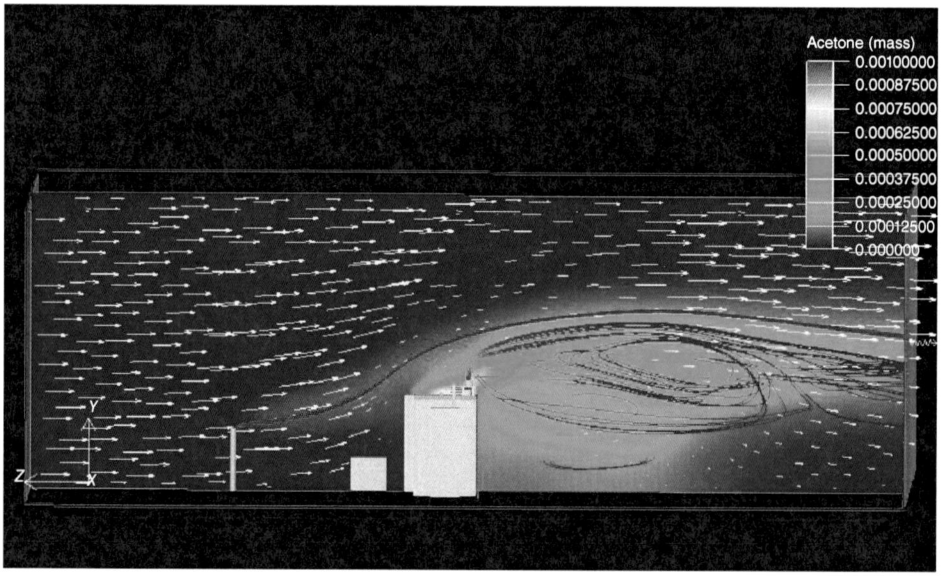

Figure 29.7 CFD study of exhaust system.

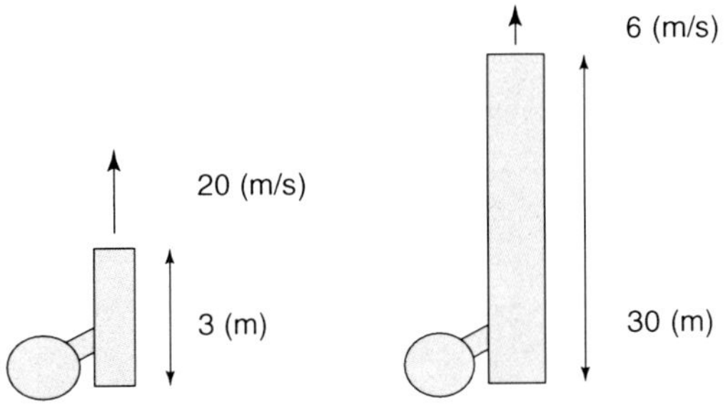

Figure 29.8 Influence of stack height.

Most of the time, the outcome of this kind of studies is that you have to discharge an amount of air with a certain velocity. The height of the extract stack can have a big influence on this and by this also on your energy bill (Figure 29.8).

We give you an example. We have to discharge 30.000 $m^3 h^{-1}$ of air from a laboratory building. We want to discharge this on the roof of the building with a stack height of 3 m above the roof. The outcome of the wind tunnel analysis gives us the result that the discharge velocity needs to be around 20 $m s^{-1}$. We could also discharge the air at a height of 30 m above the roof. In this case, a discharge velocity of 6 $m s^{-1}$ is enough to have a good dilution of the air. The first solution gives us an energy consumption of about 24.000 kWh per year. The second solution gives us an energy consumption of about 3.000 kWh per year. You save about 85% of energy.

It is possible that your architect doesnot like to put a chimney of 30 m on top of the roof. It can be still helpful to do that wind tunnel analysis. It is possible that we only need that 20 $m s^{-1}$ discharge if the wind is coming from a certain direction. If the wind comes from other directions, maybe it is possible to reduce the velocity. We have been speaking in the previous chapter that the amount of air used in a laboratory building is different from minute to minute. We come back to our example. What happens if we donot have an air consumption of 30.000 $m^3 h^{-1}$ in the lab but only 22.500 $m^3 h^{-1}$. If the wind is still coming from the bad direction and, following the wind tunnel analysis, we have to discharge the air at a velocity of 20 $m s^{-1}$, we have to take "by pass" air from the roof to discharge a total of 30.000 $m^3 h^{-1}$ to get that velocity of 20 $m s^{-1}$. But if the wind is coming from another direction, and we can discharge at a lower velocity, we do not need that by-pass air. We can discharge then at a velocity of 15 $m s^{-1}$. If we calculate the energy consumption on discharging the air with 15 $m s^{-1}$ instead of 20 $m s^{-1}$, we can save about 60% of energy (from 24.000 to 10.000 kWh per year). If we can even go lower to 10 $m s^{-1}$ we save again about 85% of energy (from 24.000 to 3.000 kWh per year)

If you need more than one fan in your extract system to discharge all the air coming from the laboratory then it is also better to give each fan its own stack outlet. If the amount of air asked by the variable air volume system from your laboratory goes down, you can shut off extract fans by still maintaining the air velocity in the individual stacks you need to have for a good dilution. In the other case, you need to take too much of by-pass air which consumes more energy than necessary.

29.4
Step 3: Install and Proper Commission the Installation

Besides a good engineering, a proper installation and commissioning is definitely as important for a good energy-efficient building. It is crucial to check the proper installation and function of all the components in the installation. It is so often the case that, due to bad installation, obstructions in ductwork or even garbage that was left in the ducts during the construction is not removed during installation. All this can lead to obstructions in the ducts with a higher resistance as result. The complete installation needs to run on a higher pressure than necessary with all wasted energy as a result of this.

This is just one example why proper installation and commissioning is so important. Unfortunately, we see in real life that often it is not done, most of the time, by time pressure in the schedule. Commissioning is always one of the last things in a building project when deadlines need to be met.

I can only address this to the endusers that a proper commissioning of a laboratory building is more important than deadlines. It is about safety for people and energy consumption for the lifetime of the building

29.5
Step 4: Maintain the Installation and Monitor

Maintenance and monitoring of the installation during the lifetime of the building is as important as all the steps before. Keeping the installation in a good performance is important for the safety of the users of the laboratory building. Temporary control of good function of fume cupboards, extract systems, and so on, is important.

Training of maintenance staff is very important. We often see that installations are built, commissioned, and then given over to the client without giving a proper training. It is also hard for the installation companies to give a proper training. For this, we advice that the maintenance engineer already is involved from the beginning of the design to understand how the system needs to be working, and during commissioning to learn how the system functions. By involving the maintenance engineer during the commissioning, gives also the advantage that he learns all the components of the installation and where they are installed.

It is also good to give the design engineers the opportunity during the first year that the installation is running to monitor on regular basis the system. Then, they can see if the installation is performing like it was designed and if not, why this could be.

29.6 Step 5: Use of Alternative Energy

29.6.1 Importance of Energy Modeling

If you want to use alternative energy sources, it is nearly a must to a have a good energy model of your building. Only then, you can make a good analysis of the systems you want to use. The "energy profile" of a laboratory building looks completely different from most other buildings since the energy consumption of such a building is driven by the high amount of fresh air you need to condition before it enters the building. For this reason, the geographical location of the building is important and therefore, solutions can be different from location to location. Solutions for Stockholm will be completely different from solutions in Seville.

Next picture gives you an example of such an energy profile of a laboratory building (Figure 29.9).

You can see that we need heating energy almost all the time of the year. Even in summertime. Cooling energy, we need only when the outside temperature is higher than our set point in our air handling unit (by example 16 °C).

If we zoom in on a time period of 2 weeks, we can see the difference in heating and cooling demand (Figure 29.10).

Heating we need during day and night, cooling only during the day.

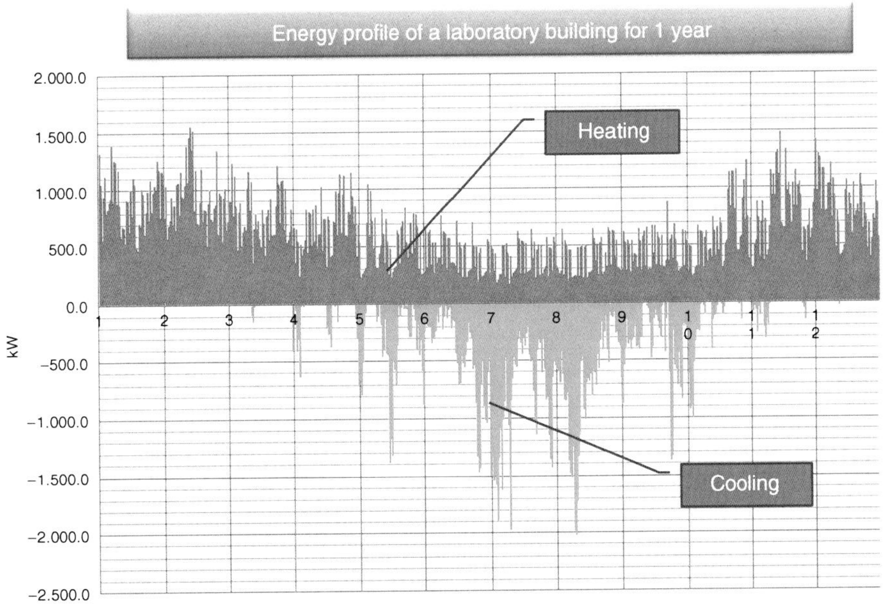

Figure 29.9 Energy profile of a laboratory building for 1 year.

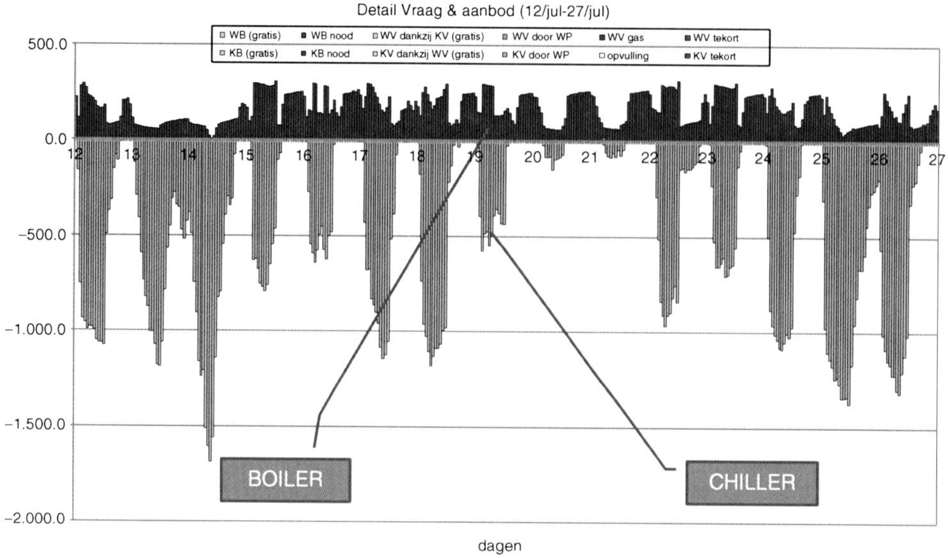

Figure 29.10 Energy profile of a laboratory building for 2 weeks.

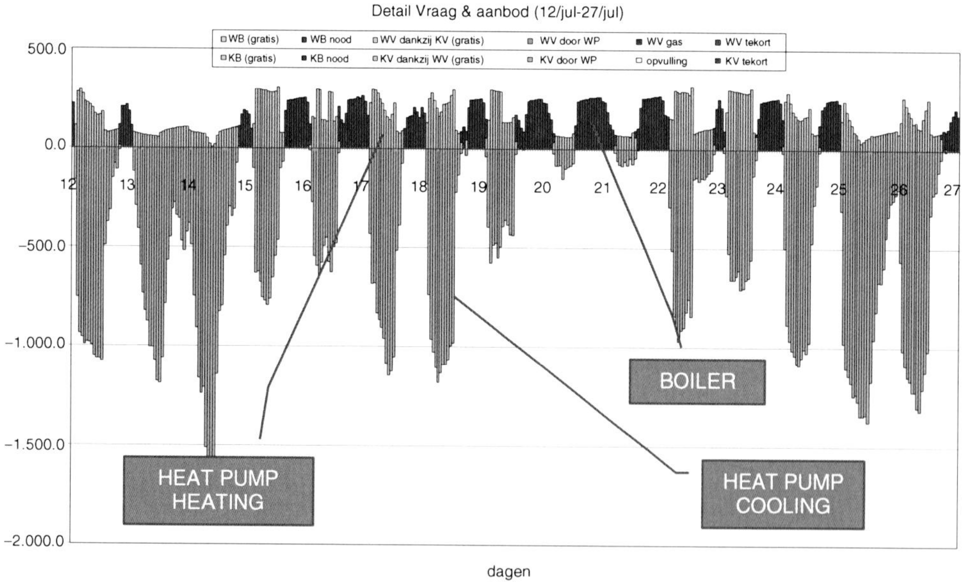

Figure 29.11 Energy profile of a laboratory building for 2 weeks using a heat pump.

29.6.2 Using Heat-Pump

If we take the energy profile from the picture mentioned earlier, and we would build "a classic" installation with boilers and chillers, we would need to run our boilers in the middle of the summer, even if the outside temperature is 30 °C.

When we would use a "heat pump" to produce our cooling energy, we can use at the same time the heat so we donot need the boiler anymore for heating (Figure 29.11).

We see that at the moment, we need cooling, the heat for the building is coming from our heat-pump. We only need the boilers during the night when we donot need cooling in the building.

29.6.3 Phase Change Materials (PCMs)

We can optimize more the running hours of the heat pump if we could also use it during nighttimes. The problem is that we don't need the cooling then. We can solve this by storing the cold energy during the night in the so-called phase change materials(PCMs) and let them give the energy back during daytime when we need the cooling. Next figure shows the influence of this on our energy profile (Figure 29.12).

The dark green color shows the heating energy from the heat pump during night. The cold energy that is produced at that moment is stored in the PCMs at that

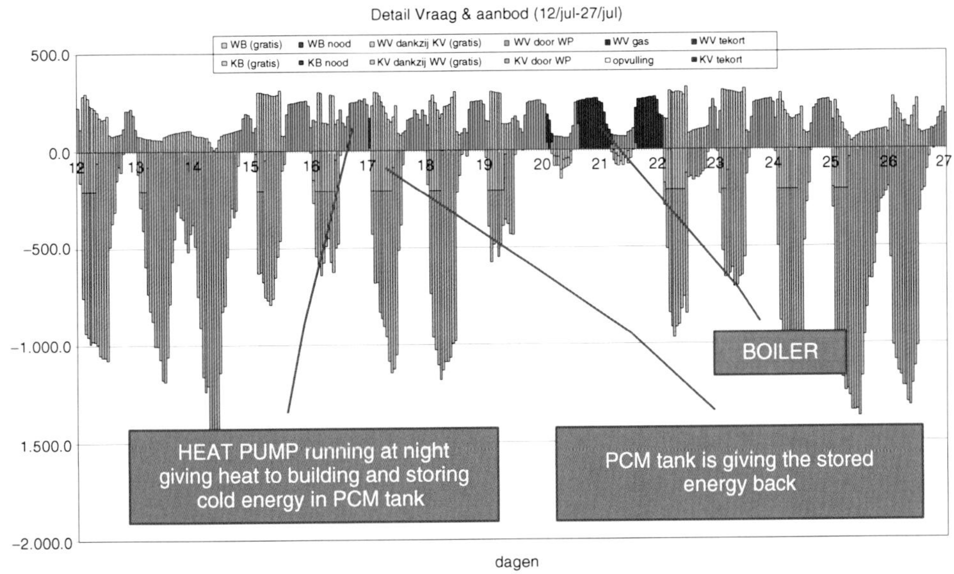

Figure 29.12 Energy profile of a laboratory building for 2 weeks using a heat pump and PCMs.

moment. The light green color shows you the energy that we get back from our PCMs when we need cooling during daytime.

29.7 Conclusion

All the different steps mentioned in the examples earlier can help us to reduce the energy demand of a laboratory building. Definitely, the energy that is needed for the ventilation, heating, and cooling can be reduced dramatically to just 20% or even less. By applying all these steps, the energy for ventilation, heating, and cooling, comes from the biggest part of the energy consumption of your laboratory building to only a minor part. It should be the goal by each design to go through all these steps to see what can be implemented.

30
Consequences of the 2009 Energy-Saving Ordinance for Laboratories

Fritz Runge and Jörg Petri

30.1
The Task Force

This report is based on the results of investigations that were conducted with the collaboration of the permanent members of the "Benchmarking – Chemistry, Pharma, & Life Science" task force in preparation for the revision of the 2009 German Energy Saving Ordinance (EnEV). At the time of these investigations, 22 experts in the design, operation, and management of laboratory buildings from a total of 12 companies in the chemical and pharmaceutical industry were active members of the task force. Bauakademie Gesellschaft für Forschung, Entwicklung und Bildung mbH, Berlin, was appointed to act as the neutral moderator and to provide scientific and organizational support, as well as data processing and analysis. It also provided the IT tools for data input and analysis.

Regular working meetings are held semi-annually. For a more in-depth discussion of the results of the analysis and to identify Best Practices, the task force has also held two additional workshops annually since 2004, focusing on topics in the area of the operation of buildings in the chemical-pharmaceutical industry. In addition to the discussion and interpretation of data analyses, the primary purpose of the workshops is to refine the methods used to collect data and to guarantee the continuous improvement of the methodology and IT tools used. The result of these many years of cooperation among experts from the 12 companies is a fully mature system for the analysis and comparison of indicators of management efficiency for office and laboratory buildings in the chemical-pharmaceutical industry. The analysis results are kept strictly confidential on account of the sensitivity of the company data on which they are based. The members of the task force decide on a case-by-case consensus basis for what purpose and in what scope the internal confidential data can be transmitted to third parties. In its own interest, the task force places great emphasis on the compilation and evaluation of data that are as complete and correct as possible.

The Sustainable Laboratory Handbook: Design, Equipment, Operation, First Edition.
Edited by Egbert Dittrich.

A particular interest of the representatives of the participating companies is the consideration and improvement of the energy efficiency of laboratory buildings. This topic has, therefore, already been the subject of several Best Practice workshops.

During the period from October 2007 to October 2009, a working group of the task force was concerned in particular with the topic "Energy efficiency of laboratory buildings," and the determination of reference values for the energy consumption of laboratory buildings.

The principal members of this working group were:

- Dipl.-Ing. Frank Emmrich, Bauakademie Berlin
- Dipl.-Kfm. Frank Hattenbauer, Bayer Schering Pharma AG
- Dr. Dietmar Kohn, Boehringer Ingelheim Pharma GmbH & Co. KG
- Dipl.-Ing. Jochen Kranz, Currenta GmbH & Co. OHG
- Dipl.-Wirt.-Ing. Ingmar Neuber, Infraserv GmbH & Co. Höchst KG
- Dipl.-Ing. Jörg Petri, Bayer Schering Pharma AG
- Prof. Dr. Fritz Runge, Bauakademie Berlin.

The work of this group can be traced back to a corresponding inquiry from ARGE Benchmark, which was tasked by the German Federal Department of Construction and Regional Planning (BBR) with the determination of energy consumption levels for nonresidential buildings. An agreement was reached with ARGE Benchmark that the energy consumption values determined by the task force would be handed over to the BBR for a decision on their inclusion in the list of reference values for the "Laboratory buildings" category.

On the basis of the investigations conducted by the task force, a study[1)] was prepared, the summary results of which are presented later.

30.2
Energy Certificates for Laboratory Buildings

The system governing the issue of energy certificates has been revised since the 2007 EnEV. Energy certificates must be issued under the conditions set forth in the ordinance for nonresidential buildings as well as for existing buildings and structures. The energy certificates can be issued on the basis of either

1) energy demand or
2) energy consumption.

1) Runge, F.; Frank Emmrich: Vergleichswerte des Energieverbrauchs von Laborgebäuden der chemisch-pharmazeutischen Industrie, Arbeitskreis Gebäudekostenbenchmarking "Chemie and Life Science," Juni 2008.

30.2.1
Issue of Demand-Based Energy Certificates

In compliance with the EnEV, the issue of demand-based energy certificates is generally required for new construction and existing structures.

The procedure for evaluating new construction or renovated existing structures in compliance with the EnEV leading to the issue of a demand-based energy certificate is as follows:

- first, the energy demand of a (virtual) reference building with standard boundary conditions is determined, and then
- the calculated energy demand of the new construction, assuming standard boundary conditions of the reference building, is compared with that of the reference building.

For the issue of demand-based energy certificates as proof of compliance with the EnEV, the following principle must be followed as part of the building permit planning process:

1) Determine the *usage profile* of the building in accordance with DIN V 1. 8599 (both for the determination of the actual values as well as for the determination of the reference values)
2) Determine the *ACTUAL values* for a planned building (in the form of ACTUAL planned values)
3) *Determine the reference values* for a (virtual) reference building containing reference equipment as required by the EnEV.
4) Create the *energy certificate* by a comparison of the ACTUAL values and the reference values, each for the selected usage profile.

As a potential usage profile for the planned building, Table 4 of Part 10 of DIN V 18599 provides a total of 33 profiles with usage boundary conditions, one of which must be selected from the list.

For each of the 33 usage profiles, 19 parameters relating to the complexes are:

- usage and operating hours,
- lighting *and*
- interior climate control and heat sources are shown.

However, a demand-based energy certificate issued on this basis as required by the EnEV is not an indication of the actual energy consumption of the planned building. This method only makes it possible to conduct a comparison with a virtual building, the selected usage profile of which is as close as possible to that of the planned building.

On the pre-printed form for the demand-based energy certificate for nonresidential buildings, it is therefore explicitly stated that, "... *on account of standardized boundary conditions ... the indicated values cannot be used as the basis for any conclusions regarding actual energy consumption.*"

However, there is currently an additional, more serious problem relating to the issue of demand-based energy certificates for laboratory buildings:

Table 4 of Part 10 does not currently contain any suitable usage profile that simulates the energy situation of "laboratory buildings"!

It is therefore an open question how one is supposed to proceed in the specific case of the preparation of a demand-based certificate for a laboratory building.

The approaches that have been attempted so far to close this gap are based on a classification of laboratory buildings according to typical sets of tasks for which laboratories are used (synthetic, analytical, physical, microbiological, biochemical, research, and technology laboratories).

However, one disadvantage of the above method, as explained in greater detail later, is that usage profiles prepared on this basis with corresponding usage boundary conditions, taking into consideration the laboratory-specific influences on energy consumption, are inappropriate because the type of laboratory usage cannot necessarily be used to determine different energy demands (Figure 30.1).

For a more realistic assessment of the energy efficiency of laboratory buildings on the basis of a demonstration of demand, it would be necessary to develop new usage profiles with corresponding usage boundary conditions that take the special features of laboratory buildings into consideration.

Based on the experience of the task force, such a differentiation of laboratory buildings on the basis of usage profiles can only be made on the basis of usage boundary conditions that primarily take into consideration the principal factor that influences energy consumption, namely the air exchange rate.

30.2.2
Issue of Consumption-Based Energy Certificates

For the issue of consumption-based energy certificates for an existing building in compliance with the EnEV, first the *energy consumption characteristics* must be determined by a consumption measurement. The three prior calendar or accounting years must be used as the basis for the determination of consumption. The characteristic consumption values based on measurements must be juxtaposed with the corresponding *reference values* provided by the German Federal Ministry for Transportation, Construction, and Regional Development (BMVBS).

The reference values are published in a bulletin[2] issued by the BMVBS supplementing the EnEV, which specifies the general procedure for the determination of characteristic energy consumption values. The annexes to this bulletin list a number of building categories, classified on the basis of their type of use.

In the edition dated July 2007, which was valid until September 2009, reference values for a total of 116 building categories were shown.[3]

2) The currently valid edition of the "Bulletin of rules for consumption-based energy characteristics and reference values in non-residential buildings" is dated 30 July 2009.

3) Thereof, 80 building categories as defined in the Structural Classification Catalog (BWZK) issued by the committee consisting of the state government ministers and senators responsible for municipal construction, construction and housing of the states (ARGEBAU), and an additional 36 types of building usage, which are not categorized according to the BWZK.

ENERGIEAUSWEIS für Nichtwohngebäude

gemäß den §§ 16 ff. Energieeinsparverordnung (EnEV)

Berechneter Energiebedarf des Gebäudes

2

Primärenergiebedarf

"Gesamtenergieeffizienz"

Dieses Gebäude: kWh/(m² · a)

CO_2-Emissionen[1]) kg/(m² · a)

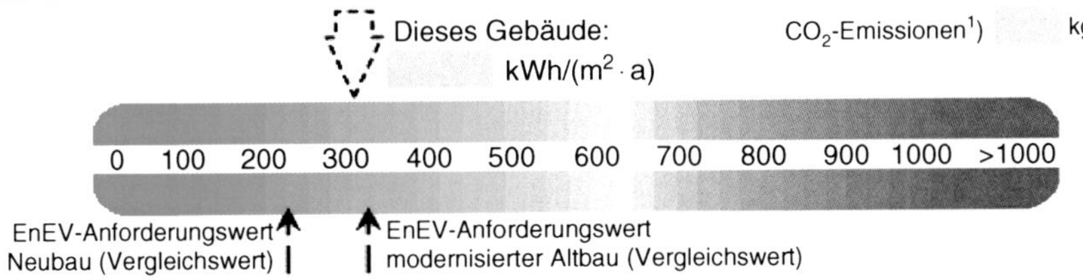

Nachweis der Einhaltung des § 4 oder § 9 Abs. 1 EnEV[2])

Primärenergiebedarf		Energetische Qualität der Gebäudehülle	
Gebäude Ist-Wert	kWh/(m²·a)	Gebäude Ist-Wert H_T'	W/(m²·K)
EnEV-Anforderungswert	kWh/(m²·a)	EnEV-Anforderungswert H_T'	W/(m²·K)

Endenergiebedarf

Energieträger	Jährlicher Endenergiebedarf in kWh/(m²a) für: Heizung	Warmwasser	Eingebaute Beleuchtung	Lüftung	Kühlung einschl. Befeuchtung	Gebäude insgesamt

Aufteilung Energiebedarf

[kWh/(m²a)]	Heizung	Warmwasser	Eingebaute Beleuchtung	Lüftung	Kühlung einschl. Befeuchtung	Gebäude insgesamt
Nutzenergie						
Endenergie						
Primärenergie						

Sonstige Angaben

Einsetzbarkeit alternativer Energieversorgungssysteme

☐ nach § 5 EnEV vor Baubeginn geprüft

Alternative Energieversorgungssysteme werden genutzt für:

☐ Heizung ☐ Warmwasser ☐ Eingebaute Beleuchtung

☐ Lüftung ☐ Kühlung

Lüftungskonzept

Die Lüftung erfolgt durch:

☐ Fensterlüftung ☐ Lüftungsanlage ohne Wärmerückgewinnung

☐ Schachlüftung ☐ Lüftungsanlage mit Wärmerückgewinnung

Gebäudezonen

Nr.	Zone	Fläche [m²]	Anteil [%]
1			
2			
3			
4			
5			
6			
☐	weitere Zonen in Anlage		

Erläuterungen zum Berechnungsverfahren

Das verwendete Berechnungsverfahren ist durch die Energieeinsparverordnung vorgegeben. Insbesondere wegen standardisierter Randbedingungen erlauben die angegebenen Werte keine Rückschlüsse auf den tatsächlichen Energieverbrauch. Die ausgewiesenen Bedarfswerte sind spezifische Werte nach der EnEV pro Quadratmeter Nettogrundfläche. Die oben als EnEV-Anforderungswert bezeichneten Anforderungen der EnEV sind nur im Falle des Neubaus und der Modernisierung nach § 9 Abs. 1 EnEV bindend.

Figure 30.1 Demand-based energy certificate (excerpt).

	Heating and WW (kWh $m^{-2}{}_{NFS}$ a^{-1})	Building category
Lowest reference value	85	Department stores, warehouses, shopping centers of more than 2000 m^2
Highest reference value	1100	Buildings for outdoor swimming pools, including outside facilities
Mean	180	—
	Electricity (kWh $m^{-2}{}_{NFS}$ a^{-1})	Building category
Lowest reference value	15	Schools, public safety buildings
Highest reference value	410	Food store up to 2000 m^2
Mean	75	—
	Heat + electricity (kWh $m^{-2}{}_{NFS}$ a^{-1})	Building category
Lowest reference value	110	Public safety buildings
Highest reference value	1380	Buildings for outdoor swimming pools, including outside facilities
Mean	255	—

Examples for reference values for heating/hot water, electricity, and averages based on the bulletin dated 6 July 2007 (averages of all building categories).

The BMVBS bulletin requires that for the creation of consumption-based energy certificates, the energy consumption for heating, hot water, cooling, ventilation, and installed lighting must be determined and indicated in kilowatt hours per year and per square meter of net floor space (NFS). According to the bulletin, the consumption determined must be split into a characteristic value for heat energy consumption and a characteristic value for electric energy consumption.

The characteristic value for heat energy consumption must include the consumption of energy, adjusted for weather, for heating (even if electricity is used for heating) and, for associated systems, the proportion of energy consumed for the central hot water supply system.

The characteristic value for electricity consumption must include at least the proportions of electric power consumed for cooling, ventilation, installed lighting, and auxiliary electrical energy for heating and central water heating.

In the consumption-based energy certificate, the *energy consumption index* determined for a building must be juxtaposed with the *reference value* appropriate to the type of usage.

These general rules for the creation of consumption-based energy certificates were valid according to the July 2007 edition of the bulletin, which was in force until September 2009, for all building categories listed therein. However, this edition of the bulletin did not include the building category "Laboratory Buildings" or the type of usage as a "Laboratory" so that no reference values for laboratory buildings could be taken from it. Consequently, up to that point, it was not possible to prepare specific consumption-based energy certificates for laboratory buildings on the basis of the above-mentioned bulletin.

Because it is well-known that laboratory buildings are among the buildings that are generally characterized by high energy consumption, the BBR[4] intended to plug this gap in the course of the revision of the bulletin for the 2009 EnEV by including the building category "Laboratory Buildings."

30.3 Special Energy Characteristics of Laboratory Buildings

On account of the work that is conducted in chemical-pharmaceutical laboratories, there is a high level of risk for employees who work there, among other things on account of toxic substances that can be released into the air. To avoid exposing the people who work in these buildings to these dangers, it is therefore necessary to adopt a series of air safety measures in laboratories. For example, to prevent the accumulation of (minor) undetected releases of gas, an air exchange rate is necessary which, according to the "Guidelines for Laboratories"[5] should be at least $25\,m^3\,m^{-2}\,h^{-1}$ in research laboratories.

That means that the ventilation system for laboratories must generally be designed and operated so that an air exchange rate of at least eight air exchanges per hour can be achieved for an effective ceiling height of approximately 3 m.

In chemical-pharmaceutical laboratories, the actual air exchange rate is frequently many times higher than eight, because in addition to the general indoor ventilation requirements, the (exhaust) requirements of the many exhaust systems required (for reasons of safety) in a laboratory must also be taken into consideration. These systems include, among other things, laboratory exhaust hoods, safety workbenches, exhaust workstations, hazardous materials enclosures, and so on, the total exhaust requirements for which, in each laboratory, can require a significantly higher rate of air exchange (Figure 30.2).

In addition to the ventilation requirements for safety reasons, additional requirements can lead to a demand for increased exhaust capacity – such as the removal of thermal loads, for example, from the electrical heat losses of large analysis instruments.

As a result, the energy consumption for the ventilation of laboratory buildings must be significantly higher than is generally the case for residential or office buildings. In laboratory buildings, therefore, significantly higher energy consumption frequently occurs, which is due less to the building than to the user devices in the building. In the chemical-pharmaceutical industry, for example, laboratory buildings are operated in which the consumption levels caused by the ventilation and exhaust systems account for 80–90% of the building's total energy consumption.

4) German Federal Department of Construction and Regional Planning.
5) BGR 120 Rules for Safety and Health Protection "Guidelines for Laboratories" August 2008 and TRGS 526 "Laboratories," February 2008 edition.

Figure 30.2 View into the mechanical equipment control room of the laboratory building.

The goal of the EnEV is to identify potential energy savings and to promote the adoption of any measures that may be necessary to improve the energy efficiency of buildings. On account of the high proportion of energy consumption required for ventilation and exhaust in laboratories, it is obvious that the very first energy-saving measures for this category of buildings must relate to the ventilation systems. The measures that are otherwise common in residential and office buildings, such as insulation of the exterior, for example, have only relatively little effect in laboratory buildings. The use of recirculation systems in laboratories is also normally out of the question for safety reasons. The measures to increase the energy efficiency of laboratory buildings must therefore be concentrated primarily on the ventilation systems. Measures of this type can include both the optimized design of the building to match the intended use as well as the sizing and operational management of the ventilation system. Nevertheless, usage-related measures to reduce the air exchange requirement can also contribute to reductions in energy consumption. However, these measures to optimize energy requirements also relate to the processes for which the laboratory itself is operated.

The characteristic values of the ventilation and exhaust systems in laboratories are the decisive control variables to ensure a proper balance between laboratory safety and energy efficiency.

Consideration must thereby be given to the fact that the type of operational management of the ventilation system of a laboratory building must always be determined on the basis of the job safety and laboratory process safety requirements. Measures to improve energy efficiency that relate to the operational management of the existing ventilation system of the laboratory building are therefore possible only in connection with detailed hazard analyses for the work to be carried out in the laboratories.

30.4 Reference Values for the Energy Consumption of Laboratory Buildings

To determine reference values, corresponding consumption data from the 3 years, 2005–2007 for a total of 85 laboratory buildings were evaluated by the "Energy

Efficiency of Laboratory Buildings" working group which is part of the "Benchmarking – Chemistry, Pharma, and Life Science" task force.

30.5 Energy Consumption Values

The EnEV requires that the characteristic energy consumption values be measured separately for heating energy and electricity. If the building is heated by electricity, the electric power required for heating is not counted in the electricity consumption but in the heating energy consumption. The energy consumption for ventilation, cooling, and lighting, as well as the auxiliary energy for heating, hot water preparation, and for auxiliary electric heaters in HVAC systems must be included in the electricity consumption. The division into characteristic consumption values specified by the EnEV for heating and electric current is not a simple matter with reference to the ventilation system in laboratory buildings, in particular because the ventilation system not only performs the exhaust and ventilation functions but generally also provides heat. The treatment of the air used for ventilation in laboratories generally also includes cooling and humidification. The relatively high treatment intensity of the incoming air is necessary not only for reasons of comfort but is also vital to the safety and reliability of the processes such as product protection, shelf life, and so on.

In the determination of the proportion of energy for the air-conditioning system, note must also be taken of the fact that in companies, the air-conditioning is supplied partly by distributed compressor units and partly by district cooling. To guarantee compliance with the EnEV for the latter case, an average COP value[6)] of 2.5 is converted into electrical energy for the determination of the reference values of the cooling energy required.

The values for the energy consumption must be indicated in accordance with the EnEV in kilowatt hours per square meter of NFS. To achieve the greatest possible correspondence in the determination of the reference values for the determination of characteristic energy consumption values according to the EnEV, the consumption data of the past 3 years was used for the reference values and an adjustment for weather was incorporated as a function of the location of the respective laboratory building.

30.6 Reference Quantities

In accordance with the EnEV and the "Bulletin of Rules for Characteristic Energy Consumption Values and Reference Values in Non-Residential Buildings", in the

6) COP value: Coefficient of Performance, ratio of useful energy (in this case: cold) to the quantity of energy used (in this case: electrical energy).

edition dated July 2007, the characteristic energy consumption values must be stated in proportion to the "energy reference area." Section 19 of the EnEV which stipulates that the energy reference area for nonresidential buildings is the NFS as defined by DIN 277.

If the usage to which a building is put is such that a clear relationship (proportionality) between the floor space in the building and the energy consumption exists, a simple reference to the NFS, in particular considering the cost of measuring all the floor space in the building, represents an altogether justifiable simplification. To that extent, the NFS can be used as a reference value for the energy assessment of residential and office buildings.

However, there are buildings in which such a "one-dimensional" reference to floor space is inappropriate for an energy assessment on account of the special use to which the building is put.

This becomes clear, for example, if we consider the energy reference values for the building category of indoor swimming pools. According to the currently valid bulletin from the BMVBS dated July 2009, the reference values indicated for heat, at 425 kWh m^{-2} of NFS, and electricity, at 155 kWh m^{-2} of NFS, are several times higher than the average of reference values for all other building categories. It is therefore reasonable to suspect that on account of the specific type of use of the building for swimming, the buildings available must be evaluated from an energy point of view in a manner that differs from that used for other buildings. Obviously, the energy consumption related to square meters of NFS is inadequate to assess the energy efficiency of an indoor swimming pool.

Reference is made to this general problem in VDI 3807,[7] for example, which states that "... *In certain cases, it may be appropriate to use other reference variables.*" In particular, VDI 3807 recommends that the reference variable used for indoor and outdoor swimming pools not be the NFS of the building, but the surface area of the water in the heated pools. A similar approach is appropriate for buildings used for industrial production. For example, in an article by Voss on energy-optimized construction in the technical Journal HLH dated July 2007, we find the following comment: "*In buildings used for production, the use of net floor space as the single reference criterion does not achieve the desired result on account of the higher ceiling heights required by the use of the structure.*[8]"

In general, it can be stated that for all types of buildings in which the processes carried out inside the buildings are the deciding factor in the level of energy consumption, a reference solely to the number of square meters of floor space is inadequate for an assessment of the energy efficiency of the building. The above statement is also true for laboratory buildings. In particular, in laboratory buildings where there are strict requirements for health protection, for the purpose of achieving potential improvements in energy efficiency, it is inappropriate to

7) VDI 3807, Sheet 1 Characteristic energy, and water consumption values for buildings – Basic principles (2007-03).

8) HLH Lüftung/Klima – Heizung/Sanitär – Gebäudetechnik, 58 (2007) No. 7 – July, p. 25, publ.: VDI Verein Deutscher Ingenieure.

use the NFS, as required in the 2007 EnEV, as the sole reference value for energy consumption.

The most important factor in energy consumption in a laboratory building is the quantity of air exchanged. This fact, which is known from experience to every laboratory operator, became clear even in the earliest assessments conducted by the task force. For example, the significance of the information derived from these statistically determined values for energy consumption differed very significantly from the reference values used. If the energy values for buildings are based on the number of square meters, the result of the calculation is no longer significant on account of the scattering of the results for buildings with a relatively high air exchange rate (ACR). The situation is different if the annual exhaust volume is used as a reference value (Figure 30.3).

For the evaluation of the energy efficiency of laboratory buildings, it is, therefore, first necessary to measure the annual exhaust volume and to use it as a reference value for energy consumption. This measurement was conducted in the framework of the task force study both by measuring the absolute quantity of air as well as the relative values in the form of the average air exchange rate.

Because no suitable instrumentation was available for measurement of the actual annual exhaust volume, an approximate value was formed for each building

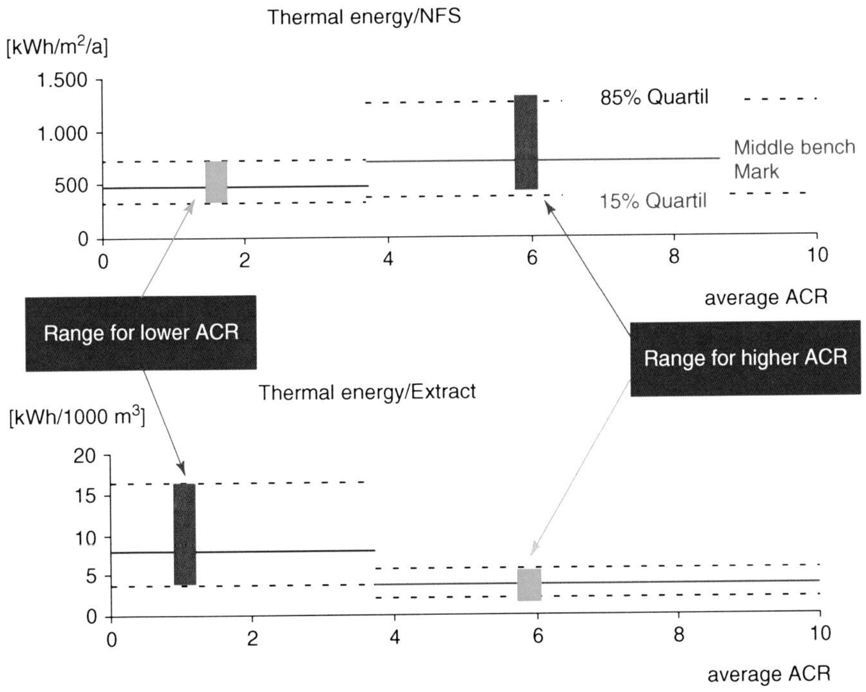

Figure 30.3 Spread of Energy Consuption depending on reference figure.

as follows:

$$V_{exh.} = L_{Nom.} \cdot h_{FL}$$

with

$V_{exh.}$	Annual exhaust volume (m^3/yr)
$L_{Nom.}$	Sum of the nominal outputs of all ventilation systems (m^3/h)
h_{FL}	Sum of the full-load operating hours of the HVAC systems (h/yr).

For the determination of the number of full-load operating hours, partial-load operating hours of the HVAC systems were converted into full-load operating hours. In addition to the annual exhaust volume, the average air exchange rate can also be used to characterize laboratory buildings. To determine the precise average air exchange rate for a specific building, it would have been necessary on account of the different air exchange rates of individual laboratories to measure all the operating modes of the ventilation and exhaust systems, broken down into days of the week and time of day for each individual room and each system over the entire year. Because that is impossible, an approximate value was determined for the average air exchange rate:

$$\mathrm{LWR_m} = V_{exh.}/(\mathrm{NFS} \cdot \mathrm{GH_m} \cdot h_{Nom.})$$

with

$\mathrm{LWR_m}$	Average air exchange rate of the building (1/h)
$V_{exh.}$	Annual exhaust volume (m^3/yr)
NFS	Floor space as defined by DIN 277 (m^2)
$\mathrm{GH_m}$	Assumed average ceiling height (m) (in this case = 3.5 m)
$h_{Nom.}$	Number of nominal hours a year (h/yr) (= 8760)

The graph in Figure 30.3 shows clearly that laboratory buildings with a low air exchange rate (low to moderate LWR) have energy conditions that are different from laboratory buildings with a higher air exchange rate. That relates to both the absolute values of energy consumption per unit of reference area as well as to the scatter around the respective average, which differs sharply in the two cases. The latter is a clear indication that the significance of the information that can be derived from the measurements is strongly dependent on the selection of the reference value for the energy assessment of laboratory buildings.

The reference value "NFS" is obviously not suited for the stricter requirements for air exchange in laboratory buildings. The use of exhaust volume in the higher ranges of air exchange rates, on the other hand, delivers good results.

Significantly, different ranges of energy consumption as a function of the reference value are used.

30.7 Groups with Homogeneous Characteristics

The laboratory buildings included in the investigation varied in size between 842 and 42 505 m^2 of NFS. The corresponding number of full-load hours of operation varied between 2753 and 8760 h. The lower and upper scatter values of the average air exchange rates were 1.6 and 24.6 $m^3\ m^{-2}$ of NFS, respectively.

The purpose of additional investigations was to determine the extent to which a variation of these parameters influences the energy consumption of laboratory buildings.

To be able to more accurately assess the degree of dependence of a measurement variable on a factor from statistically determined values, it is appropriate to divide the objects being investigated into groups (clusters) so that it becomes apparent whether the values of the measured variable differ significantly between the different groups.

For this purpose, the laboratory buildings were investigated with respect to:

- the size of the building in square meters of NFS,
- the number of full-load hours of operation (as a measurement of intensity of use),
- the presence of a heat recovery system, and
- the average air exchange rate

combined into suitable clusters with uniform properties.

If, in a cluster, the energy consumption values do not differ significantly between the clusters, it can be assumed that the factor on which the cluster is based is an inappropriate reference value for the determination of characteristics of energy consumption. On the other hand, a factor appears to be appropriate for the determination of characteristics of the energy consumption if the energy consumption values differ significantly between the clustered groups.

The results of our investigations have clearly shown that an accurate energy assessment of laboratory buildings is not a question of their division into different types of laboratories (chemical, microbiological, analytical, etc.). Rather, it was determined that it is very easily possible to carry out an energy assessment of laboratory buildings solely on the basis of the average rate of air exchange (averaged over the entire building). Figure 30.4 shows the distribution of the buildings investigated in terms of the intensity of the air exchange.

Because energy consumption, air exchange, and laboratory safety must always be considered in terms of their mutual interactions, the clustering by air exchange rates is also in line with the safety requirements, which are of decisive importance in laboratories for the specific volume of exhaust per square meter of laboratory floor space.

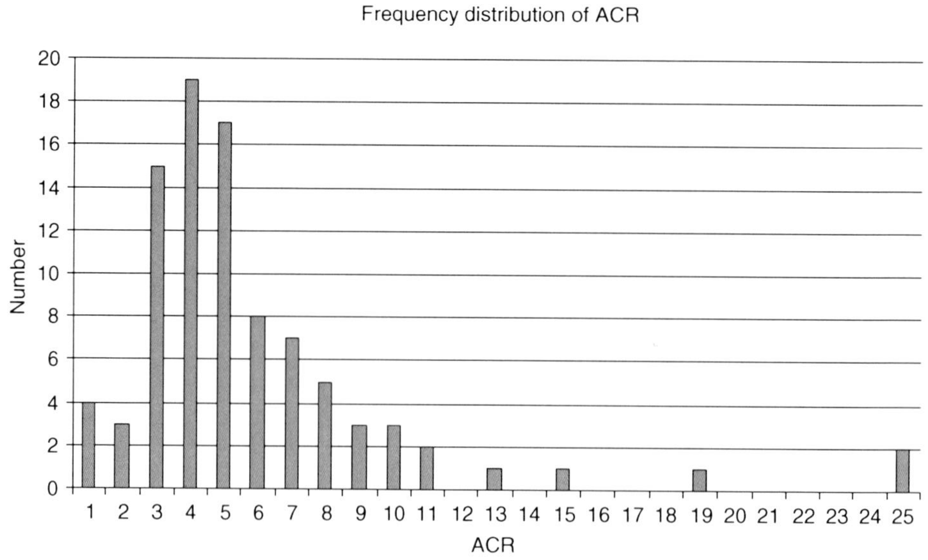

Figure 30.4 Distribution of the laboratories investigated by air exchange rate LWR.

30.8 Conclusions from the Results of the Investigations

From the usage characteristics of laboratory buildings and the intimate interrelationship between energy consumption and laboratory safety, it was determined that the energy assessment of laboratory buildings by means of consumption-based energy certification as required by the EnEV cannot be conducted in the same manner as for residential and office buildings, for example.

It has been clearly shown that reference solely to energy consumption values based on floor space in laboratory buildings is inappropriate in terms of building energy optimization. It has been determined to be significantly more appropriate to reference energy consumption to the annual quantity of exhaust air.

One problem with these results is that they are not in compliance with the requirements of the EnEV, because the energy values that must be used in the issue of energy certificates must be referenced to the NFS of the building. However, this problem was easily and effectively solved by demonstrating a linear relationship between the energy consumption reference to NFS on one hand and the average air exchange rate on the other hand, provided that the laboratory buildings investigated have an average air exchange rate greater than 2.5.

By means of a regression analysis with a suitable approach, linear relationships were established for the reference values based on NFS as a function of the air exchange rate. And that was true for both heat and electricity, and the functional relationships for heat and electricity presented in Figure 30.5 were calculated.

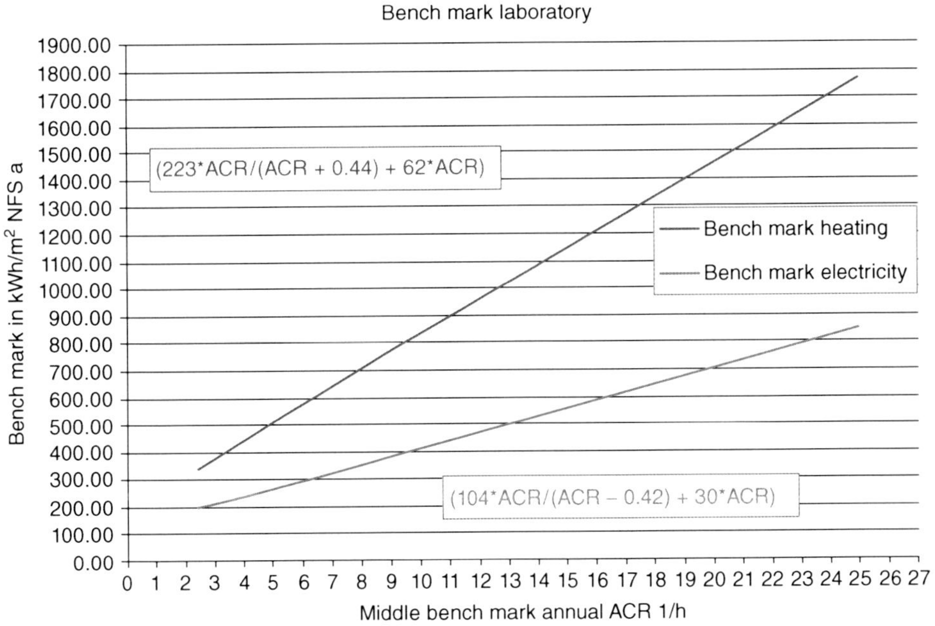

Figure 30.5 Energy consumption as a function of average air exchange rate.

The relationships determined in this manner were reflected in the revised text of the "Bulletin of Rules for Characteristic Energy Consumption Values and Reference Values in Non-Residential Buildings" dated 30 July 2009.

With the inclusion of the category "Laboratory buildings" in No. 4 of the table in the July 2009 bulletin, for the first time, constant numbers were not indicated for the reference values of heat and electricity for a building category. In place of absolute values, calculation formulas for laboratory buildings are indicated, which can be used first to determine the reference values to be used for heat and electricity for comparison with the consumption values, as a function of the specific energy situation of the building. On the basis of these calculations, the reference values "Heating and hot water" and "Electricity" that were included in the new July 2009 edition of the bulletin[9] for laboratories for private use are calculated by means of the two steps listed below:

1) *Step 1*: Determination of the average annual air exchange rate from the nominal volume flow (exhaust) of the building as well as full-load operating hours of the system in question, the NFS in square meters and ceiling height in meters.

$$n_m = \frac{\Sigma(\dot{V}_{nenn.t} \cdot t_{VL})}{A_{NGF} \cdot h_{Geschoss} \cdot 8760h/a}$$

9) Bulletin of Rules for Characteristic Energy Consumption Values and Reference Values in Non-Residential Buildings dated 30 July 2009, Chapter 7, Section 7.4.

with

n_m	Average annual air exchange rate of the building in per hour
V_{nom}	Nominal volume flow (exhaust) of the respective ventilation system in cubic meterper h
t_{vl}	Annual full-load hours of operation of the respective systems in hour per year
$h_{CEILING}$	Average ceiling height in meter
A_{NFG}	Net floor space in square meter.

If an average air exchange rate of less than 2.5/h is calculated, the calculation must be repeated with a value of 2.5

2) *Step 2*: Calculation of the reference values from the average annual air exchange rate:
   a. for heating and hot water

$$e_{Vergl.h} = \frac{223 \cdot n_m}{0.44 + n_m} + 62 \cdot n_m$$

   and
   b. for electricity

$$e_{Vergl.s} = \frac{104 \cdot n_m}{n_m - 0.42} + 30 \cdot n_m$$

   each in kWh per square meter of NFS and per year.

The July 2009 bulletin also requires that the reference values determined according to the above formulae each be reduced by 15% when the 2009 EnEV comes into force. Therefore, the calculated values must be multiplied by 0.85 for all consumption-based energy certificates issued after 30 September 2009.

30.9 Example for the Issue of a Consumption-Based Energy Certificate for a Laboratory Building

Basic information on the building used as an example:

Net floor space, laboratory building.	2600 (m^2 net floor space)
Average full-load hours of operation per day	12 (h/d)
Sum of the nominal outputs of the ventilation systems	105 000 (m^3/h)
Average ceiling height	3.5 (m)

Measured consumption levels:

Annual electric power consumption	463 911 (kWh a^{-1})
Annual heating energy (adjusted for climate)	1 646 710 (kWh a^{-1})
Annual cooling energy (electric)	37 363 (kWh a^{-1})

Reference and consumption values according to the 2009 EnEV for the building used as an example:

Sum of annual full-load hours

$\sum t_{VL} = 12(h/d) \cdot 365(d/a) = 4380(h/a)$

Sum of the nominal volume flows of the ventilation systems

$\sum V_{nom.} = 105\,000(m^3/h)$

Allocatable annual exhaust

$\sum t_{VL} \cdot \sum V_{nom.} = 4380(h/a) \cdot 105\,000(m^3/h) = 459\,900\,000(m^3/a)$

Average annual air exchange of the building

$n_m = 459\,900\,000(m^3/a)/(2600(m^2) \cdot 3.5(m) \cdot 8760(h/a)) = 5.77(1/h)$

Reference value, heating:

$e_{ref.h} = [223 \cdot 5.77/(5.77 + 0.44)] + [62 \cdot 5.77] = 564.94(kWh/m^2 \ NFS/a)$

Reference value, electricity

$e_{ref.el.} = [104 \cdot 5.77/(5.77 - 0.42)] + [30 \cdot 5.77] = 285.26(kWh/m^2 \ NFS/a)$

Consumption value, heating:

$e_{cons.h} = 1\,646\,710(kWh/a)/2600(m^2) = 633.35(kWh/m^2 \ NFS/a)$

Consumption value, electricity:

$e_{cons.el.} = (463\,911(kWh/a) + 37\,363(kWh/a))/2600(m^2) = 192.80(kWh/m^2 \ NFS/a)$

Presentation of the results in the consumption-based energy certificate in accordance with 2009 Electric Power Saving Ordinance:

Presentation of characteristic values in consumption-based energy certificate.

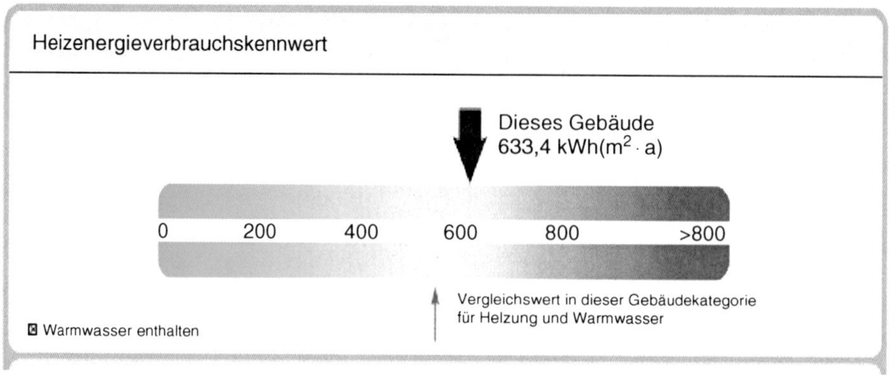

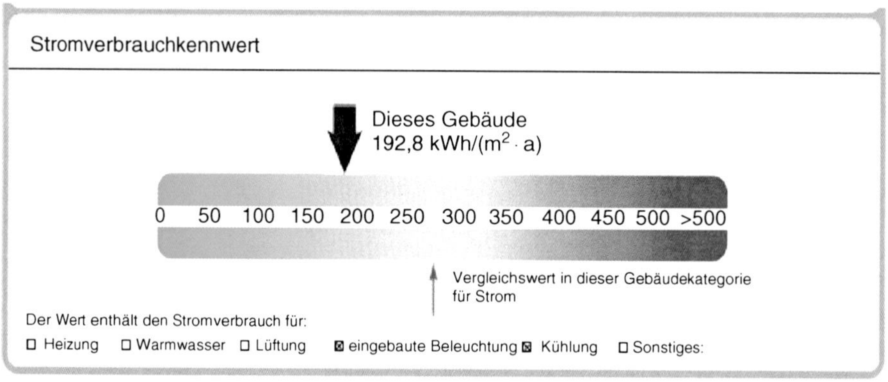

30.10 Summary

The investigations conducted by the "Benchmarking – Chemistry, Pharma, and Life Science" task force have shown that the method generally used for the issue of consumption-based energy certificates, for the comparison of the actual fixed consumption values of the respective type of building, is not suitable for the characterization of the energy efficiency of laboratories. The most important reason for this unsuitability is that in laboratory buildings, the energy situation cannot be determined simply on the basis of the characteristics of the structure of the building and the technical equipment. The requirements for process and job safety in laboratory buildings on account of the processes for which they are used are determining factors in the amount of energy required and consumed. In laboratory buildings, these requirements are primarily the result of the operation of ventilation and extract systems. The predominant portion of the energy consumption of a laboratory building is consequently a function of the purpose for which the building is used. These special requirements for laboratories must be taken into consideration in the issue of consumption-based energy certificates.

With the findings gained as a result of extensive analyses concerning the energy efficiency of ventilation conditions and the resulting terms for the calculation of the building-related reference values, it is now possible to perform the required comparison of target values and actual values for the consumption-based energy certificate for existing laboratory buildings, so that realistic assessments of the individual energy situation of a building can be made and suitable conclusions can be drawn with reference to any energy optimization measures that may be necessary.

Part V
Standards and Test Regulations

Egbert Dittrich

Everybody involved with buying process as well as user and safety managers have a considerable interest in the guarantee of the properties of all kind of laboratory equipment committed by the supplier. It must be ensured that safety-relevant products by means of a permanent functional test prove the serviceability. This is the original purpose of the global standards and regulations. The following important regulations are presented with the high impact of laboratory sections.

The Sustainable Laboratory Handbook: Design, Equipment, Operation, First Edition.
Edited by Egbert Dittrich.

31
Legislation and Standards

Burkhard Winter

31.1
Introduction

The number of government regulations that are aimed directly and exclusively to laboratories and laboratory-related facilities is still manageable. Examples of such "tailor-made" rules and regulations are the Technical Rule 526 "*Laboratories*" [1] in Germany and in the Netherlands "*Richtlijn voor de arbocatalogus – ArboInformatie 18 – Laboratoria*" [2]. Unfortunately, there is also an almost infinite number of state and under state rules with such a general field of application that these rules work and are effective for laboratories or in laboratories as well – sometimes just in parts. Examples include the *Energy Performance of Buildings Directive* (EPBD, on the European level [3]) or the *Workplace Ordinance* and the *Ordinance on Industrial Safety* on the national German level [4]. Examples of rules matching only some kind of laboratories are the *Directive 2000/54/EC on the protection of workers from risks related to exposure to biological agents at work* [5] and rules applying in part are the *Code du Travail* [6] (in France) and RITE [7] in Spain.

In order to maintain an overview in the thicket of regulations, it makes sense to sort rules and standards for "addressee":

- in Clause 2, regulations for energy efficiency are considered,
- and Clause 3 discusses the regulations for labor safety and occupational health in laboratories.

Such state regulations mainly addressed to laboratory device manufacturers will be only briefly mentioned. Examples are the *Machinery Directive* [8] (implemented in Germany by the ProdSG [9]) and the *IVD Directive on in vitro diagnostic medical devices*; both policies are more important for device manufacturers and for the organization operating a laboratory than for the laboratory planner.

At first glance, the number of European and national standards for laboratories appear even greater than the number of state and government regulations. But, most standards are dealing with analytical determination procedures that give the

The Sustainable Laboratory Handbook: Design, Equipment, Operation, First Edition.
Edited by Egbert Dittrich.

users in a variety of laboratories assistance, and they are not eligible for the scope of this handbook. Therefore, the above already introduced division also helps on here:

- standards for laboratory planning and equipment,
- standards for laboratory use (mostly analytical methods)
- standards for manufacturers of laboratory devices and equipment.

Before the rules are considered in detail, according to the above-described structure, it needs to define three terms:

- Government regulations are European directives or national laws, regulations and technical rules, for example, the *Directive concerning the minimum safety and health requirements for the workplace* (89/654/EEC) [10] transferred to national law in Germany by the *Workplaces Ordinance* [4] or the *Code du Travail* [6] in France;
- Under state regulations are very often documents of national Social Accident Insurance Organizations such as INRS in France, NHL in the United Kingdom, or DGUV in Germany;
- Standards are recommendations that are created by private sector institutions in the consensus of the parties involved; they can apply nationally (DIN or BS), European-wide (EN) or worldwide (ISO and IEC).

This conceptual understanding is also important if regulations and standards are organized into a hierarchy of legal liability.

31.2 Laboratory Planning and Building

31.2.1 General

For the planning and building of laboratories apply the same national laws and regulations as for all other buildings. Very often, the applying rules and regulations are supplemented by European or national standards giving additional technical details and recommendations.

Besides this general building legislation, specifics should be considered already in an early planning stage of the building, which arise from the function of the building or parts of the building as a laboratory or as laboratory-related rooms. Laboratories are technical facilities to gain knowledge and know-how, for example, about products or processes, or to serve the education in natural sciences and technology. Sometimes, laboratories fulfill both tasks, for example, research laboratories in universities where post-docs can be active. It is especially the laboratory infrastructure such as media supply, supply and exhaust air, and laboratory-related rooms (such as for the appropriate storage of chemicals), which need an early consideration when planning the building.

Laboratories are workplaces. Thus, the institutions operating laboratories need to observe regulations such as the European Council Directive 89/654/EEC *concerning the minimum safety and health requirements for the workplace* [10], respectively, the transfer of this directive to national law in the EU member states. This and the related regulations take effect on the workplace design in the lab rooms, on the air ventilation of the whole building, and on the media supply including supply tubing, and need to be considered when planning and erecting the building and before installing the rooms with laboratory equipment.

Unfortunately, only few countries offer a general view or complete compendium of lab planning-related technical protection measures to be applied and observed in the country. For The Netherlands exists the "*Richtlijn voor de arbocatalogus – ArboInformatie 18 – Laboratoria*" [2] and for Germany the "*Guidelines for the Laboratory*" BGI/GUV-I 850-0 [11] where planning-related protection measures are treated in clause 6 of the guideline.

A planning aid, especially for the complicated air ventilation of lab buildings, lab rooms, and lab-related rooms is at the time being only available in Germany (DIN 1946-7 [12]). An appropriate European-wide document is under consideration and development in the Technical Committee CEN/TC 332 "Laboratory equipment." The same Technical Committee has already published the European standard EN 14056 [13] giving recommendations for the arrangement and installation of lab furniture such as benches, fume cupboards, and microbiological safety cabinets.

31.2.2 Regulations for Energy Efficiency

31.2.2.1 Legislative Requirements

In respect of planning and erection of sustainable laboratories, the European Directive for *Energy Performance of Buildings* (EPBD [3]) seems to be the most important regulation at first glance. All EU member states have transferred this directive to national legislation, for example, in Germany by the *Energy Savings Law* (EnEG [14]) and the *Energy Savings Ordinance* (EnEV [15]). But there are two reasons why the sustainability of lab buildings is very incompletely covered by the EPBD and its national conversions which, however, must be observed.

One reason is that the EPBD applies for all newly constructed buildings in the European Union as far as they are heated or cooled (conditioned) by means of energy; in addition, the EPBD applies for new building's equipment serving heating, cooling, ventilation, and illumination. Moreover, the directive applies for all modifications, enlargements, and upgrading of existing buildings and their technical equipment if the useful floor area is larger than 1000 m^2. According to EPBD, a classification of buildings in appropriate groups shall be done in national responsibility of the EU member states, and in most countries, lab buildings belong to the group of non-residential buildings. But lab buildings are only a very minor part of the non-residential buildings because there are much more office buildings, shop buildings, schools, hotels, parking buildings, and many others than lab buildings.

The second reason is that the EPBD focuses very much on the energy loss via the building's walls, which may be appropriate for most non-residential buildings but not for laboratories. In laboratory buildings, there are two other parameters which have a much higher influence on the energy balance than the outer building walls. One parameter is the processes operated in laboratories which often need much energy for heating and cooling. The second parameter is the indispensable air exchange needed in laboratory rooms to satisfy (legislative) requirements of labor safety and occupational health (for details see Clause 3). Very often, this air exchange can only be performed by "forced ventilation" with circumferential technical ventilation equipment and supply and exhaust air devices. The energetic optimization of these two parameters usually contribute much more to the sustainability of a lab building than the reduction of energy loss via the building's walls as stipulated by the EPBD, which more easily can be reached by use of today's construction materials and state-of-the-art architectural technology.

Putting the EPBD into practice is far from trivial and needs technical know-how in building construction engineering, building material sciences, and in circumferential calculation methods for all kinds of energy. Thus, the project group CEN/TC 371 and related Technical Committees have developed a large set of EN standards to give the user guidance how to apply and interpret the EPBD and to perform necessary calculations. The umbrella document CEN/TR 15615 [16] explains the general relationship between various European standards and the EPBD. The Technical University of Denmark maintains on the internet an information center with many valuable documents to be found with *www.iee-cense.eu*.

Finally, it needs to be mentioned that the EU policies intend to erect only "climatic neutral" buildings starting from 2018. This would mean that new buildings gain as much energy from regenerative sources as they "consume" for heating, cooling, conditioning, and illumination. Thus, laboratory planners and institutions erecting labs should be prepared for major changes in the EU and national legislation and the related standardization during the next years.

31.2.2.2 **Voluntary Certification**

Many builder-owners of lab buildings and institutions operating laboratories use, when erecting laboratories, materials and technologies exceeding legislative and normative requirements. It is often considered whether the higher investment could be honored by more than only the decreasing operating cost, for example, by energy savings.[1)]

Private certification institutions have realized this demand and they offer the certification of sustainability or environmental efficiency after evaluation with a very circumferential set of rules and requirements. Usually, certificates are issued in "Bronze," "Silver," "Gold," or "Platinum" depending on the requirements met by the building during the evaluation. To date, three building certification systems are established on the market: LEED (Leadership in Energy and Environmental

1) Comment Dittrich: The German government has implemented BNB (Bewertungssystem Nachhaltiges Bauen). This is more or less mandatory if the builder wants to raise public funds.

Design), BREEAM, and DGNB (Deutsche Gesellschaft für Nachhaltiges Bauen e.V.). The three systems consider the special situation of lab buildings more or less in different extent and way. All the three systems have much more experience when certifying office, administrative, or bank buildings, but LEED and BREEAM already have certified a small number of research buildings.

The internationally widest spread certification system LEED was established in 1995, in Washington by the US Green Building Council (USGBC) with the intention to build single buildings and even assemblies of buildings with environmental and social responsible care. Today, there are LEED certifications in Europe and Asia, and there are European consultancy and engineering offices offering support and service when certifying with LEED. Information on the Internet can be found with *www.usgbc.org*, among others a case study for a research building with the Lasry Center for Bioscience at the Clark University in Massachusetts. This certification was performed in the stage "Gold." A certification in "Platinum" was granted to the Arizona State University for its Biodesign Institute – Building B.

The oldest certification system for ecological and social-cultural building's planning, erection, and enlargement was established in the Great Britain in 1990, see *www.breeam.org*. BREEAM indicates more than 110.000 certified buildings in the whole world and has specific certification "schemes" for training schools and for hospitals. Laboratory buildings can be certified in accordance with the certification scheme BREEAM Bespoke. The "Veterinary Laboratories Agency" initiated the certification of its "The Mills Building" in Weybridge in 2006 and gained a BREEAM Award for the realization of an especially ecological concept.

The German federal ministry for traffic, construction engineering, and city development set up in 2009 the "Deutsches Gütesiegel Nachhaltiges Bauen" (German Quality Certificate for Sustainable Construction, DGNB) for a holistic evaluation of buildings. The evaluation is performed at the time being by the "DGNB" (*www.dgnb.de*) and by the "Österreichische Gesellschaft für Nachhaltiges Bauen." The certificate is mainly known in German-speaking countries, even when the "Austrian-House" at the Olympic Winter Games in Vancouver/Canada gained a pre-certificate in "Silver." DGNB certifications require with German thoroughness an analysis of the environmental impact of the materials used for the building and its technical installations over the complete life cycle including overhauling and modernization works. EN ISO 14040 [17] and EN ISO 14044 [18] may be applied for these purposes. DGNB Working Groups develop with support by EGNATON (European Society for Sustainable Laboratory Technologies, see *www.egnaton.com*) special certification rules and criteria for lab buildings.

This article is not the right place to introduce the exhaustive and circumferential sets of rules used by the three certification systems for evaluation of sustainability; instead, we would like to refer to the internet addresses already mentioned earlier when introducing the organizations. But, it should be noted which additional cost may be expected when intending such a certification: the DGNB indicates about 2 and 5% of the investment cost for the building. It is recommendable to involve the certifying organization as early as possible in the planning process of the lab building. Thus, different concepts and layouts can be evaluated early for sustainability criteria, and the circumferential documentation for the certification can be incorporated and embedded in the planning process.

31.3 Regulations for Labor Safety and Occupational Health

In most countries, there is a whole bundle of state and under state regulations for safety and health at the workplace which gear and complement into each other. Although most of these regulations are addressed toward the company or organization, which operates the laboratory, many requirements regarding technical provisions must be considered early during the planning stage and effect the erection of the lab buildings and the installation of lab rooms, for

Table 31.1 Example for the transfer of European directives to national legislation.

EU directive	Designation	German law or regulation	Actual version [a)]
Directive on the energy performance of buildings (EPBD)	2002/91/EC	Energieeinsparungsgesetz (EnEG)	7/2013
		Energieeinsparverordnung (EnEV)	11/2013
Minimum safety and health requirements for the workplace	89/654/EEC	Arbeitsstättenverordnung (ArbStättV)	7/2010
Safety and/or health signs at work	92/58/EEC	Arbeitsstättenverordnung (ArbStättV)	7/2010
Protection of workers from risks related to exposure to biological agents at work	2000/54/EC	Biostoffverordnung (BioStoffV)	7/2013
Machinery directive	2006/42/EC	Gesetz über die Bereitstellung von Produkten auf dem Markt (Produktsicherheitsgesetz – ProdSG)	11/2011
Low voltage directive	2006/95/EC	Gesetz über die Bereitstellung von Produkten auf dem Markt (Produktsicherheitsgesetz – ProdSG)	11/2011
ATEX directive	94/9/EG	Betriebssicherheitsverordnung (BetrSichV)	11/2011
Classification, packaging, and labeling of dangerous preparations	1999/45/EG	Gefahrstoffverordnung (GefStoffV)	7/2013

a) Stage September 2014.
Sources see Table 31.2.

example, the layout of emergency routes and exits. In many countries, there are no "tailor-made" regulations for laboratories, but general regulations for safety at the workplace and for occupational health of employees apply for laboratories as well.

To date, there is no harmonization of the occupational health legislation in Europe. The European Union has issued some directives in this field (see Table 31.1), but these directives lay down so-called "minimum requirements" for labor safety and occupational health. The EU member states are obliged to transfer these minimum requirements to national law and they are authorized to keep already existing regulations and to embed the minimum requirements in the existing regulations as far as not already covered, and they also may add additional provisions and requirements if deemed necessary. This means that laboratory planning needs a sound knowledge of the occupational health regulations applying in the country where the lab will be erected.

31.3.1 Minimum Safety and Health Requirements

Minimum safety and health requirements at the workplace are laid down in directive 89/654/EEC [10]. The directive applies for workplaces "used for the first time" (see Annex I of the directive) and for workplaces "already in use" (Annex II) when the directive was published in November 1989. Minimum requirements are specified for the building and its (electrical) installations, for emergency and traffic routes, ventilation, fire detection and fighting, and for different kind of rooms. When hazards cannot be avoided or adequately reduced by technical measures, safety and/or health signs at work shall be provided in accordance with directive 92/58/EEC [19] (amended 2007 and 2014). These signs used in the building and rooms with places of work must comply with the general requirements given in directive 77/576/EEC [20]. Consolidated versions of all directives can be downloaded in many European languages from *www.eur-lex.europe.eu*. For more information sources see Table 31.2.

There are a few European standards issued by CEN/TC 332 which ease the work of lab planners. EN 14056 [13] giving recommendations for the arrangement and installation of lab furniture such as benches, fume cupboards, and microbiological safety cabinets applies in more than 30 European countries and in countries associated with CEN, such as Turkey and Ukraine. For the positioning and installation of fume cupboards, CEN/TS 14175-5 [21] has been published. The positioning of fume cupboards needs special attention, because the function of this very important safety device in laboratories can be affected by disturbing air drafts generated in the laboratory room. Also, the installation of local exhaust devices such as capture devices with articulated extract arm can need early consideration in lab planning, which is described in CEN/TR 16589 [22].

Table 31.2 Sources for legislation and regulations.

Regulation	Internet URL	Publisher
European legislation [a] (directives and regulations)	*http://eur-lex.europa.eu*	European Union, Publication office
National legislation in EU member states (links to national databases)	*http://eur-lex.europa.eu/n-lex/index_en.htm*	European Union, Publication office
Legislation in France	*http://www.legifrance.gouv.fr*	République Francaise, Service public de la diffusion du droit
Legislation in The Netherlands	*https://www.overheid.nl*	Dutch government
Legislation in United Kingdom	*http://www.legislation.gov.uk*	Ministry of Justice of the United Kingdom government
Legislation in Spain	*http://www.boe.es/legislacion*	Gobierno de Espania, Ministerio de la Presidencia
Legislation in Germany	*www.gesetze-im-internet.de*	Federal Ministry of Justice of the German government
Technical rules in Germany (TRBS, TRBA, TRGS)	*www.baua.de*	Bundesanstalt für Arbeitsschutz und Arbeitsmedizin
Germany: BGI/GUV-I 850-0 (in German and English)	*http://bgi850-0.vur.jedermann.de/index.jsp*	BGRCI (*www.bgrci.de*)
Info about EPBD	*http://www.epbd-ca.org/index.cfm?cat=home*	European Commission, Directorate-General for Energy
Info about EPBD	*www.iee-cense.eu*	Technical University of Denmark
Info about sustainable laboratories	*www.egnaton.com*	EGNATON e.V.
Info about European standardization	*www.cen.eu*	CEN; Brussels
Standards (all)	*www.beuth.de*	DIN e.V. (Beuth Verlag)
Standards and info about laboratory planning and installations	*www.fnla.din.de*	DIN Standards Committee on Lab Equipment (FNLa)
Occupational health in France	*www.inrs.fr*	Institut National de la Recherche de Sécurité
Occupational health in United Kingdom	*http://www.hsl.gov.uk*	Health and Safety Laboratory
	http://www.hse.gov.uk	Health and Safety Executive

a) All legislative documents downloadable free of charge in most European languages.

31.3.2 Chemicals and Hazardous Substances Regulations

Very often, laboratory work includes the handling of hazardous substances. Thus, laboratory personnel and the institutions operating laboratories need to observe national regulations restricting and controlling the use of hazardous substances. Usually, these national regulations go (sometimes far) beyond the more general requirements of the European Directive 1999/45/EC *concerning the approximation of the laws, regulations and administrative provisions of the Member States relating to the classification, packaging and labeling of dangerous preparations* [23]. Dangerous substances (preparations) in accordance with Article 2 of this directive are among others toxic, carcinogenic, mutagenic, flammable, and explosive chemicals and chemicals dangerous for the environment. Many of these substances often need to be handled in laboratories (Table 31.3).

An example for the far-beyond-going transfer of this directive to national law is in Germany the *Chemicals Law* (*Chemikaliengesetz*, ChemG) and the *Hazardous*

Table 31.3 EN standards for laboratory installations.

European standard	Edition	Title
EN 12128	1998	Containment levels of microbiology laboratories
EN 12469	2000	Microbiological safety cabinets
EN 13150	2001	Workbenches for laboratories – Safety requirements and test methods
EN 13792	2002	Color coding of taps and valves for use in laboratories
EN 14056	2003	Laboratory equipment – Recommendations for design and installation
EN 14175-1	2003	Fume cupboards – Vocabulary
EN 14175-2	2003	Fume cupboards – Safety and performance requirements
EN 14175-3	2004	Fume cupboards – Type test methods
EN 14175-4	2004	Fume cupboards – On-site test methods
EN 14175-6	2006	Fume cupboards – Variable air volume fume cupboards
EN 14175-7	2012	Fume cupboards – Fume cupboards for high heat and acidic load
EN 14470-1	2004	Fire safety storage cabinets – for flammable liquids
EN 14470-2	2006	Fire safety storage cabinets – for pressurized gas cylinders
EN 14727	2005	Storage units for laboratories
EN 15154-1	2006	Emergency safety showers – Plumbed-in body showers
EN 15154-2	2006	Emergency safety showers – Plumbed-in eye washes
EN 15154-3	2009	Emergency safety showers – Non plumbed-in body showers
EN 15154-4	2009	Emergency safety showers – Non plumbed-in eyewash units
CEN/TS 14175-5	2007	Fume cupboards – Recommendations for installation and maintenance
CEN/TR 16589	2014	Laboratory installations – Capture devices with articulated extract arm

All EN standards are available in English, French, and German (*www.beuth.de*).

Substances Ordinance (*Gefahrstoffverordnung*, GefStoffV, [24]). The law and the ordinance are backed up and concretized by two more regulations which are TRGS 526 [1] and BGI/GUV-I 850-0 [11]. Advantage for the laboratory planner is that the last mentioned document is also available in English language and describes in Clause 6, in a very comprehensive way, the technical measures needed to fulfill the more general requirements of the law and the directive. Another example of national transfer is in The Netherlands *"Richtlijn voor de arbocatalogus – ArboInformatie 18 – Laboratoria"* [2].

To date, the REACH regulation (EC No 1907/2006) does fortunately not seem to effect laboratory planning and laboratory work in practice and daily work, despite that the "use" of the regulated chemical substances and the protection of human health and the environment belong to the scope of this regulation.

31.3.3
Biological Agents and Safety

Laboratories where the lab personnel could be exposed to pathogenic microorganisms must observe EU directive 2000/54 EC *on the protection of workers from risks related to exposure to biological agents at work* [5] respectively, the national legislation transforming this directive. The directive classifies biological agents in four risk groups and laboratories handling microorganisms of a risk group must have the related "containment level." Annex IV of the directive specifies minimum safety measures for each of the four containment levels. The European standard EN 12128 [25] describes more detailed technical devices and equipment necessary for each containment level. Especially, the technical equipment and safety measures including air filtration needed for containment levels 3 and 4 need an early consideration during the planning stage of the lab. A later upgrading of an already existing laboratory is not possible. The directive applies also for genetically modified microorganisms.

References

1. Ausschuss für Gefahrstoffe - AGS-Geschäftsführung - BAuA (2008) Germany: TRGS 526. Technische Regel für Gefahrstoffe 526 — Laboratorien (Technical Regulation for Hazardous Substances No 526 — Laboratories).
2. The Netherlands: Richtlijn voor de arbocatalogus – ArboInformatie 18 – Laboratoria.
3. EPBD (2002) Directive 2002/91/EC of the European Parliament and of the Counsil of 16 December 2002 on the energy performance of buildings. *Official Journal of the European Communities*, **1**, 65–71.
4. Betriebssicherheitsverordnung - BetrSichV (2014). Verordnung über Sicherheit und Gesundheitsschutz bei der Bereitstellung von Arbeitsmitteln und deren Benutzung bei der Arbeit, über Sicherheit beim Betrieb überwachungsbedürftiger Anlagen und über die Organization des betrieblichen Arbeitsschutzes (Germany).
5. EC (2000) 2000/54/EC, Directive 2000/54/EC on the protection of workers from risks related to exposure to biological agents at work. *Official Journal of the European Union*, **L 262**, 21–45.

6. France: Code du Travail par Décret n°2008-244 du 7 Mars 2008.
7. Spain: RITE, Reglamento de Instalaciones Térmicas en los Edificios (approved by REAL DECRETO 1027/2007)
8. EN (2006) 2006/42/EC, Directive 2006/42/EC of the European Parliament and of the Council of 17 May 2006 on machinery, and amending Directive 95/16/EC (recast) (Text with EEA relevance). *Official Journal of the European Union*, **L 157**, 24–85.
9. Produktsicherheitsgesetz – ProdSG (2014). Gesetz über die Bereitstellung von Produkten auf dem Markt (Germany).
10. 89/654/EEC, Council Directive of 30 November 1989 concerning the minimum safety and health requirements for the workplace (first individual directive within the meaning of Article 16 (1) of Directive 89/391/EEC).
11. BG RCI/DGUV (2014) BGI/GUV-I 850-0. Working Safely in Laboratories – Basic Principles and Guidelines (Sicheres Arbeiten in Laboratorien — Grundlagen und Handlungshilfen), Available in English and German, see *http://bgi850-0.vur.jedermann.de/index.jsp* (accessed 28 November 2014) (Germany).
12. Germany: DIN 1946-7, *Raumlufttechnik — Teil 7: Raumlufttechnische Anlagen in Laboratorien*, Available in English and German, see *www.beuth.de* (accessed 28 November 2014).
13. British Standards Institution EN 14056. *Laboratory Furniture – Recommendations for Design and Installation*
14. Energieeinsparungsgesetz – EnEG (2014) Gesetz zur Einsparung von Energie in Gebäuden (Germany).
15. Energieeinsparverordnung – EnEV (2014) Verordnung über energiesparenden Wärmeschutz und energiesparende Anlagentechnik bei Gebäuden (Germany).
16. CEN CEN/TR 15615. *Energy Performance of Buildings – Module M1-x – Accompanying Technical Report on Draft Overarching Standard EPB.*
17. ISO EN ISO 14040:2006 (2006). *Environmental Management – Life Cycle Assessment – Principles and Framework.*
18. ISO EN ISO 14044:2006 (2006). *Environmental Management – Life Cycle Assessment – Requirements and Guidelines.*
19. EU (1992) 92/58/EEC, Council Directive 92/58/EEC of 24 June 1992 on the minimum requirements for the provision of safety and/or health signs at work (ninth individual Directive within the meaning of Article 16 (1) of Directive 89/391/EEC). *Official Journal of the European Communities*, **L245**, 23–42.
20. EU (1977) 77/576/EEC, Council Directive 77/576/EEC of 25 July 1977 on the approximation of the laws, regulations and administrative provisions of the Member States relating to the provision of safety signs at places of work. *Official Journal of the European Communities*, **L229**, 12–21.
21. CEN/TS 14175-5 (2012), *Fume Cupboards – Part 5: Recommendations for Installation and Maintenance.*
22. CEN/TR 16589 (2013), *Laboratory Installations – Capture Devices with Articulated Extract Arm.*
23. EU (1999) 1999/45/EC, Directive 1999/45/EC of the European Parliament and of the Council of 31 May 1999 concerning the approximation of the laws, regulations and administrative provisions of the Member States relating to the classification, packaging and labelling of dangerous preparations. *Official Journal*, **L 200**, 0001–0068.
24. Gefahrstoffverordnung – GefStoffV (2014). Verordnung zur Anpassung der Gefahrstoffverordnung an die EG-Richtlinie 98/24/EG und andere EG-Richtlinien (Verordnung zum Schutz vor Gefahrstoffen) (Germany).
25. EN 12128 (2012). *Biotechnology – Laboratories for Research, Development and Analysis – Containment Levels of Microbiology Laboratories; Areas of Risk, Localities and Physical Safety Requirements.*

32
Examination, Requirements, and Handling of Fume Cupboards

Bernhard Mohr and Bernd Schubert

32.1
Introduction

In the laboratories of industry, research institutes, universities, schools, and educational institutions all kinds are fume cupboards are used extensively to protect those people working in the laboratory from the harmful effects of the chemical or physical processes that operate within the fume hoods.

The current European Standard DIN EN 14175-2 demands as the primary objectives of the use of a fume cupboard:

- Operator protection from harmful erupting gases and vapors or dusts
- Retention of flying particles that may arise in case of an unforeseen event
- Efficient transfer of gases or vapors to avoid a dangerous or explosive atmosphere inside the fume cupboard.

This objective is consistent with the requirement of supervisors and safety engineers in all areas of working life and education for increased safety at work and in the training space.

The reason for such demands is the realization that even very small concentrations of certain substances in the air can lead to high health risk. This led in recent years to a strongly increasing number of activities of protective work space of fume hoods for experimentation and production. It was also the major motivation for the development of the European standard series EN 14175th which deals with fume cupboards. The series was published in March 2003 and is valid in the entire European Community area.

While the protection against flying particles is relatively easy to ensure by the closing of the study area with a window that is constructed in accordance with and made of suitable materials, the necessity of holding back of noxious gases causes considerable problems.

Just the asset of the fume cupboard to contain existing or resulting gases and vapors in its interior is to be regarded as its main task.

For decades, the determination of the flow of laboratory air through the hood had been regarded as adequate for the protection of the user.

The Sustainable Laboratory Handbook: Design, Equipment, Operation, First Edition.
Edited by Egbert Dittrich.

However, this cannot claim that security concerns have been met. An accurate determination of the harmful gas retention is necessary, as described in DIN EN 14175.

32.2 Principle of Operation

A fume cupboard is an up to the front on all sides enclosed work area, where for the user harmful gases or vapors may be developed by chemical experiments.

In order to prevent these substances from getting into the breathing zone of the user, air is drawn through the front opening of the laboratory over the work surface and removed along with the harmful substances through an extract air line from the laboratory building.

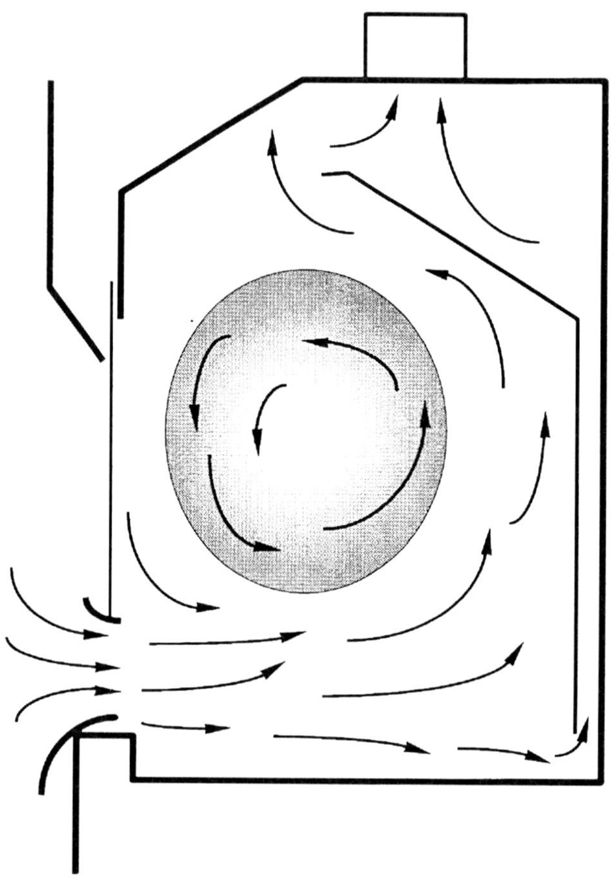

Figure 32.1 Fume cupboard with horizontal air roll.

This is done by a duct connection on the roof mostly over a side extract nozzle fan or, increasingly in recent years, through a central ventilation system, which provides the required pressure difference.

Different principles apply in stabilizing the flow through the inside. Usually, the flow is guided through a baffle at the back of the fume cupboard so that a horizontally positioned vortex rises up as shown in Figure 32.1.

The laying air cylinder is driven by the incoming air. So the pollutants are concentrated in the center of the roll, but they are diluted at several extraction points. The vertical extension of the roll for a given fume cupboard construction depends on the opening height.

To avoid excessive inflow velocities at the smaller front opening, a similar principle is implemented as in the so-called bypass fume cupboard. In the upper front area created by the closing window is another opening through which also air flows. From Figure 32.2, it is clear that this may provide an additional vortex, which reduces the original one and shifts it to the withdrawal center.

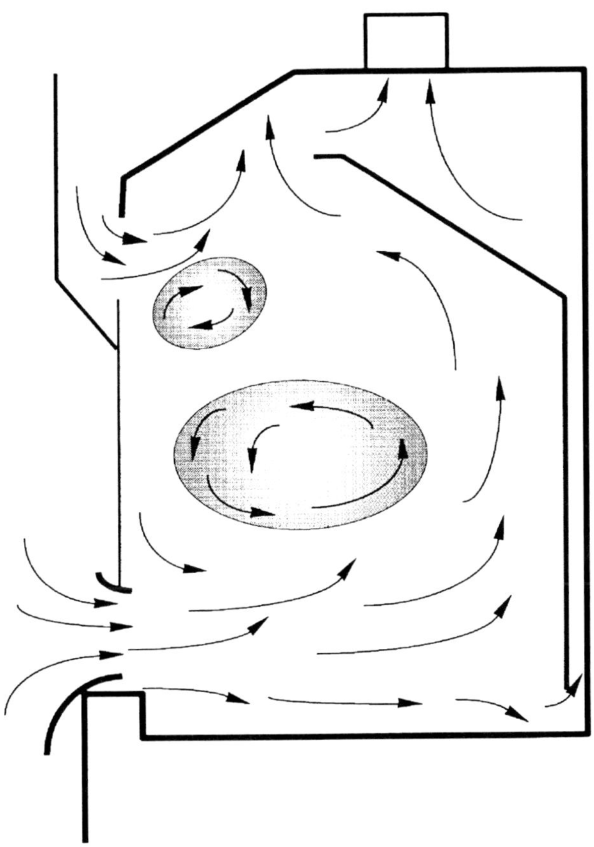

Figure 32.2 Fume cupboard with horizontal air roll and bypass.

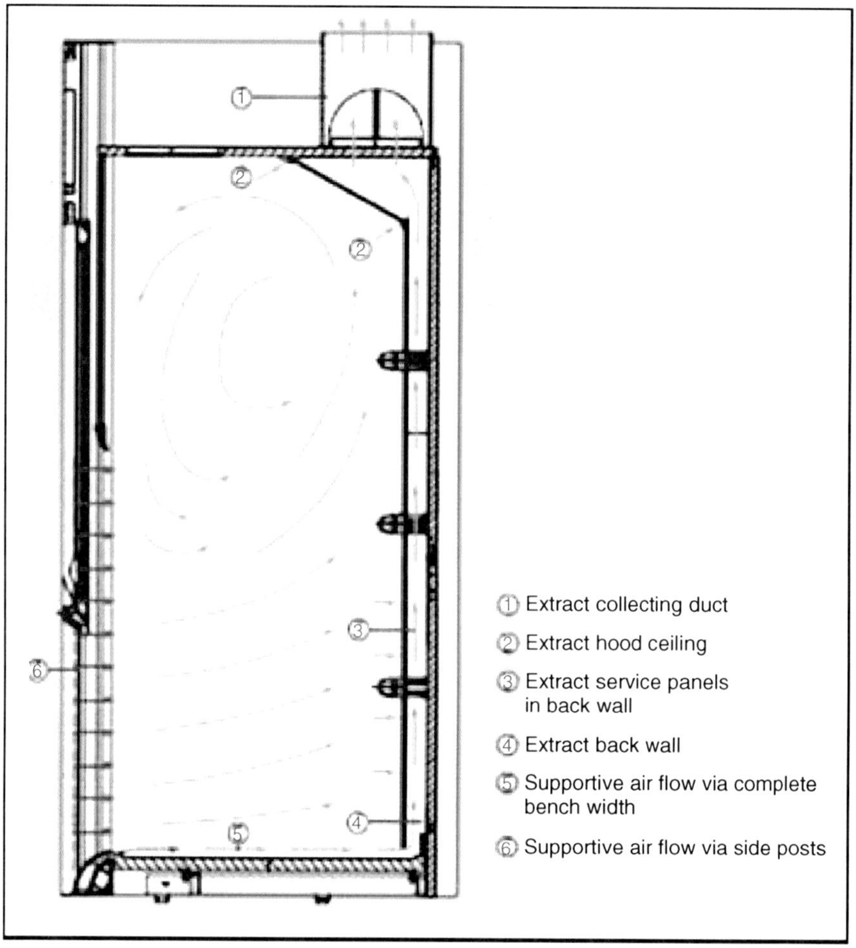

Figure 32.3 Fume cupboard with support air.

For large-scale extraction, for example, by many rows of holes on the back walls, some designs try to move the flow pattern in the direction of a plug flow. However, this is only possible in the respective front opening; in the upper hood space the air continues to move in the characteristic vortex (Figure 32.3).

The current trend is the use of additional air injection in the region of the front opening. This type of support air tries to avoid vortex formation at the inflow edges, or at least shift it further into the interior. Alternatively, one can also reach this goal with wide side posts, as is common in the English-speaking world. The aim of is still to offer sufficient protection against the escape of contaminants despite a lower total extract volume.

According to local development, the following are the typical features of fume cupboards:

- *Type A:* The table fume cupboards have the front slightly inclined backward. The side columns are relatively narrow and aerodynamically styled. They have usually only one vertical sliding window. Usually, their widths are 1.2 and 1.5 m; they are mainly used in the Scandinavian countries.
- *Type B:* The fume hoods have a lower interior height. The sides are a double wall construction with surfaces tilted to the interior on which the media operating units sit. Because of the higher amount of air, they usually have a bypass in the upper region, which is open only when the window is closed. Large air guide plates are mounted on the table edge. In most cases they have only a vertical sash. The widths can reach up to 2.4 m. These are mainly used inthe United Kingdom, United States, and Australia, and also in Asia (the Far East, India).
- *Type C:* The bench fume cupboards have narrow side posts and additional aids in the front. The systems can be active (an electric blower brings air through nozzles on the wall) or passive (baffles are directly at the front edge or they bypass the opening parallel to the front opening) or a combination of both. These fume cupboards are for small amounts of extract air, but they also need a minimum of room air disturbances. The sash consists of a vertical slider with horizontal split panes. Their distribution is currently mainly in Central Europe (German-speaking regions).
- *Type D:* The bench fume cupboards are similar to the table top fume cupboards, with single wall, but wide front posts. They are a combination of types B and C, the air baffles or inlet areas are narrow and simple. The extracted air quantities are higher than for type C, which makes the type more stable against external influences. In most cases simply, but practically designed with one vertical sash. Their distribution is in the Mediterranean, Western Europe, and the Middle East.

32.3 Types of Fume Cupboards

32.3.1 Standard Fume Cupboard

The most common type is the table fume cupboard. It provides the chemist a work surface which is available, depending on the version, in 900 mm (standard) or 500 mm (deep table) elevation. The interior height varies depending on the model and manufacturer in the range 1.1 – 1.5 m in the standard version. For rooms with a lower interior height, the fume cupboards are available specially designed with a low ceiling hood with split sash.

The usual external width is 1200, 1500, or 1800 mm, with a spacing of 300 mm, but also some smaller or larger sashes are available.

32.3.2
Walk-In Fume Cupboards

Walk-fume cupboards are found in the chemical industry especially for very high experimental setups, for example, for use in distillation columns. This type of fume cupboard takes its name from the height of the work surface, which is identical with the floor level. The interior height of these fume cupboards is thus increased by 900 mm above normal fume cupboards. The requirements for the retention of DIN EN 14 1753 also apply to this type.

It is not possible to achieve the required containment with the similar air flow rates as used for standard fume cupboards. Especially with larger opening heights of the most multi-divided front window, much higher volume flow rates are needed even when the type test has been performed with a low flow rate at the only 500 mm test opening. With large opening height the face velocity drops in the opening area where already small perturbations in the large room have mostly negative effects on the retention of the trigger.

32.3.3
Fume Cupboards for Thermal Loads

The EN 14175 Part 7 applies to fume cupboards with thermal loads and/or special acids. Special application fume cupboards are used for open digestions, for example, digestions carried out with perchloric acid, sulfuric acid, and/or hydrofluoric acid. For this, special demands are made for hood designs, which are based on the dangerous aspects of the work carried out there. Digestion of metals is frequently carried out on an open flame or on heating baths. The resulting thermal loads affect the air flow inside the hood, as also the containment of pollutants. Therefore a type of test is prescribed, which must also satisfy the same demands on the containment efficiency when tested with such loads in addition to the requirements (without thermal loads) laid down in Part 1 of the standard.

Experimental results have confirmed that the containment of a fume cupboard for normal use substantially fails when there is a buildup of thermal loads, which causes disturbances in the flow field.

In addition, open chemical digestions are often associated with dense smoke. Therefore, the residence time of the smoke inside the fume cupboard should be as short as possible to achieve the best containment with the thermal loads as also the benefit of a clearer view.

32.3.4
Special Constructions

32.3.4.1 Hand Over Fume Cupboards

In science classrooms, primarily in general education schools, hand over fume cupboards are used as special hood types, which differ mainly because they have an additional slider instead of the rear wall of the usual fume cupboards. These fume cupboards are mounted between the preparation room and the classroom. However, they are always used only from one side. In the DIN 12924 Part 3, published in April 1993, this is supported by a visual alarm that sends an alert in case of insufficient closure of the opposite side. On the basis of experience gained in the type tests with normal fume cupboards, the requirements for the containment capacity compared to EN 14175-3 have not been changed or even decreased, but only extended such that the requirements must be met for retention on both sides.

Owing to incorrect air mass balances of supply and extract air, a pressure difference between the preparation room and the teaching space may arise. This causes pollutants are discharged from the hood by uncontrolled air currents at the hand over fume cupboard. This is not included in the DIN 12924 Part 3, because the standard does not cover the installation of the fume cupboards in the building or the space requirements for ventilation. The planner for air equipment must therefore carefully design and control the air volume balance to cover this possible fundamental element of danger.

32.3.4.2 Pharmacy Fume Cupboards

According to the pharmacies operating order, a fume cupboard is required in every pharmacy in Germany. In cooperation with organizations representing the German pharmacists, therefore, Part 4 of DIN 12924 was developed for fume cupboards in pharmacies. While extensive structural changes are permitted to the base standard, in the interest of the safety of the user, no modification of the requirements for the containment compared to the standard is made.

32.3.4.3 Fume Cupboards for Radioactive Substances

These fume cupboards are provided for the handling of unsealed radioactive substances to protect against radiation exposure. Because of possible hazards to employees handling radioactive materials, compared with general fume cupboards, additional technical precautionary measures must be taken. These are described in DIN 25466 and involve decontamination of the entire structure, bearing capacity of the table construction due to the weight of additional lead shielding and equipment for extract air cleaning and contamination-free filter change. While the previous standard of 1985 demands an air flow rate of $540\,m^3\,h^{-1}\,m^{-1}$ frontage to prevent "irrotational entry," the requirements of EN 14175 in terms of containment for these fume cupboards took the standard without restrictions, only adding an additional test with the typical lead barrier inside.

32.3.4.4 Safety Benches with Air Recirculation

In EN 14175, these devices are clearly separated from the general fume cupboards. In addition to their dimensions, they differ significantly in that the extract air flow is not removed by separate pipes from the building, but purified by filtration plants and brought back into the laboratory room. They are recirculatory hoods, which work without a connection to a ventilation system. These are designed with secure arrangements so that no hazardous substances – or at least, quantities within safety limits – may pass through the recirculated air.

Depending on the specified user, responsible manufacturers investigate the retention by possible filters of future chemical substances accurately and include a corresponding warning system.

Security is solved technically, but problems arise under the strict test conditions. No operator in any laboratory can foresee and guarantee for the future, the nature of the hazardous substances used or created and also the exact amounts of these. Even for a small change in use, effective protection of laboratory personnel concerned can no longer be assumed. The hazard potential of the filter changes and disposal of the contaminated filter materials must be checked and minimized. It is understandable, therefore, that these types of extracted chambers can be recommended only in exceptional cases as equal protection against exposure to toxic substances.

32.3.4.5 Individual Constructions

In addition to the standard fume cupboards offered, the larger laboratory equipment manufacturers offer some specialized companies special constructions that are partially built only once. In addition, for these fume cupboards, the rules of EN 14175 may be used. But in order to keep the amount of testing within reasonable limits, Part 4 of the series of standards offers the ability to perform the test on the already built-in hood on site. Depending on the task, but also on the actual work, adapted test methods are used to show the reliable protection of the working persons.

32.3.4.6 Fume Cupboards with Auxiliary Air

A special design of the fume cupboards presented earlier is that of the so-called auxiliary air fume cupboards. In the 1980s and 1990s, the auxiliary fume cupboards were favored, especially by planning engineers, because of the expected advantage of savings from recycled extract air, which is taken from the laboratory. The ventilation principle of these fume cupboards was to replace the extracted volume flow normally by the laboratory input air by directly supplying air inside the hood. Thus, primarily, an energy saving was expected. First, there was the idea that for air replacement, untreated or at least less expensive pretreated outside air could be used. Filtering the incoming air to protect the lines from contamination and warming or dehumidification for safe dew point was necessary in any case. This resulted in considerable extra costs because of separate wiring and special hood structures that had to be considered in the identification of possible savings. These considerations set supply air to extract air ratios of 70–90%, however, which proved to be neither practicable nor safe. An energy saving through the use of auxiliary air fume hoods cannot be achieved for safety as well as for reasons of comfort. Today, such fume cupboards are hardly used for the above reasons, but are still very much in operation.

32.3.5
Control Systems

While both the extract and the tracking of the room supply air was kept constant in earlier times, it is now state of the art to adapt the fume cupboard's extract flow air to the circumstances. Such systems are called *fume cupboard with variable air flow*, also internationally known as the VAV (variable air volume) systems.

The simplest solution is to use the "On" or "Off" switch in the fume cupboard, so that the resulting pollutants are discharged through the extract during work in the fume hood. This solution, however, has limited safety, as possible contaminants in the hood or extract duct connected can get deposited. After switching off the suction, hazardous substances can diffuse back from the deposits into the room. In addition, it is necessary to consider the input air and how the extract of the room is to be controlled in order to ensure room air balance.

When the sash is closed, even a small amount of extract air provides good protection against contamination outbreak. Thus, it is only logical to install a suitable device that controls the amount of extract air as a function of the position of the window: window closed – reduced air flow; window open – normal amount of extract air. This two-step system has the advantage that a reduction in the indoor air can be tracked in corresponding stages.

The next stage is a continuous variation of the extract amount according to the actual opening area. The control information comes by way of the length of the signal from a sensor, which detects the opening of the height or width or a reference level sensor, such as an anemometer that measures the face velocity.

The level of complication involved in the control of the ventilation system depends on the number of fume cupboards standing in the room or are connected to the same line. As a reward for the effort, an optimized system may save about 50% of the energy of the room air, thus increasing the safety significantly by proper indoor air replacement.

Before a control is used, the following questions should be answered:

- Does the fume cupboard operate safely with a VAV?
- Which are the volume flow rates that are necessary?
  For this purpose, Part 6 of the series of standards EN 14175 states, "hoods with variable air volume flow." The methods are described for the hood with the VAV (Variable Air Volume) system itself as well as for examining the containment of the fume cupboard equipped with the VAV. The measurement results show the limits for the operation of the fume cupboard with the corresponding control. The manufacturer the fume cupboard can provide this information.
  Variation in the volume flow in the fume cupboard with special functions (e.g., using thermal loads) should be avoided, in principle.
- Is a VAV system in the given space situation useful?
  Depending on the size of the room, it may be useful to suck out the air constantly from the area, because only then the sufficient and necessary air exchange can be ensured.

The following questions can arise for the supply air:

- Which outlet elements are suitable?
- How can one avoid flow noise, for example, caused by excessive pressure in the duct system?
- Is it always necessary to maintain a pressure difference between the laboratory and adjacent spaces?
- What are the operating costs, including maintenance, compared to the actual savings?
  If the payback period of the cost of the additional components and their installation are above the usual warranty period, the subsequent costs must be taken into account.
- How is the actual user behavior?
  In laboratories with only part-time operation, such as during a practicum, an easy on/off system may be the best solution. During operation, the employees are standing at open fume cupboards, so the cupboards can be shut down and cleaned again during the rest period.
  In contrast, major research laboratories with, for example, two hoods per employee make use of VAV systems as usually several tests are run simultaneously.

Only intensive training of users with the use of controlled hoods ensures the rational use of this technique. A key point is the understanding of the relationship between energy savings and safety.

Both laboratory manufacturers and specialized suppliers of ventilation components offer various systems. They all consist essentially of the following components:

- *Throttling devices for the extract*
  This can, for example, be a damper with electric or pneumatic actuator. Other designs, such as a pressure plate in a Venturi tube or chambers with rubber bellows have not become widely accepted in Europe
- *Detecting unit for receiving a measured value, or a signal for the setpoint*
  For the detection of the window position, the simplest form is a switch at the window. The so-called draw-wire sensor continuously detects the position of the sash.
  From the front opening and the air drawn by the volume flow (= extract) a corresponding inflow results into the hood. Since the pressure differences in the workspace are relatively low, one can measure the corresponding velocity in openings in the side wall or the ceiling. For this, a velocity sensor is mostly used. Its signal is processed simultaneously for control and actual value information.
- *Display/operating unit as a machine–human interface*
  It shows the user the operating status of the fume hood function by means of a display – in order or malfunction. Some manufacturers offer additional detailed information, such as extract air flow or inflow velocity. The user must, however, know the exact minimum requirements in order to derive this information or deal with disorders.

The use of an on/off-switch is controversial, as it carries a high risk of malfunction.

- *The control electronics*
  It collects information such as air flow, air face velocity, window or damper position, and links them logically. By controlling the throttle unit, the values are kept within specified tolerances.

Usually, interfaces to the BCS or for setting the target values are also provided. Depending on the manufacturer, the connection is via a bus system or a special adapter, for example, in the control panel.

The control principle is relatively simple. The current actual state is detected from the detection unit and compared to the electronic control system for the desired value. If there is a difference, a control signal is applied to the throttle device until the deviation is within the tolerance range. An essential feature of the whole system is that it must be relatively fast in increasing the amount of extract air. It is the only way to ensure that the user is not exposed in any manner to the highly concentrated polluted air from the interior, while opening the sash. On the other hand, the process of air reduction should be rather slow, so that, in most cases, the slower systems use for room air control can, at the same or at least similar time, can reach the room pressure.

In practice, it takes about 3 s for the closed sash to be opened. In this period, the extract air volume is increased from the minimum value to at least 80% of the target value for the window. Depending on the type of fume cupboard, the minimum and target air levels are set in this time but it may take longer. The decisive factor is the pollutant retention capacity of the extract air, which is tested according to EN 14175.

Part 6 of EN 14175 describes the requirements, the examination procedures, and the on-site inspections of fume cupboards with variable air flow, allowing the purchaser to select a fume cupboard with the requirements regarding containment, air exchange capacity, and so on, with a suitable control (VAV system).

32.3.6 Window Closing System

Further accessories have been added to the fume cupboards in recent years, among which is the automatic window closing system. It consists usually of a presence detector, a light barrier, and a drive. The presence detector is, for example, a proximity switch that is detects the presence of people. If no movement is detected for a longer period before the fume cupboard, the system assumes that the fume cupboard is not being used, and tries to close the sash via the drive.

A light barrier below the sash ensures that any object extending into the opening area stops the sash. Since this system is a motorized unit in which there is risk to the user, the entire hood is covered by the Machinery Directive and must meet much stricter conditions.

Targeted training of users to keep the sash closed whenever possible may be a viable alternative. However, it predicts autonomous conscious work of the fume cupboard's user. Experience shows that, unfortunately, in many cases, this cannot be assumed.

32.4 Standards

The following standards are known:

- ASHRAE 110-1995, USA (ASHRAE 110-2005 Entwurf)
- EN 14175, Part 1–7, Europa
- XPX 15-206, France
- AS/NZS 2243.8:2014, Australia, New Zealand
- DIN 12924 Teile 2–4, Germany
- China

32.4.1 U.S. Standard (ASHRAE)

The American Standard 110 describes measurement methods for testing the performance of laboratory fume hoods. The 1995 version is still valid, with a revised version of 2005, which is commonly used in draft form., No limits are specified in the standard, but typical values are partly given in the informative annex.

Generally, the standard distinguishes three different types of test:

- Measurement of the hood by the manufacturer.
  This measurement is similar to an examination in which a hood is tested under ideal space condition.
- Measurements during commissioning.
  This serves to document the functionality of the hood in the installed state and in conjunction with the on-site air conditioning.
- Measurements of the hood in operation.
  Here, the influences are taken into account on site such as from obstructions or with heat loads.

In the normative part of the standard, the following tests are described.

32.4.1.1 Flow Visualization

- *Examination with great output volume*: Here in the workspace, a larger amount of smoke (fine particles) or fog (water droplets) is generated and observed when something comes out of the fume cupboard.
- *Test with small output volume*: With smoke or fog lines, the flow is visualized in the work area of the hood to identify dead zones or return flows.

32.4.1.2 Measurements of Air Velocity

The inlet velocity of air in the front opening is measured at standard well-defined points. Depending on the design of the hood, this may involve different window positions. The average over all the points and the points with the highest and lowest values are noted.

32.4.1.3 Measurements with Test Gas

For testing the containment, a specific gas is released inside the hood and then the gas that comes out is measured.

As a tracer, pure sulfur hexafluoride (SF6), a synthetic gas which is relatively safe and can be measured even in small concentrations, is used. The release takes place through a nozzle-tube grid construction and the gas and the air inside the fume cupboard mix uniformly.

The measurement in front of the fume cupboard is made in the breathing zone of the laboratory technicians, which is simulated by a mannequin. During a measurement, the static window remains open. In a further test, the influence of change in the front opening is shown by the gas test.

In addition, the front area is scanned around the window opening to identify any leaks.[1)]

32.4.1.4 Additional Measurements for Fume Cupboards with VAV Systems

For fume cupboards where the extract air is altered by a control system depending on the front opening, additional measurements are described below:

- Determining the reaction time of the system upon opening
- Determining the stability
- Checking the repeatability.

Either the flow rate or air speed at a reference position is recorded over time and displayed in response to the changes of the front opening.

32.4.2 European Standard

The European Standard EN 14175 consists of several parts, which were developed by representatives of the various countries of Europe from 1995 to 2003. The essence of this product standard as opposed to a safety standard is that limits to the results described in the standard tests of the national safety/health authorities are required. As the various countries have different rules about the use and operation of the equipment, the current version a compromise agreed to by all participating countries.

The following parts of the series of standards are currently valid:

1) Comment of Egbert Dittrich: It is not clear the use of SF_6 is still the state of the art. There is nothing in the market to substitute even the CO_2 if equivalent of 23.800 for the time being. Some research organizations are working on this issue.

- Part 1 Definitions
- Part 2 Requirements for safety and performance
- Part 3 Type test methods
- Part 4 On-site test methods
- Part 5 Recommendations for the installation and maintenance
- Part 6 Volume flow-controlled fume cupboards
- Part 7 Fume cupboards for high thermal and acid loads.

In Part 1, the terms used are explained and translated into a list in the world's most commonly used European languages.

Part 2 lists the minimum general requirements for safety, material, or lighting. Here, the reference to the aerodynamic function test is carried out under Part 3 and 4.

Part 3 is used for the examination of a fume cupboard in a laboratory. It describes in detail the individual test and measurement methods that are carried out on the fume cupboard.

The list is structured as follows:

- Measurement of the air entry velocity at the window opening.
- Determination of containment by releasing a test gas and measuring the concentrations occurring before the fume cupboard. In detail, the European procedure is distinguished from the American method by the use of only 10% sulfur hexafluoride in nitrogen, whereby the density of the test gas is rather similar to that of air. Furthermore, the test gas is distributed over nine ejectors in the interior and the possible leakage determined integrally at several measuring points in front of the fume cupboard.
  - Determination of the robustness against disturbances in the room air. Since the measurements take place in a standardized test room under ideal conditions generated by a moving plate, creating a defined disorder before the fume cupboard, robustness is measured in terms of the resistance to interference.
  - Determination of the purge factor in the interior as a measure for possible enrichment of an explosive mixture.
  - Determination of additional data such as pressure loss, luminance, or force required to move the window.

Part 4 is an enumeration of the possible measuring methods that are to be used in built-in cupboards on-site. A distinction is made between non-type-tested and type-tested fume cupboards; in the second case, the test is again divided into a commissioning test and a periodic test.

Part 5 is only a technical recommendation and contains general information about the fume cupboard, for example, its placement in the room. This part is partially covered in different countries by other directives

Part 6 includes an additional requirement on the fume cupboard and also control system, if this changes the extract air volume in dependence of the front opening. This control system and fume cupboard can be tested independently.

Part 7 is a listing of the additional testing of fume cupboards intended for work with increased thermal loads and harsh chemicals.

32.4.3 Other Standards

32.4.3.1 France

The XPX 15-206 is a French experimental standard which has as its basis the EN 14175. This mainly gives notes regarding installation and air ducting. An important point is the setting also of a limiting value for the containment test according to EN 14175-3: it is 0.1 ppm for the mean value of the measurement at the inner grid.

32.4.3.2 Australia/New Zealand

The latest version (2014) of the Australian/New Zealand Standard AS/NZS 2243:8 describes the general requirements for laboratory fume hoods, their installation, and operation. The guidelines are based on the principle no longer relevant to the state of knowledge that the inflow velocity must be at least 0.5 m s^{-1} for safe operation of a fume hood.

32.4.3.3 Germany – Standard for Special Fume Cupboards

In addition to the EN 14175 standard fume cupboards, the DIN 12924 Part 3 applies until further notice in Germany for pass-through fume cupboards and the DIN 12924 Part 4 for pharmacy fume cupboards. In both of these, the aerodynamic function test is indeed referred to in accordance with Part 1 of the old standard, but has been replaced by EN 14175. The same applies until further notice for the DIN 25466 of October 1995, as it mainly concerns the technical description of the fume cupboards for each particular application to the above standards; an analog to EN 14175 applies to the examination of the ventilation part

32.4.4 Comparison

There are currently only two major global standards in the field of laboratory fume cupboards: the EN 14175 and ASHRAE Standard 110th.

Throughout Europe, the existing EN standards are applicable as the only standards. It is therefore recommended to use the EN 14175. This standard is also currently the most advanced and recognized by standard in recent years throughout Europe.

32.5 Safety Criterion

Before the introduction of fume cupboards in laboratories, harmful gas removal was carried out through a chimney stack although not of the high safety standards of today. Safety requirements were not applicable at that time. In the situation of vapor formation, the windows were merely opened. Fully enclosed working spaces with natural extract facilities, as seen for the first time in Liebig's analytical laboratory in 1839, corresponded to the usual fume cupboards today using the same fundamental objective of fume cupboards, namely, the removal of emerging pollutants. Later, the thermal buoyancy flow was supported by a burning flame within the chimney. The principle of the fume cupboard, as we know it today, originated first with the use of a fan, which could be adjusted to define the setting of the extract air volume flow.

Initially, flow through the working space in the fume hood was adequate as a safety criterion, but this led to later investigations of the performance under various conditions and the demand for certain levels of extracted volume flow, which would cause flow velocities perpendicular to the front plane of the opening. By specifying certain values of air flow rate and/or the effective face velocity, an indirect criterion was given, which ensured the containment of harmful gases in a fume cupboard (Figure 32.4).

A comparison of the concentrations leaking out from inside and measured in front of the fume cupboard, with different front openings and entry speeds makes it immediately clear that the inflow is not sufficient criterion for the safety of personnel, and consequently for the performance of the fume cupboard. With the first description of a method of measurement of the pollutant retention capacity of a fume cupboard in 1978, the basis was created to test this primary characteristic of a fume cupboard directly. This meant not only a fundamental change in the safety requirements for the satisfactory operation of a fume cupboard but also a corresponding adjustment of previous safety methods. The standards available up to 1986 took this fact only partially into account. In most standards committees,

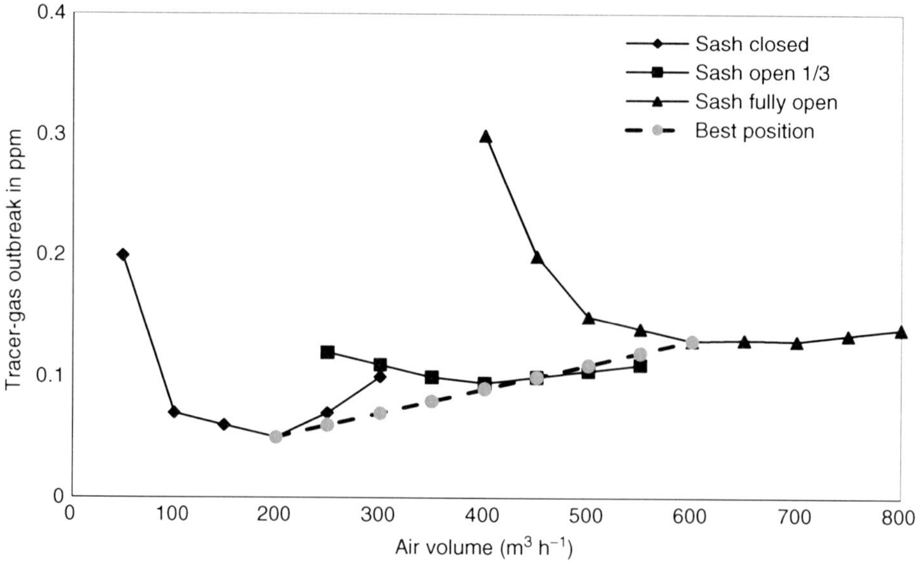

Figure 32.4 Dependence of outbreaks of harmful gas from the extract air quantity on face velocity.

the principle of direct verification of the safety requirement of a fume hood, the containment, was adopted later on.

32.6 Fume Cupboard Testing

Since the beginning of systematic testing of laboratory fume hoods about 30 years ago, not only the performance, hence the containment of fume cupboards, but also energy efficiency has improved dramatically. As part of these tests, it has been made clear that it is primarily necessary to ensure that a model, that is, the prototype of a fume cupboard should be able to meet the requirements. This resulted in Europe in the standards Part 3 of DIN EN 14175, namely, the type examination. For fume cupboards with variable flow, DIN EN 14175 Part 6 applies.

However, it could also be shown that the ambient conditions in the laboratory have a significant impact on the performance after the installation of one or more fume cupboards in the lab. Hence the requirements of DIN EN 14175-4 deduce that a selection of test methods provides for the examination of different requirements that are generally found in Part 2. The given methods should allow comparability of results and, although very broad, ensure that the application of appropriate test methods excludes unsuitable audit approaches in reverse.

32.6.1 Type Test

In order to produce the same boundary conditions and thus a fair comparison of results of different fume cupboard types (type), the type tests shall be performed in a standardized laboratory. For the user of the tests, it is only important that it is proved that the values originate in such a laboratory.

The user or buyer has the opportunity of founded values to be obtained by submitting a test report which qualifies his fume cupboard. But as the European Standard provides no performance requirements, each user must decide if the test results comply with or satisfy its requirements. The statement "tested according to DIN EN 14175" does not admit any conclusions about the performance of the fume cupboard. Only the results stated in the test report show this. This important fact is often overlooked.

Important properties of a fume cupboard can be taken from the report, not the certificate, as described below.

32.6.1.1 Containment

By placing a certain amount of test gas, a situation is simulated inside the fume cupboard. It is measured how high the concentration of test gas is before the fume cupboard, that is, in the zone of movement of a user. This value is a measure of the outbreak of substances from the fume cupboard interior. The escape of contaminants is measured in a plane 50 mm in the front of the fume cupboard and/or directly in the plane of the front opening again, as a function of the given extract air volumes. The smaller the amounts reported in the test report of concentration values, the better is the retention capacity at the corresponding extract air flow. Thus, the performance can be assessed in relation to the safety of a user in the laboratory.

By similar measurements behind a moving plate, which simulates disturbances in the laboratory in front of the fume cupboard, statements about the robustness of the containment can be made. Again, the smaller the measured concentration value, the better the containment.

32.6.1.2 Other Requirements

The standard asks for more properties of fume cupboards, which are of importance for its quality:

- The inflow into the fume cupboard at 500 mm opening height of the sash at given air flow rate is determined directly in the plane of the front opening. The more uniform the distribution of the measured values is, the lower the risk of pollution outbreaks in the laboratory will be.
- The pressure drop of the flow through the fume hood is a measure of the energy losses during the flow through the device.
- The sash is examined for its suspension, the necessary force when moved, its sash-stop at a given opening height, and the protection provided against splashes coming out.
- The examination of the monitor system of the fume cupboard unit is of high importance as it indicates unsafe extract condition of the fume cupboard to the user.
- Construction, materials used, and the lighting are also subjects of tests.

All results can be found in the test report. This document is an important part of the information about a fume cupboard and must already exist in the phase before an order is placed.

32.6.2 Examination in the Laboratory

The responsibility for on-site testing of fume cupboards lies with the laboratory operator. Notes thereto give in particular the policies of the local authorities. The European Standard contains a collection of useful test methods under which the standard user can select the suitable test methods. The details are given in the EN 14175 Part 4.

After considerable deliberation, a distinction has been made between a test during commissioning (initial assessment) and periodic testing. The first ensures that type-certified qualified hoods do not drop in their performance too much as a result of the local situations. Periodic testing at regular intervals should show the results verified in the previous test have not fallen below the required levels due to changes in the technical environment.

32.6.2.1 Commissioning Testing

Such studies are rather recent in Germany; however significant benefits usually justify the financial expenses.

The verification of the ability of a new installation to function is an opportunity to determine whether the fume cupboard has the promised properties regardless of work of the manufacturer, contractor, or supplier.

In such acceptance tests, related errors during production or installation are often identified and can be eliminated immediately. In addition, inadequate performance of other components such as ventilation can be identified as being the reason for detected deviations in the performance of the fume cupboards. The results can serve as the basis for safety risk assessments, as they provide the necessary data, especially the containment test data.

According to the principle for the selection of tests and their methodology as specified in the European standard, the comparison must be with the boundary conditions defined in the type test.

If the installed hoods type is approved, the complex and expensive containment tests in most cases are not necessary. The control of the face velocity at the same volume flow may be sufficient, possibly even the determination of the pressure drop, which is a specific value for each fume cupboard type. One should, on the other hand, be aware that with the reduced tests, other influences from the location of a fume cupboard may not be noticed.[2)]

The containment values in the type test of other extract values than in the examined real laboratory are not comparable and therefore practically worthless. They should be repeated in the initial test. The standard refers to this as the commissioning test.

It is undisputed that the planning of the commissioning tests must be carefully considered and be strictly adapted to the specific circumstances. It shall also be the basis for subsequent periodic testing in the form of then comparable results.

32.6.2.2 Periodic Inspection and Testing

Periodic inspection is based on the results of the commissioning test. The idea here is that it shall be proven that even after definite time of use, the performance once found in a fume cupboard has not dropped. If the values of extract air flow, face velocity, pressure loss, or retention, identified during an initial test, are found again with small deviations, the system can be considered of equal performance.

The more accurate and/or larger the database from the initial test is, the less the expense for the repeat test will be. This, of course, is also reflected in the financial expenses. The pleasant side effect of such tests is to be able to find faults in the vicinity of the actually examined fume cupboard, for example, faults in the ducts of the ventilation system or defective fire dampers, fans, or air volume controller, whose functions do not indefinitely remain the same.

2) Comment of Egbert Dittrich: This aspect is very much needs to be considered as on-site circumstances constantly vary and despite the results of type tests, external influences may cause outbreaks at a high level.

The methods to be used for these examinations can easily be taken from the EN 14175-4. The reason that this is recommended is to avoid basic measurement error. However, trained and proven specialists must also be used.

32.7 Influences of Real Conditions

Idealized conditions defined for the type test for reproducible and comparable analysis of fume cupboards of different fluid mechanical layout and various designs for specific application in chemical laboratories are rarely given in actual installations in industry or research institutes. The reasons for this lie in constraints such as the architectonic conditions of existing buildings or planning traditions and preferences in design of new buildings. However, they are also in the needs of future users, who are attached to their traditionally trained and conventional ideas in the preparation of fume cupboards in relation to foot traffic routes or other functional assemblies as well as in the handling of the fume cupboard. In addition, on presentation of clear measurement results, a longer experience or, perhaps also an educational process, will probably be needed to make the safety capabilities of modern hoods in real life laboratory fully usable.

In the extensive literature in the field of laboratory fume cupboards, these changes in reliability under actual user conditions are often discussed only in a general way. Thus, from the point of view of the negative effects of construction and given room flows induced by the user, and disturbance on account of the behavior of users, the operation of good fume cupboards designed by fluid mechanical knowledge under ideal conditions is in jeopardy. Detailed information about the positioning of fume cupboards in laboratories are published in EN 14175, Part 5: Recommendations for installation and maintenance. Knowing the aforementioned resistances and complex interactions of many influencing parameters merely suggestions and not rules are given for distances of a fume cupboards from frequently used traffic, connecting and escape routes, laboratory benches, walls, other fume cupboards, or columns and doors.

In recent years, performance of fume cupboards has been analyzed by increasing the influence of the user. The results are limited to either single problems or lead to general advice on compliance with the various influencing factors. The information that follows is the result of the first attempts for traceable documents for the correct installation of a fume cupboard in a room, the necessary relation of optimized room inlet and extract air; it shall deliver the customized behavior of users in daily laboratory operations. The results are part of inspections of fume cupboards that have been carried out on behalf of companies and government organizations.

32.7.1
Changes to Fume Cupboards

For reasons of organizational or experimental necessities, existing fume cupboard constructions are often changed on behalf of the users or by the users themselves, without in each case, a reevaluation of the function of the fume cupboard. The need for such renewed tests is apparent from the following examples.

The examined object is a 1500 mm wide fume cupboard with an air flow of vertical vortices, which is installed at the head of the laboratory space. It may thus be expected that there is no major influences through doors, windows, and traffic areas. The fume cupboard is on the side facing the room provided with a sliding side window to serve the passage of cables or wires for the supply of test facilities. This change of the original design had already been applied by the manufacturer. With the closed side window, measurements showed satisfactory containment behavior up to a sash height of 500 mm. But with a completely, that is, 900 mm open sash, a leakage shows up to 5 ppm; thus, under these conditions, it cannot be said that there is an adequate protection factor.

If the side windows open only one-third of the possible extent, the harmful gas containment behavior continues even with front window completely closed to periodic fluctuations. If the side window is completely, that is, 500 mm opened, the retention capacity of the fume cupboard degrades drastically. It turns out that only the side window is responsible for the high pollutant gas outbursts. The influence of the side window is of the same order of magnitude as that of the totally open front window, although the resulting surface open toward the laboratory is substantially smaller than the corresponding area in front.

Even with the opening intended by the manufacturer for the cables, a significant reduction of the harmful gas retention can occur, particularly when the fume cupboard is operated on the border of its proven capabilities. In practice, often changes, similar to those described above, can be found in the proper design and operation of fume cupboards, such as hand-through openings or direct connection to adjacent fume cupboards or changes to the supply or extract air conditions. Such geometric changes affect the aerodynamic design specified by the manufacturer and described in the type test flow situation inside the fume cupboard. It cannot be assumed that the containment results observed in the type test are valid any longer and further examination on-site is necessary in such cases.

Access doors in the immediate vicinity of a fume cupboard often have a significant influence on its containment. Each opening operation of the laboratory door shows with only a small delay, a short but significant concentration of harmful gas escape. Since often the number of the access door openings is quite high, despite the absolutely small amounts of harmful gas escaping into the laboratory, a non-negligible total amount will feed on potentially harmful substances per day in the staying area of employees. Periodic outbreaks are not uncommon, indicating an unstable state of the fume cupboard flow. Thus, the cumulative effect is amplified further. Therefore, an influence of these emissions on the limited emission of MAK values must be assumed, even if a functioning ventilation system is available.

It is often found that when fume cupboards are next to each other, their supply air is compensated by direct supply air from the same line. This shows that the uncontrolled switching on or off of other fume cupboards or other ventilation technology components in a room can interfere with the stable air flow in the laboratory or in the ventilation system such that the occurring oscillations can lead to substantial harmful gas outbreaks. Here the control of the ventilation system can help, which takes into account the time constants of the individual components.

However, there is general experience about the influence of the arrangement of fume cupboards in the laboratory by manufacturers, designers, and especially users of fume cupboards. These experiences are consistent with the experimental results. But because of the complex interrelationships of other influences on the harmful containment capacity of a fume cupboard, a general statement that could lead to a rule in the standardization purposes does not appear possible. For this reason, only few published proposals for the positioning of fume cupboards in the laboratory can be found. It will always be noted that the recommendations are provided only as a guide for the planning of laboratories, to prevent interference in the performance of fume cupboards.

It must not be doubted that concrete and detailed instructions for installation in the laboratory can be given, because this problem area may provide important clues for the facilities planner. It merely is one of the influence of size on the performance of a fume cupboard. It is quite possible that other parameters change this influence in both positive and negative directions. This means that the laboratory planner needs to become familiar with all sorts of influences and take them into account in their entirety. Only in this way, the optimum performance of the fume cupboard can be ensured under the most varied circumstances.[3]

32.7.2 Ventilation System

The ventilation system that is connected to the fume cupboards to supply the necessary air makes a significant contribution toward the proper functioning of the cupboard or, conversely, is often the cause of malfunctions. The need for adequate inflow in rooms equipped with air conditioning should be stressed here. Too high a negative pressure, which is detectable by difficult to open doors, affects the performance of fume cupboards. Another difficult aspect is the variable partitioning of extract flows for several fume cupboards on a common extract manifold, which can be the reason for inadequate containment of otherwise first-class fume cupboards. It must always be remembered that changes, such as readjustment of the hood on the extract side, are most likely to affect the extract situation of the other fume cupboards. The extract values of the other cupboards must always be calculated.

3) Comment of E. Dittrich: The conclusions clearly show the necessity for close collaboration of all the planners involved to take all the parameters into account.

Large changes, but sometimes even small ones, in the ventilation system require the extract side validation of fume cupboards and/or the supply air.

32.7.3
User Behavior

Any other uncontrolled change in the inflow in front of the fume cupboard can, like the room flow, directly have a significant impact on the containment of the fume cupboard. Such disturbances are, for example, movements by laboratory personnel in front of the fume cupboard. Movements transverse to the main inflow direction, especially, affect the inflow into the fume cupboard considerably as the air velocities so caused can easily reach values of about 1 m s^{-1}. In contrast, the typical air inlet velocities perpendicular to the front opening of the hood are at 0.1 m s^{-1} in the case of a fully open sash (900 mm opening height) or about 0.3 m s^{-1}at one-third sash height. Only when the sash is closed to stop the inflow, the velocities at the windows are at 1.0 m s^{-1} and thus in the order of traffic movements in front of the window.

Measurements of the containment that is caused by this disturbance show no differences in the undisturbed case, as long as the sash is closed. However, even at an opening of one-third (300 mm), depending on the speed of movement, test gas outbursts with an increase by a factor of 1/40 can be found.

Although such movements are naturally unavoidable because of work processes and connections to different work areas (scales, sinks, and the like), the result must take into account. This can be done both through work instructions to staff through appropriate planning and suitable structural arrangements.

In studies in different laboratories, it has been repeatedly found that significant adverse effects on the extraction capacity are caused by disorders of the inflow into the fume cupboard. So if the lab technician should lean against the edge of the table, this may generate a negative pressure area which produces a direct inflow into the breathing zone. For lack of suitable placement options, other disturbances such as large gas cylinders are frequently placed at the front edge of the hoods. For this, similar conclusions apply (Figure 32.5).

In reality, the work surfaces are often fully utilized inside the fume cupboard. Owing to the obstruction of the working space, which should also serve the trouble-free flow for the discharge of gases, this flow can be modified to the extent that areas with reverse flow or turbulence zones occur which transport harmful gases from the fume cupboard inside the laboratory away from the work area. An extreme but not untypical example is the fixed installation of a bench in a walk-in fume hood. Very good retention values can be seen in the measurement protocol of investigations on a 1800 mm wide walk-hood in a standard laboratory indicating that harmful gases are still quite well retained when the sliding window is closed. However, any opening of the sashes leads to a leakage of harmful gases in extremely high concentrations.

Very often, sand or water baths are found on the benchtop, which do not permit an unobstructed flow over the work surface. This creates a flow vortex near the

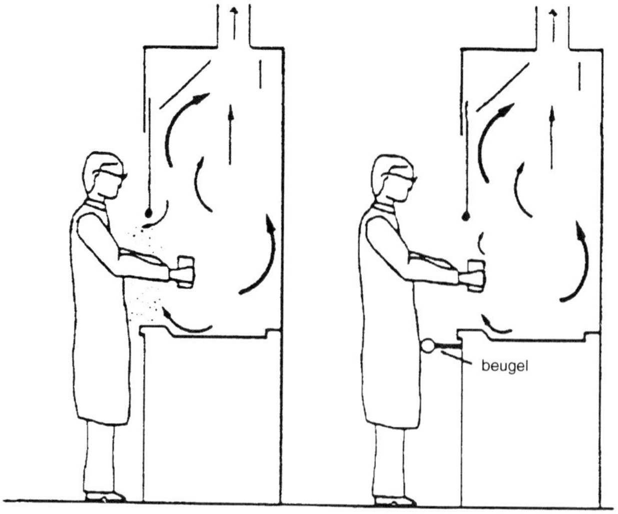

Figure 32.5 Influence of distance between user and hood front.

working surface moving existing harmful gases outward, as a drain up to the interior of hood is no longer possible. This situation can be easily remedied through the establishment of a possibility for a flow below the object, such as small bases or feet. Manufacture of corresponding devices taking this fact into consideration has increased, but is not yet fully complied with in all cupboards.

Many other obstructive installations in the fume cupboards make it difficult for the undisturbed flow in a potentially contaminated work area. Shelves for storage of chemicals often installed as retrofits by the users are the reason for harmful gas outbreaks. Apart from the fact that chemicals must be stored in designated storage cabinets because of their potential to create danger in case of fire or explosion, the impairment of the flow can be kept low through the use of perforated boards (with shelves) or by a free distance of 1 – 2 cm from the bench surface (instruments).

This applies in particular to experiment setups within the fume cupboard, which directly influence and alter the internal flow. A fairly example is the use of a hair drier as a heater or as a drying apparatus. It is obvious that the so-induced uncontrolled flow has a strong negative influence on the containment of the fume cupboard.

Part VI
Safety in Laboratories

Egbert Dittrich

Major devices of the engineering building services are entirely used for OHS (Occupational Health and Safety) purposes. Thus, any change of standards and technical devices requires preceding risk assessments. Different conditions and the numerous substances and chemicals in use squeeze consultants, experts, and planers to really take care of the individual subjects. At the end of the day, the lab planner must acquire skills from many contrary disciples inter alia OHS basic knowledge.

The requirements articulated in this piece to achieve sustainable buildings must be considered with keeping in mind that OHS issues beat everything. One must know that this is no contradiction with sustainability. It is a good deal more of an intrinsic value of sustainability.

No question that collaboration with occupants will support solution, that is, being effective, ecological, and safe. Like already addressed, these requirements of sustainability must be brought in balance.

With respect of the permanent change of the laboratory processes and the relevant technologies, the chance to realize sustainability appears even more often in a laboratory.

The Sustainable Laboratory Handbook: Design, Equipment, Operation, First Edition.
Edited by Egbert Dittrich.

33
Health and Safety – An Inherent Part of Sustainability

Thomas Brock

33.1
Scope

In the past, laboratories were often simply designed rooms whose only ventilation measure, for example, was an open window, and perhaps also a fume hood with natural draught powered by a gas flame. Health and safety was only of minor importance – after all, it was considered honorable to risk one's health in the interests of research. As the famous chemist, Justus von Liebig[1] once told August von Kekulé,[2] "If you want to become a chemist, you have to ruin your health; he who does not ruin his health through study will not get anywhere in chemistry." Certainly, even in those days, no one risked his health on purpose, but even risks taken unconsciously can have fatal consequences. In the modern world, it is therefore not acceptable to lose one's health or even one's life, while risking the results of the work as well as material property, and causing extra costs, which might be quite considerable in some cases. Health and safety thus pays off. According to an investigation by the DGUV (German Social Accident Insurance) in cooperation with the Justus Liebig University Giessen and spanning 19 countries, an average return on prevention of 2.2 is achieved by implementing such measures.

Health and safety in laboratories is the result of a range of coordinated measures. The success of these measures depends on various stakeholders, in particular the user of the laboratory, the planner, the equipment supplier, and the building operator. They can be divided into four large groups:

- technical measures such as the proper design and construction of the building or the design of laboratory equipment and devices,
- organizational measures such as the provision of informative documents on the building or the training of laboratory staff,

1) Justus von Liebig was a nineteenth century chemistry professor at the universities of Gießen and Munich, where he developed a very modern concept of research and teaching laboratories, contributed important findings to organic and agricultural chemistry and invented the Liebig Extract of Meat.
2) Friedrich August Kekulé von Stradonitz, chemistry professor at the universities of Heidelberg, Ghent and Bonn and principal founder of the theory of chemical structure (the benzene ring, e.g.).

The Sustainable Laboratory Handbook: Design, Equipment, Operation, First Edition.
Edited by Egbert Dittrich.

- personal measures such as the use of personal protective equipment or occupational medical prophylaxis,

and, of course, also necessary in the laboratory are:

- checking the possibility of using alternative substances or methods if feasible and if this actually results in a reduction of risk.

In this interconnected system, laboratory buildings fulfill a range of different purposes. As well as allowing the accommodation and operation of equipment and devices and thereby the execution of a wide range of activities in order to gain knowledge, they also allow the people employed there to carry out these tasks in a way that ensures the highest possible level of protection against accidents and illness. In addition, they prevent material damage as well as harm to the environment.

Sustainability cannot solely be seen as a concept for preserving environmental resources or saving costs. Ethical basic principles relating to people at the workplace must also be taken into consideration. As such, one protection measure may not be taken at the expense of another. Although prioritizing assets to be protected certainly offers a subject for a philosophical discussion, for our purposes, it is sufficient to keep in view the multiplicity of the problem.

A concept encompassing the intrinsic safety of buildings, facilities, and the people working there, consisting of technical (including structural), organizational, and personal protective measures serves this purpose. To meet this end, it is not sufficient to simply add such safety measures to an otherwise fully planned and coherent concept. Rather, health and safety requirements must be incorporated at a very early stage and thus integrated as seamlessly as possible. Only by taking this approach can the loss of time and costs for improvements, which very often only result in second-class overall concepts, be avoided. Otherwise, organizational or personal protective measures can only resolve deficiencies in technical safety to a limited extent posing problems to a lesser or greater degree. One example of this is when carrying out reactions under pressure. If the appropriate facilities (e.g., autoclave rooms with the required walls and protective equipment for flying debris and escaping gases or flames) are not taken into account during planning, the execution of such reactions will always be strictly limited as secondary measures such as protective walls or fume cupboards and personal protective equipment can only eliminate hazards to a very limited extent. This work is, therefore, drastically limited or even excluded altogether. The retrofitting of the corresponding equipment is very often associated with considerable cost and yet, still does not always result in satisfactory solutions.

The recommendations in the following should therefore not be viewed as legal constraints that arise as a matter of course, but rather as sensible requirements. For laboratories, this system of rules and additional publications has been tried and tested for more than 60 years. The incidence of accidents and illnesses in laboratories is exceptionally low. However, this level is not given; it is only kept low with continuous effort. Despite this, incidents still occur, some of them serious. In order to avoid these, it is necessary to continuously develop and improve the field of occupational health and safety.

33.2 Legal Foundations

The Federal Republic of Germany has a number of parallel legal systems (some of which supplement or overlap each other), which are also influenced or even determined by EU regulations to an ever greater degree. These systems are principally divided into the system of national law, which includes any regulations made by the federal states in Germany, the autonomous bylaws of the German statutory accident insurance organizations (which possess legislative competence under state control as public institutions within the scope of their statute), technical rules (in particular standards), and rules agreed within the scope of relationships under civil law (e.g., the rules and regulations of general insurers). In addition, there are also the interpretations of judicial decisions. All four areas can contain rules relevant to health and safety at the workplace.

Occupational health and safety mainly relies on the standards describing safe devices and equipment. To this end, standardization from the field of statutory accident insurance is supported and, depending on the availability of resources, active involvement in drawing up standards is provided. Agreements concluded within the scope of relationships under civil law and associated rules are also not covered in this chapter.

It is necessary to deal with the areas of state occupational health and safety and statutory accident insurance in greater detail. The German central government and federal states are responsible for the field of state occupational health and safety, while statutory accident insurances fall within the competence of the DGUV institutions for trade and industry, the public sector and the agricultural sector.

With regard to German Basic Law, the state is responsible for regulating the safety of employees, in this case in the laboratory. The German constitution, the Basic Law, stipulates that the state is to perform this duty in a suitable, appropriate, and proportionate way (Art. 20 (3) of the German Basic Law). To this end, it has the right to pass laws and provisions which can be backed by accompanying rules and regulations. These are addressed to the employer and superiors, that is, the line managers. However, in order to enable them to fulfill their responsibility, third parties such as architects or laboratory planners must know what is required from the operators. Responsibility for quality or suitability may also be placed on architects, planners, and equipment suppliers. This applies to the same extent to the statutory accident insurance regulations, which will be discussed later.

The state regulations are now based to a great extent on European guidelines and are subject to their obligational transposition into national law. With regard to occupational health and safety, they represent minimum requirements, the level of which can be raised within the scope of their transposition into national law but never fallen below. Up to 1 December 2009, these were based on Art. 137 of the "Treaty establishing the European Economic Community" (Treaty of Rome), and since then, on Art. 153 of the "Treaty on the functioning of the European Union" (Lisbon Treaty). On the other hand, movement of goods, which also affects aspects of health and safety, was based on Art. 95 of the Treaty of Rome and has now been replaced by Art. 114 of the Lisbon Treaty. No national derogation is permitted here.

European regulations, on the other hand, apply directly to the entire European Union without transposition into national law. Examples of this are the REACH regulation, which covers the registration, evaluation, and authorization, as well as the limitation of chemicals in the EU, and the European CLP regulation for the classification, labeling, and packaging of chemicals (the implementation of the Globally Harmonized System – GHS – for the worldwide standardization of these regulations, in particular for simplifying the global movement of goods). Such regulations also have an impact on construction and equipment, for example, when considering how to label pipes. The transition period for the labeling of pure substances ended on 1 December 2010; for mixtures, such as solutions, it expires on 1 December 2015.

In the field of health and safety in laboratories, the following laws and regulations are especially relevant:

- Occupational Health and Safety Act
- Chemicals Act
- Genetic Engineering Law
- Ordinance on Hazardous Substances
- Ordinance on Biological Agents
- Ordinance on Industrial Health and Safety
- Ordinance on Workplaces
- Genetic Engineering Safety Ordinance
- Occupational Medical Prophylaxis Ordinance.

All of the above include requirements that affect planning, construction, and equipment. The sometimes rather abstract requirements of these laws and regulations are in many cases interpreted and specified in the form of technical rules, which are drawn up by the bodies of the responsible Federal Ministries. These include the "Technical Rules for Hazardous Substances (TRGS)," the "Technical Rules for Biological Agents (TRBA)," the "Technical Rules for Workplaces" (ASR, for example, ASR A2.3 "Escape routes, emergency exits, escape, and rescue plan") and the "Technical Rules for Operational Safety" (TRBS, for example, TRBS 1201 "Inspections of working materials and technical equipment subject to mandatory surveillance").

While the TRGS and the TRBA have a longer tradition and are therefore available in a relatively complete form, the rules for operational safety are still under development. In addition, all of these rules are subject to a more or less continuous reviewing process.

33.3
Laboratory Guidelines

For over 60 years, a DGUV committee made up of laboratory specialists under the auspices of the German Accident Insurance and Prevention Institution for Raw Materials and the Chemical Industry (BG RCI) has updated the laboratory guidelines (BGI/GUV-I 850-0 Working Safely in Laboratories) as the standard set of rules in Germany for laboratory safety. The committee's diverse composition of members from industry, universities, research institutions and companies, standardization, laboratory construction, trade unions, and other stakeholders allows it to keep up with legal and technical developments and react to events such as accidents or unexpected issues without delay as well as to offer advice. New findings are published on the Internet at short notice and the laboratory guidelines and other relevant publications are updated (*www.guidelinesforlaboratories.de*). The committee ensures conformity with the applicable legal situation so that users can generally assume that they comply with legal requirements relating to hazardous substances in the laboratory by observing these guidelines.

TRGS 526 "Laboratories" is of particular importance for laboratories. This technical regulation was drawn up as a result of cooperation between the state and DGUV in the Laboratories Workgroup along with the revision of the laboratory guidelines in coordination with the Hazardous Substances Committee of the German Federal Ministry of Labour and Social Affairs. It is, therefore, also focused on the elaborations regarding laboratory guidelines. TRGS 526 contains the same wording as the laboratory guidelines. This joint drafting and additional publication by the ministry gives the laboratory guidelines additional legal weight; after all, the employer should be able to assume that by applying the Technical Rules, they are complying with the requirements of the regulation. This creates additional legal certainty.

On the legal level, the requirements must be met in their respective scope of application. These are laws, regulations, and provisions laid down by providers of statutory accident insurance. The Occupational Health and Safety Act, the Ordinance on Hazardous Substances, and the Prevention Principles are examples of this. Users must check, in particular on the basis of the scope of application and the definition of terms, whether this provision applies. If so, adherence to these guidelines is mandatory. Exceptions are occasionally formulated that may abrogate the obligation. If no such exceptions are present and the person subject to the regulations still wishes not to comply, an exemption request must be submitted to the relevant authority or the statutory social accident insurance institution. Otherwise, fines or even punishments (under criminal law) may follow, and in the event of damage recourse would also result. An exemption can be denied following a review of the facts, in which case the requirement must be met. After all, one will generally not receive exemption from observing speed limits on the road for one's own vehicle, as finding a reasonable justification will prove difficult. Failure to observe this rule would then have well-known legal consequences and may also result in insurance institutions refusing to provide cover. It is therefore advisable to observe laws, even if the content of these regulations may not always reveal itself immediately. It may be necessary in this case to obtain advice when one's own specialist knowledge is not sufficient.

Rules and publications below this legislative level are handled in a less restrictive way. These include the Technical Rules of the state committees and the rules and publications of the statutory accident insurance providers. These are drawn up by broad-based specialist committees. Deviation is possible here (if these rules are not made mandatory through other provisions or agreements), although it is then necessary to prove that one's own regulation for the respective individual case represents a solution that guarantees at least the same degree of safety. One example of this is the obligation in the laboratory guidelines to equip laboratories with fume cupboards. These are intended to protect people in the laboratory from (inadmissibly high) exposure to hazardous substances when breathing air, to provide protection (within sensible limits) from explosions, and also from splinters and splashes coming from the fume cupboard. If the operator wishes to deviate from this, for whatever reason, it must first be checked whether there are any hazards which could be contained by a fume cupboard. For planning, it is also important to know the intended use and whether changes to that use, that may lead to mandatory retrofitting can be foreseen. Such a change in use is of course only possible when the necessary measures have already been taken at a previous stage and to the required extent. In our example, this would mean that work that – for the first time – requires a fume cupboard cannot be performed until a fume cupboard is fitted (if this is possible at all). It goes without saying that this may present a major problem, as important work is hindered as a consequence, research results delayed or development work held back. In both cases, this can be extremely damaging when a competing research group is able to claim leadership, subsidies cannot be applied for in time, or a competitor is able to bring its development to market maturity quicker. If this predicament results in the work being carried out without a "functional and tested fume cupboard," this may bring about legal consequences, in particular if it results in an accident.

On the basis of a sensible assessment of use, planners and operators are both charged with the task of defining the number, construction type, and spatial arrangement of the fume cupboards which is appropriate for current and anticipated utilization purposes and determining which precautions must be taken for unforeseen changes. These form part of a package of measures made up of technical, organizational, and personnel-related steps, including technical reserves (fume cupboards, ventilation capacities, surfaces, energies, and more) and an operator's package of measures, which in the event of a change must be implemented if that change results in limits being exceeded when the existing infrastructure is used. This in turn comprises a new risk assessment taking into account the new basic conditions and the specification of measures, for example, retrofitting procedures or – possibly on a temporary basis – different scheduling and spatial distribution of work. This is in order to better utilize existing infrastructure, potential expansion of protective equipment for a limited time, as well as a requirement to train and often also motivate all staff concerned, whose work structures change to a greater or lesser degree as a result. It is therefore important to achieve a sensible balance between individual requirements within the scope of the legal regulations.

If, however, the corresponding hazards are indeed absent, the protective measure can be eliminated with a clear conscience. By definition, a protection target is also achieved when no protective measure was taken in the absence of a hazard to be protected. Yet, a recommendation to dispense with fume cupboards in laboratories cannot usually be made, as hazardous substances are handled – albeit quite different in type and scope – in virtually all laboratories, with the exception of facilities that are also called *laboratories* but are not within the scope of this publication. However, hazardous substances are used even in an electronics laboratory, a sleep laboratory, maybe even in a "theater laboratory"; one only needs to take a look at the range of cleaning fluids used. Naturally, these generally do not necessitate the use of a fume cupboard. Such cleaning media are certainly designed in such a way that they can be used without the need for a fume cupboard, although depending on the composition, gloves and protective goggles may be worn as a necessary and sufficient safety measure.

Another way to meet the requirements is to take measures that guarantee at least an equal level of safety. Using our example of the fume cupboard, this would mean other technical protective measures or suitable laboratory procedures that can replace a fume cupboard. As such, instead of a fume cupboard, a glove box, for example, can be installed to entrap the hazardous atmosphere or a laboratory device can be closed in such a way that no gases, vapors, mists, or dust can escape, for example, during operation under vacuum. However, these alternatives also have their limits: working in a glove box is significantly more time-consuming than working in a fume cupboard, the vacuum technique described here does not work well with substances with a very low vapor pressure, and the flexible extraction arm can only represent a sensible alternative when hazardous substances are sufficiently captured. If there are no type-approval tests for alternative equipment or devices that demonstrate sufficient effectiveness under the basic conditions applicable for the devices, the effectiveness of the measures must be checked on a case-by-case basis. This can be done in an individual check, verification by calculation or a conclusion by analogy. In each case, this must be performed by persons with the required level of specialist knowledge and experience. TRGS 402 is a good reference for hazardous substances.

Although it is not mandatory to obtain an approval to derogate from these rules (n.b. that it is mandatory for derogations from acts and ordinances), it is, however, advisable that internal staff members (including the safety officer, occupational physician, and works or employee council) notify the responsible representatives of the authorities and the statutory accident insurance institution, and that they document this.

33.4
Hazardous Substances

The national law covering hazardous substances is now almost completely based on European Union law. A central pillar of occupational health and safety is the Ordinance on Hazardous Substances (GefStoffV), currently in its 2013 version. A new version is planned for 2015 with the addition of risk-based limit values for carcinogenic, mutagenic, and teratogenic substances. The Ordinance on Hazardous Substances has a broad scope of influence, and as such, only few areas or activities with substances and materials do not fall under the Ordinance in one way or another:

- Substances and preparations in accordance with Section 3a of the Chemicals Act (ChemG) or other chronically harmful substances and preparations
- Explosive substances and preparations
- Substances, preparations, and products that can produce an explosive atmosphere
- Other hazardous chemical agents that present dangers such as risks of suffocation, damage through heat or cold or also through working in humid conditions, or harm caused by the degreasing of skin.

This means for the laboratory that alongside the "classic" laboratory chemicals with their orange hazard symbols (in future the new, red-framed pictograms as per the new identification regulation of the GHS), also contamination, reaction products, (including undesired ones) or liquid nitrogen must also be taken into consideration. The substances used are handled at laboratory workplaces in accordance with the guidelines of the Ordinance on Hazardous Substances depending on their toxicological properties. Laboratories operated in line with the laboratory guidelines generally allow for work with toxic materials and substances that cause respiratory sensitization, and with some organizational alterations also with CMR substances (activities with carcinogenic, mutagenic, and reprotoxic substances, whereby the teratogens are covered under the maternity protection law for legal reasons). The protective measures that derive from the toxicological properties of the substances are currently found in Sections 8–10, and those resulting from other substance properties (named "physical–chemical" properties in the Ordinance) are contained in Section 11. These mainly encompass fire and explosion hazards, but also others, such as risks of suffocation through cryogenic liquid nitrogen. Other concepts for protective measures, such as those based on internal pharmaceutical guidelines, can be integrated to advantage here.

33.5
Biological Agents

Similar to hazardous substances, there are rules in place regulating work with biological agents. In this connection, there are regulations both for genetically modified and nongenetically modified biological agents. Biological agents are:

- Microorganisms (bacteria, viruses, and fungi)
- Human pathogenic endoparasites (for technical reasons some ectoparasites are also included in the DGUV regulations as biological agents, for example, jigger fleas or anopheles mosquitoes)
- Causes of the transmissible spongiform encephalopathy (TSE, BSE; this is actually an effect of exposure to hazardous substances through the autocatalytic folding of a protein, but which manifests itself in the manner of an infection).

Cell cultures are also covered under biological agents in the BGRCI regulations. As well as so-called targeted activities, the far more common nontargeted activities in which the biological agent is an undesired factor, for example, mold in archives or microorganisms in air conditioning units, are also grouped within this category.

Key provisions here are the Ordinance on Biological Agents and the Genetic Engineering Safety Ordinance, which also contain requirements for laboratories depending on the risk group (1, 2, 3, 3**, or 4) of the biological agents. The Ordinance on Biological Agents also includes the TRBA. TRBA 100 (laboratories) and 120 (keeping of animals for experiments), as well as 460–466 (BG RCI classification lists for biological agents, which are also integrated in national law within the scope of the cooperation model) are also important for the laboratory. BG RCI information sheet B 002 provides further information on the construction and operation of such laboratories. Requirements relate, among other things, to floor design or permanent negative pressurization with double door systems.

In all cases, the provisions for other hazard areas should also be taken into account: for example, chemical hazards that regularly occur in microbiological and genetic engineering laboratories are covered under the laws relating to hazardous substances. The laboratory guidelines should therefore also be observed. If this is forgotten, the unfavorable situation may arise in which, when setting up a building for the purpose of researching genetic engineering, although safety cabinets were intended, no fume cupboards were planned for activities with hazardous substances that result in gases, vapors, or dust being produced (e.g., toxic vapors from trichloromethane from a phenol/chloroform solution).

33.6 Other Hazards

The risk assessment must also include all other areas that present potential hazards such as electrical dangers, risks through radiation, or noise as well as mental stress. Naturally, this can already have an impact as early as the planning and construction stage, but certainly also when equipping laboratories. If the separation of sources of noise (for example, mechanical workshops or vacuum pumps) from work areas is not taken into consideration during the planning phase, this can affect work performance and well-being and in the worst case, may result in hearing damage. If in the planning phase unfavorable locations have been chosen for office and writing areas, the long distances to be covered by staff will probably lead to such work being carried out inside the laboratory itself instead. On the other hand, if these areas are poorly integrated into the laboratory, problems then arise relating to ergonomics and staff wishing to remove their protective goggles in the writing area. Even today, the integration of computers in the laboratory continues to present problems and is occasionally poorly executed. Information sheets A 016 and A 017 as well as BGI 850-1 (Risk assessment in the laboratory) from BG RCI provide help on the overall topic. Information sheet A 017 presents a general list of risk and stress factors in all workplaces. BGI 850-1 takes this up and systematically assigns the measures contained within the laboratory guidelines and other selected sources to the individual factors.

33.7 Occurrence of Accidents and Illnesses

For many years, safety technology in German laboratories has been highly advanced. As proven and tested practice, this is integrated as early as the planning and design phase, thus giving laboratories a considerable degree of tolerance toward user errors as well as unforeseen or even objectively unforeseeable events and accidents. The incidence of occupational diseases in laboratories is therefore fortunately low-key, and in terms of accidents, too, laboratories are very safe workplaces. However, the occurrence of accidents can never be ruled out completely and sometimes with very tragic results. Even today, people lose their lives in the laboratory, usually as a result of mistakes against which insufficient precautions were taken. The reduction of space or energy input must under no circumstances lead, for example, to running autoclave reactions (to an uncontrollable extent) taking place (or having to take place) in the laboratory instead of in a suitable facility (autoclave room) or, due to the absence of a (mini) pilot plant, activities with quantities of hazardous substances that cannot be controlled in a normal laboratory but take place there nonetheless. A bursting autoclave can propel debris with a mass of several kilograms several meters at high momentum, and excessive quantities of solvents in apparatuses that barely fit in a fume hood can lead to fire and explosion with terrible consequences for laboratory staff. Increasing the risk of accident or illness by simply reducing the ventilation must never be tolerated – intelligent and feasible concepts for alternatives are required here. Planning based on reactions on the milligram scale for reducing risks is inappropriate in a research area in which syntheses are regularly performed over many stages and must therefore start with relatively large quantities. N. b. the construction of autoclave facilities needs in-depth expert knowledge.

33.8
Risk Assessment and Measures

Risk assessment is an important instrument for controlling the selection of the correct protective measures. Activities may only be undertaken following a risk assessment and after the required protective measures have been taken. As such, the risk assessment systematizes processes that would otherwise run on an intuitive basis. This systematic approach forces those involved to also deal with issues with which they are not familiar. Another factor playing a role here is the so-called organizational blindness, which means that aspects from personal day-to-day work are blanked out, thus allowing even major errors to be overlooked.

Unfortunately, checklists in laboratories can only be of limited help. In cases where the matters in question are relatively simply structured and only change slightly, checklists can result in at least the most important points being taken into consideration. In laboratories, with their highly complex and flexible framework of building, people, equipment, and chemicals, as well as fast-changing (particularly in the area of research) tasks and objectives, it is only possible to perform rudimentary risk assessment tasks using checklists. In general, important aspects, which in the individual case may even be of decisive importance, cannot all be contained within the checklist. By relying solely on the checklist, critical problems may be overlooked while enjoying a false sense of safety. In this way, details that are not always of general interest but which can in individual cases lead directly to a catastrophe may simply be missing in such a list for reasons of length. For example, the use of standard checklists would not be adequate for performing safety assessments of autoclave rooms in the condition in which they were frequently found by the author, which would present a significant hazard for the building and the people within it, due to the resulting pressure wave, propelled debris, and escaping hazardous substances (alongside fire and explosion hazards this is also a question of toxicological properties) in the event of an accident.

According to the Occupational Health and Safety Act (Sections 5 and 6), the assessment of hazards in every workplace is mandatory for the employer and management staff. Risk assessments must fulfill various requirements such as:

- Determining the hazards
- Assessment
- Defining preventive measures
- Employment limitations
- Documentation.

For the purpose of assessment, groups of comparable workplaces can be formed. This is a living system that needs to be updated continuously. If used correctly, there is a chance of detecting hidden risks from systematizing intuitive processes and preventing them from occurring.

For this purpose, the operator or user generally needs to build on the protective measures offered by the building and its equipment. If they later recognize that the building, its infrastructure (which he or she generally does not know in detail and does not wish to know) or the laboratory equipment does not permit an effective measure, the dilemma is obvious. If the capacities for incoming and exhaust air are not sufficient for building additional fume cupboards, but these are urgently needed, then work might be disrupted. If there is no space for setting up a compressed-gas cylinder cupboard, it is generally necessary to remove the cylinders from the laboratory and preferably feed in the gases via pipes, which would be accompanied by extra costs and delays. If on the other hand compressed-gas cylinders have to be constantly fetched and removed again in order to perform the work, the risk of an accident during transportation rises and the procedure becomes an additional burden. If the cylinders remain in the laboratory, this will lead to significant risks in the event of a fire due to bursting cylinders and their released content.

The fact that risk situations change fast in laboratories due to changing materials and procedures means that the otherwise conventional method of determining protective measures on a case-by-case basis depending on the properties of substances and the work involved is often not suitable for laboratories. Safety in laboratories is therefore primarily determined by the building, equipment, procedures, and operation as well as the training of the laboratory personnel. By combining technical, organizational, and personnel-related measures, risks relating to the work in laboratories are minimized. Building and equipment therefore play a central role in determining the activities that can be performed. Because it is not possible to completely eliminate accidents due to a naturally incomplete knowledge of the properties of new substances or research chemicals, laboratories must use this intrinsic safety concept to control accidents within certain basic conditions in such a way that there are no unacceptable consequences for people and the environment or for the building and equipment. Such basic conditions (according to which a laboratory that is built, equipped, and operated in compliance with applicable laboratory guidelines will provide sufficient controllability) can be found in the laboratory guidelines. If these basic conditions are exceeded because, for example, significantly larger quantities of solvents need to be used, additional measures must be taken, which could also include structural work.

Laboratory operators can generally expect that the exposure of hazardous substances is within limits when specialist and reliable staff work in accordance with the applicable guidelines using the latest technology and in particular, adhere to this rule and standard laboratory conditions (suitable work procedures, generally relating to fume cupboards and the adherence to quantity limits for hazardous substances).

Laboratories handling hazardous substances are generally exposed to the following risks:

- Risk of fire and explosion through flammable solid, liquid, and gaseous substances
- Risk of damage to health caused by solid, liquid, and gaseous substances
- Hazards from unknown, severe, or continuous reactions as well as risk to eyes and skin through corrosive and irritating substances

- People working in laboratories are often burdened or put at risk from:
- Insufficient lighting or lighting that is inadequate for the visual task to be performed
- Unfavorable indoor climactic conditions
- Hazards from containers subject to overpressure or vacuum
- Hazards from hot or cold surfaces and media
- Noise from devices and equipment, mechanical hazards from devices and equipment
- Risk to skin through working in wet conditions, in particular when wearing gloves
- Risk of slipping due to wet surfaces
- Risk of stumbling
- Burdens on the musculoskeletal system due to repetitive activities or constrained postures
- Psychological stress from repetitive activities, time pressure, isolation, high concentration requirements
- Stress on staff due to personal protective equipment.

Further risks arise, for example, from:

- Ionizing radiation
- Electromagnetic fields
- Optical radiation (UV, laser, IR) and
- Biological agents.

The risk assessment includes compiling all required information, in particular with regard to the properties of substances used and of those produced (in the form of main products or by-products, contamination, also if the reaction process does not proceed as planned or when the substances age). Furthermore, information is required on the devices used as well as the laboratory building and its equipment in order to allow effective protective measures to be sensibly defined. It is vital that the effectiveness of measures is checked. Failure to identify a lack of effectiveness or ignoring such a fact creates hazards.

Further, elements of the risk assessment are the determination of exposure, which in the laboratory only needs to take place through an exposure measurement as and if necessary. In general, it can be said that when the specialist and reliable staff work in line with the applicable provisions using the latest technology and in particular work in accordance with this rule and under normal laboratory conditions (see Section 3.3.3 of the laboratory guidelines), it can be assumed that they will not be subject to impermissibly high levels of exposure. Exposure can be determined by measuring hazardous substances, performing model calculations or through analogies. As such, one can typically expect that limit values will not be exceeded when work is performed in line with the applicable provisions, the general rules for health and safety (in particular the laboratory guidelines) and using procedures developed by the specialist and responsible staff.

Connected to this, however, is the requirement that buildings and infrastructure are appropriate for this purpose and are consistently kept at this level of functioning. This can be problematic in an age in which competencies and responsibilities are often divided. It is crucial that users and operators work together with planners to draw up specifications in order to avoid failures arising from not knowing correlations, effects, and task distribution. For example, if the building operator and the planner are not aware that the user must also carry out reactions under pressure, they will not provide an autoclave room or even a separate autoclave building right from the outset. However, if it turns out that such a room or building becomes necessary, more or less unsafe makeshift solutions may be implemented. If the building operator performs work on the laboratory water supply without first consulting the user, it could happen that in the event of an accident, victims may find themselves standing under a dry emergency shower that nobody had felt responsible for checking.

The level of training of the laboratory staff is an important component in the safety concept, as the technical measures can only work seamlessly when used and operated in an appropriate and responsible manner. After all, what use are fire doors when they are held open with door stoppers? However, in such a case, it may be asked whether it was clear as early as the planning stage that the movement of people, combined with the transport of chemicals, devices, compressed-gas cylinders, and so on, would make it impractical to constantly have to open heavy doors. The use of a device for holding open the doors connected to the fire protection system may have been eliminated for cost reasons, thus endangering the building with the people and work results within it in the event of fire. Whether this can be counted as planning of effective protective measures is certainly questionable. And how are the chemicals and compressed-gas cylinders actually transported between floors if there are no suitable lifts (goods lift, external control)? Due to the lack of alternatives, in a standard lift accepting the risk that in the event of a leakage any help may come too late, or by using the stairs, where the risk of falling is involved. However, one will always have to make compromises; it cannot be a matter of introducing luxury solutions on the grounds of the safety argument. On the other hand, the specified minimum level must be attained.

"*Normal laboratory conditions*" are defined as framework conditions for the effectiveness of the measures contained within the "laboratory guidelines" and are mainly based on using suitable apparatuses and equipment as well as the limitation of quantities used. When diverging from these framework conditions, the dangers that increase as a result must be assessed separately and covered by protective measures. It is therefore not necessarily true that these quantities cannot be handled in laboratories, but rather that they can require corresponding additional measures. If these additional measures are not taken, however, the work cannot be carried out in the laboratory. Further aspects of the safety concept are rules on emergencies and faults, on the ingestion of hazardous substances through injuries, on activities of third parties in laboratories, on special activities that cannot be fully secured by technical protective measures, on the reaction process and properties of new substances, on employment restrictions and on the documentation of risk assessment.

Technical, organizational, and personal protective measures frequently interact in these safety concepts. Shortcomings in one measure can only be compensated by others to a limited extent. This is why careful planning of the initial situation is so important. If there is no pilot plant or overnight room for working with larger quantities of solvents or for experiments over a longer period of time, altered working conditions can only compensate this to a limited degree.

In laboratories, there are only very limited possibilities for gathering all necessary information within the scope of a risk assessment, assessing it and then performing the actual required measures. The high level of complexity of changing risk situations usually demands a pragmatic approach. It has therefore paid off to equip laboratories to a minimum standard that permits current laboratory activities in accordance with regulations without having to constantly think about any retrofitting or modification that may have become necessary. This is also in line with TRGS 400 "Risk assessment," which permits the user to deploy ready-made risk assessments in conjunction with the respective measures. Due to the large diversity of laboratories, which can never be fully assessed in advance, alternative measures can still be taken that offer the same level of protection and are equally effective. There will always be cases in which a different protective measure may even be better suited for the case in question.

The laboratory guidelines allow this type of approach, in that, Sections 1–4, 6, and 7 describe those measures that are required in laboratories or that one can expect to be necessary. For example, very frequent activities with flammable liquids or toxic substances are possible within the limits specified in Section 3 without further measures generally having to be taken. It is therefore only necessary to check whether all activities performed are covered within this framework or whether there are special substances or procedures that necessitate an additional assessment. Additional measures can be taken on a modular basis from Section 6 where necessary, which means that activities with carcinogenic substances, too, are possible without further limitations. Of course, the obligation remains to, if necessary, take stricter or additional measures than those described in the laboratory guidelines, but on the other hand, also allow the freedom to omit measures in individual cases if the risk assessment objectively permits this. It is therefore certainly possible for laboratories, or at least individual laboratory rooms, to manage without a fume cupboard or emergency showers. However, one also has to accept certain limitations in the work that can be performed. If at short notice work needs to be done that requires such equipment, either another equally safe solution must be found or the work will be held up due to the necessity of an – expensive and time-consuming – retrofitting process. Missing ventilation equipment or capacities regularly present a major problem. This begins with the question of where the currently required fume cupboard can be located, continues with the increasing number in analytical blank values caused by the formation of dust when making openings in walls in order to lay new ducts, and ends with the question of how to create the missing infeed and exhaust air capacities for ventilation. During the planning phase it is, therefore, recommended to consider very carefully with laboratory planners and architects, safety specialists and in particular the users of the laboratory, which measures can really be omitted and which may also be necessary for the individual cases above and beyond the guidelines of BGI/GUV-I 850-0.

These deliberations may stand in the way of the goal of enabling a laboratory to be set up quickly with as few costs as possible and with minimized energy requirements. If the starting point here is defined in such a way that one side of this conflict is favored too strongly to the detriment of the other, subsequent problems are present from the outset. The art of laboratory planning will therefore focus in future even more on bringing together demands that are sometimes difficult to harmonize to produce forward-looking solutions. For this purpose, it is far more necessary than previously that well-qualified and experienced specialist planners intensively involve all stakeholders, in particular, the users and the operators of the buildings and their infrastructure. Upon commissioning, a principal should ensure that the potential contractors possess the required know-how and experience in order to suitably take health and safety matters into consideration. A common language that allows the demands of all parties to be defined and discussed, as well as the knowledge of all relevant framework conditions is vital here. We hope that this book can make a contribution to this.

References

1. Bräunig, D. and Kohstall, T. (2013) *Calculating the International Return on Prevention for Companies: Costs and Benefits of Investments in Occupational Safety and Health*, DGUV, Berlin, *http://publikationen.dguv.de/dguv/pdf/10002/23_05_report_2013-en--web-doppelseite.pdf* (accessed 28 November 2014).
2. Lenk, H. and Maring, M. (2003) *Natur – Umwelt – Ethik*, Münster.
3. Brock, T.H. (1997) *Sicherheit und Gesundheitsschutz im Laboratorium*, Springer, Berlin.
4. Adelmann, S. and Schulze-Halberg, S. (1996) *Arbeitsschutz in Biotechnologie und Gentechnik*, Springer, Berlin.
5. SG Laboratorien der DGUV (2014) BGI/GUV-I 850-0. *Sicheres Arbeiten in Laboratorien* (the "laboratory guidelines." Available in English as BGI/GUV-I 850-0e, "Working Safely in Laboratories"), *www.laborrichtlinien.de* and *www.guidelinesforlaboratories.de* (accessed 28 November 2014).
6. BGI 850-1. *Gefährdungsbeurteilung im Labor*, BGI 850-2. *Abzüge*, BGI 850-3. *Besondere Schutzmaßnahmen* (in preparation), DGUV-I 213-853. *Handhabung von Nanomaterialien in Laboratorien* (available in English as DGUV 213-854); further information available at *www.laborrichtlinien.de* (accessed 28 November 2014).
7. Information sheets and other publications of BG RCI: see *http://bgrci.shop.jedermann.de/shop/* (accessed 28 November 2014).
8. State Technical Rules: see Bundesanstalt für Arbeitsschutz und Arbeitsmedizin (BAuA) *www.baua.de* (accessed 28 November 2014).

34
Operational Safety in Laboratories

Norbert Teufelhart

34.1
Safety Principles

The safety in laboratories is, on the one hand, affected by the construction and infrastructure, as well as the available resources and devices but, on the other hand, by the experimental procedures and specific operation, and is last but not the least, significantly influenced and co-determined by the skills of the employees. Hence, the construction of laboratories according to the appropriate regulations and state of the technology is a fundamental prerequisite for operational safety that can only be guaranteed in combination with an operation that conforms to the regulations. In this context, structural, technical, and personal expenses should always be in reasonable proportion to the objective.

That is why the objective of a reliable operational concept for laboratories in a research building is to guarantee at least appropriate protection for the employed personnel and their fellow human beings in the surroundings, depending on the respective hazards and risk potentials (protection levels). For this purpose, it is necessary to fulfill fundamental operational and safety principles of corresponding lab areas. This starts with the compliance with the regulations and the fulfillment of the state of science and technology, and continues with guaranteeing the most efficient and low-risk possible work sequences, considering the results of a hazard assessment (workplace and activity-related protective measures), safeguarding the most hazard-free possible maintenance and upkeep, and extends not in the least to the definition of preventive measures (hygiene plans, release certifications, screenings, etc.).

In order to guarantee the operational safety in the laboratory, generally applicable principles and rules must be observed on the one hand, and additional protective and safety measures must be implemented on the other hand, depending on the respective type of laboratory and the activities associated with it. Primarily, the general measures conduce to cleanliness and hygiene, and the specific safety measures guarantee the required product, personal, and environmental protection.

The Sustainable Laboratory Handbook: Design, Equipment, Operation, First Edition.
Edited by Egbert Dittrich.

34.2
Safety Management

Since the German Occupational Safety and Health Act came into effect in the year 1996, the basic duties of the employer include only maintaining safety and the protection of health of the employees through appropriate measures at work. In fact, the employer is requested to establish a suitable occupational health and safety organization which guarantees the planning and implementation of these protective and safety measures. Furthermore, the employer is obligated to make arrangements for these measures, while included in the operational management structures, are adhered to in all activities and the employees can fulfill their obligations to cooperate.

34.2.1
Occupational Health and Safety Organization

An extensive and efficient occupational health and safety organization is a basic prerequisite for effective occupational health and safety. This organization is all the more significant for operators and managers of laboratory buildings based on the given increased hazard potentials.

In this regard, a functioning basic structure is characterized by the fact that the employer implements at least the organizational requirements of the Occupational Safety and Health Act. The necessary operational and organizational structure is guaranteed by naming and appointing, in writing, the essential occupational safety and health officers – specialized personnel for occupational safety, company physicians, and safety officers of the different laboratory and task areas – as well as providing them with the necessary abilities, authority, and financial freedom of action.

People in charge – employers and managers – and occupational health and safety officers are appointed to a parent body, the so-called occupational health and safety committee. This is where the tasks and goals of occupational health, in general, and safety, in particular, are communicated and coordinated. The required measures are deliberated and determined and, last but not the least, their implementation and effectiveness is reviewed.

The occupational health and safety committee is established by the employer, holds meetings at regular intervals, at least once per quarter, and creates a written report and/or prepares decision memos for the management. By-laws are advisable to clearly regulate individual processes.

34.2.2
Occupational Health and Safety Management System and Audits

Based on the increased and considerably differentiated hazard potential in laboratories, special management structures and activities are necessary that will make sure that protective measures are notably taken into account for all hazardous activities and procedures. In order to meet these increased requirements, the introduction of a systematic occupational health and safety management system has proven to be useful. As for the safety, it is of inferior significance whether this voluntary management system is operated as an independent occupational health and safety management system or as an extension of an existing quality or environmental protection management system. In any event, it clearly contributes to an increase of the level of occupational safety since, on the one hand, the individual responsibility is observed to the necessary extent and, on the other hand, the efficiency of trade is optimized as a result of dealing completely with the subject on a high level. The system includes organizational structures, responsibilities, strategic planning, and test instruments as well as methods, procedures, and resources to develop, fulfill, and assess, as well as maintain the specified protection and safety objectives in the different lab areas. The system is characterized by the systematic planning and organization of occupational health and safety protective measures.

Planning includes,

- the determination of appropriate safety objectives and priorities
- the description of the activities associated with it
- the definition of periods to be observed
- the course of action in case of arising changes and malfunctions of the processes.

Organization is implemented through the

- goal-oriented selection of personnel and resources
- determination of responsibilities and skills
- systematic identification and assessment of risks at the workplace
- determination of protective measures and control mechanisms as needed
- scheduled inspection of individual work areas and consistent execution monitoring
- systematic monitoring of dates of safety-relevant measures (review of working materials and resources)
- management of relevant safety field books
- preventive occupational medical care
- preparation for hazard and emergency situations
- regular reporting to document activities
- continuous briefing and qualification of executives and personnel through training and instruction.
- recording, reporting, and analysis of accidents and unexpected events.

The voluntary implementation of system and compliance audits is a useful addition to increase occupational safety in the laboratory. The primary purpose of these audits is to reveal organizational weaknesses and to identify defects related to safety or relevant to health at the workstations and in experimental procedures. Fundamental optimization potentials are identified quite often in the process. Self-monitoring that is perceived with discernment not only improves the transparency of structures and processes in the laboratory but also promotes understanding and acceptance for safety measures which, at first glance, appear to be obstructive and inconvenient or even superfluous. If the monitoring is planned well and systematically implemented on one's own authority, the activities in terms of occupational safety are not only limited to specially sensitive lab areas – laboratories of high protection and safety levels – but apply in the entire operation. Practical experiences prove that, by implementing laboratory safety audits as a result of improved internal communication, an interdisciplinary transfer of knowledge is set in motion that is also beneficial to safety.

34.2.3 Hazard Assessment

The hazard assessment is an effective and central instrument for the largest possible targeted occupational safety and health protection of the employees in the laboratory. It is an essential part of a concept of systematic prevention of occupational accidents and work-related health risks, including ergonomics at the workplace. The goal-oriented safety measures are based on the assessment of the hazards associated with the different laboratory activities. Based on professional and medical care expertise about causes, types, and effects of hazards, this instrument is used to systematically determine laboratory-specific hazard factors at the workplace and to specifically assess the hazard potentials on the basis of criteria that depend on the procedure and activity. The frequency of the occurrence of damage must be estimated in connection with the anticipated extent of the damage. The nature, duration, and extent of the exposure play a decisive role. Goal-oriented and need-driven safety measures must be defined and taken, and their implementation consistently monitored in accordance with the result of the assessment.

Last but not the least, the hazard assessment is useful for the managerial ability of the people in charge since it helps to set priorities and make the necessary decisions. The hazard assessment must be performed prior to accepting the tasks and in case of operational changes. Additionally, it must be repeated and updated at appropriate and regular intervals, but in any event, at least every three years. Where possible, the rule of substitution must be observed, that is, replace hazardous procedures and hazardous materials with less hazardous ones.

An extensive, possibly complete collection of information is an essential basis of the assessment. Tools and procedural guidelines of BG-RCI such as hazard catalogs "hazard assessment – check lists, hazard and stress factors" (bulletin A 017) and "hazard assessment in the laboratory" (BGI 798) and BGI 850 "working safely in laboratories," not the least also other bulletins on the topic, for example, the B series "safe biotechnology" have proven to be particularly helpful in this context. The respectively applicable technical rules for instance for hazardous materials (TRGS) and biological agents (TRBA) provide concrete information.

Due to the great multitude of laboratories and fields of research, and an even larger number of procedures and activities with the most diverse hazardous working materials – dangerous substances, agents, biological agents, radionuclides, and so on – and alternative hazards such as optical radiation and electromagnetic fields, the otherwise common case-related approach often cannot even be put into practice. Nevertheless, to achieve maximum safety during common laboratory operation, the Brains Trust defined basic conditions to work safely, which are to be observed in the laboratory. These basic conditions include the following points:

- construction and infrastructure according to BGI 850 and the relevant regulations
- use of skilled personnel
- working according to the relevant rules and state of technology
- working on the laboratory scale (quantities common for a laboratory according to BGI 850)
- working according BGI 850.

A safety concept is given under these conditions, the consistent compliance of which no longer explicitly requires further hazard assessments for every single attempt. It is hereby considered a so-called procedure-specific criterion (VSK) for which it can be assumed that all necessary safety measures, which would emerge from a hazard assessment according to the hazardous substances ordinance, are present. However, any deviation from VSK, on the other hand, makes a detailed single-case assessment imperative. Activities with large quantities of hazardous substances are for instance not covered by the VSK. As a result, additional measures become often necessary that clearly go beyond the requirements of BGI 850. Only the hazard assessment to be performed shows the actual need for action.

34.3 Regulation of Internal Processes

34.3.1 Skilled Trained Personnel

Qualified and trained personnel are an essential prerequisite to maintain an orderly and safe laboratory operation. From this point of view, the selection of competent and reliable employees, on the one hand, and the work and operation-specific briefing as well as regular technical further training, on the other hand, are of special importance. The employees must be informed about possible hazards and the safety measures to be taken from the start and be kept up to date about safety-relevant information, innovations, and expertise. In this context, the rotational health and safety briefing plays an important, if not the most important role. It is thereby crucial that the training is provided by a competent party, the safety rules are communicated in an interesting and credible manner, and the superiors of course play a true role model. The communicated theory and content must be "experienced" in practice and are considered absolutely binding. The safety awareness of every individual must constantly be resharpened so that the latent, virtual human risk of organizational blindness cannot escalate, because negligence is the greatest enemy to safety.

34.3.2 Laboratory Rules and Regulations and Safety Information

General laboratory rules and regulations – general operating instructions for laboratories – are very useful for efficient occupational health and safety protection. These govern internal processes with a hazard potential and essentially define the required organizational measures. In this respect, these cover all areas of a laboratory operation in which hazardous working materials are handled, supplied, and stored. These generally applicable regulations must be observed by all employees and adjusted to the respective circumstances in relation to the location and activity depending on the requirement of the different workplaces and procedures. Essentially, the laboratory rules and regulations govern and describe the following aspects:

- hazard areas and access authorization
- work restrictions and prohibition from work
- working hours
- work and protective clothes for different safe areas
- general behavioral and hygiene measures
- orderliness at the workstation
- application of safety systems and due diligence
- behavioral patterns in case of hazardous work and in hazardous situations
- general handling of hazardous working materials (hazardous substances, gases, biological agents, etc.)
- handling of storage equipment
- collection and disposal of hazardous waste (waste disposal concept)
- people in charge of occupational health and safety protection (expert for occupational safety, company/occupational physician, etc.).

(Annex: "sample of laboratory rules and regulations").

34.3.2.1 Work and Operating Instructions

These general operating instructions are supplemented by differentiated work and operating instructions that accommodate the respective work and safety areas as well as their specific hazards caused by devices, procedures, and substances. The form and classification as regards to content are largely standardized and describe the following aspects in detail:

- name of the activity(ies)
- hazards for personnel
- protective measures and code of conduct
- behavior in a hazardous situation and first-aid

- measures upon completion of activity(ies)
- disposal measures.

Depending on the respective laboratory workstation, a series of different instructions must be created accordingly, kept available on location for the employees, and be updated promptly as needed (see the draft of operating instructions). This includes, for example,

- work instructions for safety and disposal systems such as hoods, safety cabinets, autoclaves
- operating instructions for hazardous substances
- operating instructions for biological agents
- work instructions for devices with a hazard potential such as centrifuges and vacuum apparatuses
- and many more.

(Annexes: "sample of operating instructions").

Generally applicable skin protection and hygiene plans are created and posted in the laboratory to observe the required skin protection and hygiene conditions (see later).

34.3.2.2 Working Substance Registry

In order to maintain an overview of the inventory of all hazardous working substances – hazardous substances, biological agents, radionuclides, and so on – these are systematically recorded in substance registries such as a hazardous substance registry or biological agent registry according to the regulations. These registries are updated at regular intervals of time. This ensures that all organizational measures associated with it – operating instructions, occupational medical care – are observed.

34.3.2.3 Outside Company Coordination

Written release regulations are required for several safety areas at least as of protection level 2 in order to minimize a laboratory-specific hazard for external people, particularly maintenance and service personnel. The use of a so-called release certification has proven itself in practice. This certification describes, on the one hand, the safety measures to be taken by a person in charge in the laboratory as a mandatory prerequisite to access and/or start the service work and, on the other hand, the safety instructions to be observed by external personnel in the area. At the same time, it is used to brief people and to document the content and measures taken. It must be signed by the parties involved. Something similar applies for the safety briefing of cleaning personnel of an outside company. It must be ensured that the assigned company has the particular expertise and experience that is necessary for the activities and has been briefed about the operational hazards and specific code of conduct prior to starting the work. Services and safety measures must be clearly regulated as per a contract with the contractor. The contractor will then have the duty of briefing his employees accordingly. The so-called complementary safety monitoring is incumbent upon the laboratory operator, that is, he intervenes immediately in case of apparent safety violations and guarantees maximum safety in this manner. As needed, the operator appoints a coordinator prior to accepting the outside activities and initiates a hazard assessment that is specific to the activity. Laboratory operators and contractors work together and make mutual arrangements with respect to the required protective measures.

34.4
Functional Efficiency of Systems and Equipment

The reliable and certain use of resources, systems and devices implies its complete functional efficiency. On the one hand, this is guaranteed by proper commissioning and, on the other hand, by regular maintenance measures and upkeep.

In either case, all laboratory-specific safety equipment must be subjected to extensive visual tests prior to commissioning a new laboratory. The coordination of inlet and outlet air and/or high and low-pressure ratios of the ventilation system, the correct setting of control and monitoring units and not the least, the determination of the effectiveness of all alarm systems as well as the systems and equipment in terms of fire safety are at least equally important.

Maintenance and service work is just as indispensable to maintain performance. The resources and equipment are recorded in a comprehensive testing medium registry to safeguard the regular tests. It is advised to differentiate between physical structures and infrastructure, general working materials and specific safety systems. The testing medium registry defines the required and/or legally prescribed extent of the test according to the state of technology in connection with the testing institute, authorized person, monitoring body permitted for this purpose and determines testing periods based on the present operating instructions or applicable regulations. In particular cases, it can be useful to record the following additional data and to compose documentation and keep it at hand:

- manufacturer's instructions
- location
- operating instructions
- maintenance recommendations
- maintenance contracts
- documentation of deficiencies
- error rates.

These documents and information are not only very helpful when assessing future test periods but are also partially a prerequisite. The compliance with the test periods is governed by law.

(Annex: sample testing medium registry).

In addition to the recurring tests, the commissioning test is of particular importance for more complex and safety-relevant equipment and systems such as a ventilation system or autoclave (see earlier).

34.5
Occupational Medical Care

In either case, the laboratory operator provides appropriate occupational medical care. Essentially, the purpose of this care is to recognize and prevent work-related diseases, including occupational diseases prematurely. Occupational medical care is also used for the individual assessment and consultation of employees with respect to interactions of hazardous activities with individual health.

Every employee is subject to an initial examination performed by a qualified specialist of occupational medicine or an industrial physician. The medical determination that there is no sanity objection to the exercise of the intended activity is a prerequisite to take up the work. In addition, the laboratory managers ensure that all employees who perform activities with hazardous working materials, such hazardous substances receive general occupational medicine and toxicological consultation prior to the start of the laboratory work. Generally, this consultation is provided as part of the initial briefing. The employees are informed about optional and mandatory check-ups just as much as they are advised of particular health risks caused by certain hazardous substances. The consultation includes the involvement of the physician in the event that this has to be indicated for reasons of occupational medicine.

34.5.1
Preventive Measures

The occupational medical care includes all occupational medical measures that are required to prevent work-related health risks. As for activities with hazardous substances, this includes in particular the occupational medical assessment of health hazards caused by hazardous substances and in relation to the activity, including the recommendation of appropriate protective measures, special occupational medical care check-ups for the premature recognition of health disorders and occupational diseases, as well as justified recommendations in terms of occupational medicine to review workstations and to repeat the hazard assessment.

34.5.1.1 Optional and Mandatory Medical Care

Essentially, the so-called optional and mandatory check-ups must be differentiated. This differentiation results from the view of the employee. Both forms of screening are offered by the employer. The employee can elect to accept or reject the optional check-up. The mandatory check-up, on the other hand, must absolutely be observed. A mandatory check-up on a biological agent can only be rejected when the employee is already sufficiently vaccinated, as has been proven.

(Annex: "screenings in case of laboratory activities").

The different necessary occupational health screenings are authorized by the manager as needed. These screenings are performed in the following according to regulations:

1) initial examinations prior to taking up a hazardous activity,
2) subsequent examinations at regular intervals during this activity,
3) subsequent examinations upon completion of this activity,
4) subsequent examinations in case of activities with carcinogenic or mutagenic substances of categories 1 and 2, also upon completion of the work,
5) examinations for a particular reason.

Generally, the screenings include,

- the inspection or acknowledgement of the workstation by the physician,
- the occupational medical questioning and examination of the employee,
- the assessment of the condition of health of the employee taking into account the workstation circumstances,
- the individual occupational medical questioning and documentation of the examination results.

The practical occupational medical examinations are performed purposefully according to the rules established by the occupation cooperative, the so-called occupation cooperative principles. These propose examination content and extent of the examination for specific hazardous substances and activities and, in this respect, they are more or less a binding guideline for the physician and employer. Subsequent examinations are consistently performed according to the terms established therein.

Biological monitoring is part of the occupational medical screenings as far as recognized procedures are established for that purpose and values are available for assessment, particularly biological threshold values.

34.5.1.2 Vaccination

If applicable, an available vaccination constitutes an additional useful preventive measure when handling potentially infectious agents of risk group 2 and higher. Depending on the agent, this vaccination prophylaxis can be done voluntarily or is a prerequisite for the scheduled activities and thus, prescribed and binding. In any case, effective vaccines are made available to the employees who are exposed or possibly exposed to a biological agent but are not yet immune and/or a corresponding vaccination offer is submitted. The employees are informed about the advantages and disadvantages of vaccination and/or non-vaccination in the process.

34.5.2 Health Monitoring

If an employee contracts an infectious disease that can be attributed to occupational exposure, the competent physician will offer to monitor the health closely and will do so upon request for other employees who are exposed in the same manner.

If a screening shows that an employee has a medical objection to the further exercise of the activity based on the workplace conditions, the laboratory operator will immediately take additional required protective measures to the point of reassigning the employee to an alternative activity for which there is certainly no hazard caused by further exposure.

The employer keeps a screening file for the employees examined by a physician. This file contains particular information on the risk of exposure as well as the results of the occupational health check-ups. The file is kept in such a way that it can be expediently analyzed at a later point in time, if necessary.

In order for the company physician to fulfill his medical tasks and duties to the fullest extent, he/she is provided with all the necessary information about the workplace circumstances, particularly also the results of the hazard assessment, and is given the opportunity to regularly inspect the workstations, preferably in close consultation with the occupational safety expert.

34.6 Employment Restrictions

Based on an amplitude of legal provisions and regulations, the protection of health against incalculable risks, excessive demands, and the impact of hazardous substances at the workplace is guaranteed.

34.6.1 Protection of Minors and Mothers

The consistent compliance with employment restrictions and prohibitions in laboratories for identified groups of people such as minors or expectant and breastfeeding mothers contributes to their safety. Minors and expectant or breastfeeding mothers are only kept busy with activities with biological agents provided that this is compatible with the stipulations of the Young Persons Protection of Employment Act and its related ordinances, particularly the Maternity Protection Guideline Ordinance. General and individual employment restrictions are differentiated and must be observed.

For example, the laboratory workstation of an expectant or breastfeeding mother is designed in such a way that the life and health of the mother and the child is by no means jeopardized by the occupational activity. This is accomplished by an individual hazard assessment associated with corresponding activity restrictions and protective measures. Essentially, overtime and activities with an increased accident risk are prohibited. This also applies for handling very toxic, toxic, harmful or otherwise chronically damaging hazardous substances for people, for example, chemicals, cleaning agents, and disinfectants when their threshold is exceeded. Expectant mothers must not be exposed to carcinogenic, teratogenic or mutagenic hazardous substances at all. Working with substances, preparations or products which, from experience, could transmit pathogens depending on its nature, and/or when the employees are exposed to the pathogens, is just as fundamentally prohibited as lingering in a restricted access area of ionizing radiation or handling of open radioactive substances.

In either case, appropriate impermeable protective gloves must be worn and, as needed, additional eye/facial and breathing protection must be worn in addition to the protective coat when working with skin-resorptive substances. It is prohibited to handle cutting, stabbing, breakable or rotating devices and instruments since this work equipment and resources impair the effectiveness of the personal protective equipment or can even neutralize it completely. The required personal protective equipment is made available by the laboratory operator. In this context, it speaks for itself that the generally applicable hygiene regulations (see later) must be observed.

If the workplace assessment shows that there is a remaining health hazard for expectant or breastfeeding mothers. despite complying with all stipulated measures, a change of workstation is advisable or, in particular cases, an exemption from work due to a basic employment ban. An individual employment ban applies accordingly when the life and health of the mother or the child is jeopardized while the work continues according to a medical report.

34.7 Access Regulations and Protection against Theft

34.7.1 Identification and Access Control

Common access regulations and restrictions such as a constantly occupied entry door and the labeling of publicly accessible and publicly inaccessible areas has proven to be sufficiently effective in a laboratory building with a highly differentiated and interdisciplinary spectrum of use, on the one hand, and public foot traffic, on the other hand. It is for this reason, that access to the actual lab areas is reserved exclusively to qualified personnel. Visitors and service personnel only have access when escorted. An additional provision of services via side entrances, for example, for deliveries, is not given on operational and safety grounds. Side doors are used exclusively as escape routes and are equipped with a hand knob on the outside and panic lock on the inside accordingly.

The different lab and safety areas such as genetic engineering facilities, radionuclide or laser laboratories are labeled with information and prohibition signs according to the regulations. These labels such as "bio-hazard" or "caution laser" are used, on the one hand, as a clear and significant designation of the different hazard areas and, on the other hand, contribute substantially to the occupational health and safety protection of the employees. As a result, every employee recognizes the specific occupational hazards and can adjust his/her behavior to the respective hazardous situations. For reasons of protecting products and people, as well as to protect heavily invested equipment, these general organizational measures have been reasonably supplemented, as needed, for special use and safety areas on the second level by additional technical access control mechanisms such as card readers or transponders. Access to a selected, yet necessary group of people is restricted here and personally documented at all times via logs and/or automatic electronic records.

As for different special use areas with an elevated hazard potential, such as biological laboratories of protection level 3 or radionuclide areas of protection category 2 and higher, special emphasis is put on controlled incoming and outgoing traffic of people and material. This safety measure is not only specified but also effectively prevents the carry-over of contaminations. Area-specific personal protective equipment is put on and taken off in the man locks. Required hygiene and disinfection measures are implemented here by the employees just as much as potentially contaminated disposable protective gear is properly collected and disposed of. The designation of black and white areas increases the barrier aspect of the locks and contributes effectively to the protection of the employees and environment.

Personal or camera and/or video monitoring of the entrances and exits of high security laboratories as well as their work areas generally increases the occupational safety and, additionally, allows for employees to work in isolation in a controlled manner in a hazardous working environment.

34.7.2 Protection against Burglary and Theft

Special security measures to protect against burglary and theft are taken into account for the preservation of toxic substances and anesthetics and for a radionuclide area. Toxic substances are generally kept in locked safety storage cabinets of the type 90 so that these are only accessible by the respectively authorized group of people. The same applies for drugs that are subject to the Narcotics Act. These are preserved in a specially certified safe with resistance grade I or higher according to EN 1143-1 which is bolted in conformity with DIN.

Depending on the nature and extent of the narcotic handling, a structural shield and electronic monitoring via a burglary alarm system are required in addition to the mechanical security as determined by the Federal Opium Agency.

34.8 Cleanliness and Hygiene

Cleanliness and hygiene play a crucial role for the product and personal protection in research laboratories. The pollution and uncontrolled contamination carry-over must not be tolerated for the experimental tests and for the occupational health and safety protection. Valid analyses and research as well as the handling of health hazardous substances require maximum cleanliness and hygiene. Hence, a comprehensive hygiene concept is essential for a controlled and safety research operation. Hygiene includes all precautionary measures to keep people healthy in general and to prevent infections and diseases of the employees at the workplace in particular. Essentially, it's a matter of keeping contamination with hazardous substances and biological agents at the workplace to a level that is acceptable in terms of health and/or to reduce it to that level. For this purpose, organizational measures are supplemented with technical and personal safety measures that meet the respective degree of hazard – for example, improbable risk of (infection) in the case of biological agents of risk group 1 to high risk for risk group 4.

34.8.1 Minimum Hygiene Standards

34.8.1.1 Cleaning and Hygiene Measures

In order to fulfill the operational necessities, practice has shown that a series of organizational measures have proven to be of value in addition to generally common, regular cleaning measures – wiping work of surfaces and floors, window cleaning, and so on, with disinfectant cleaning agents according to a cleaning regime – regardless of the nature of the laboratory. These measures are summarized in technical jargon under the term "basic rules of a good laboratory practice." It is strictly prohibited to bring foodstuffs, drinks, and tobacco of any kind into the laboratories. Just as well, food and beverages must not be stored with chemicals, for example, in cold storage rooms. Cosmetics must not be used in laboratories either. Due to the given likelihood of confusion, chemicals must not be prepared and stored in common food and beverage containers.

Additionally, the minimum hygiene requirements include mainly the following points:

- hands must be washed before the start of a break and upon completion of the daily work when leaving the laboratory
- appropriate means are made available (see sample skin protection plan) for the hygienic cleaning and drying of the hands and to protect and care for the skin.
- appropriate facilities and common areas are provided for the purpose of storing lunch/snacks separately from the working materials and also to consume food and beverages in a hazard-free manner.
- work clothes and personal protective equipment are stored separately from the regular clothes
- workspaces are cleaned regularly and as needed with appropriate methods
- avoid entering the break and lunch rooms with severely contaminated work clothes
- biologically contaminated waste is collected in appropriate receptacles and transportation containers
- fundamental means of wound care are kept at hand at the appropriate location
- personal protective equipment is made available as needed and used and/or worn by the employees based on the needs.

In either case, the workstations are kept free from contaminations via appropriate hygiene measures. Inadvertently released chemicals, for example, through spillage or spraying, are absorbed immediately and completely with appropriate chemical binding agents and professionally disposed of. When wearing protective gloves, it must be ensured that contaminations are not unconsciously spread in the laboratory, for instance on keyboards, telephone receiver, or door handles. For this purpose, the protective gloves are regularly changed and immediately removed in case of an obvious contamination. Likewise, laboratory coats are changed and cleaned.

34.8.1.2 Microbiological Requirements

These general hygiene measures are supplemented with the so-called basic rules of good microbiological technique (see "basic rules of good microbiological technique") in biological work areas.

Taking into account the said points of view, the following design characteristics are of vital importance when constructing biological laboratories:

- dressing facilities separated from the workplace (at least a wardrobe for regular clothes)
- appropriate washing facilities
- easy to clean surfaces of floors and working equipment (e.g., machines, installations) in the work area as far as these are within the scope of the operational possibilities
- technical measures and accommodations to avoid and/or reduce aerosols, mist, and dust.

(Annex: "basic rules of good microbiological technique")

34.8.2 Disinfection Measures and Hygiene Plan

If necessary, work areas and used devices must be subjected to appropriate disinfection measures on a routine basis for the purpose of sterilization and/or reduction of bacteria before the start and/or upon completion of the activities with (potentially) infectious biological agents. Simple, chemical disinfection procedures via spraying and wiping are basically used to treat surfaces and, in particular cases, disinfection via submerging is applied especially for working materials and more complex devices.

Chemical disinfectants consist of and/or contain substances that kill and/or deactivate bacteria, fungi or viruses. Accordingly, they often feature certain toxicities for human and animal tissue that can be expressed in intolerability for the skin and mucosa – especially in concentrations. Contact with the skin or mucosa should be avoided as much as possible with the exception of products that are permitted explicitly for skin disinfection according to the safety data sheet. Occasionally, employees must take appropriate occupational safety measures, such as wearing protective gloves or goggles, when manufacturing conventional thinners and when implementing disinfection measures.

Alcohol (e.g., 80% ethanol, 70% isopropanol), aldehyde (e.g., formaldehyde, acetaldehyd), or quaternary ammonium compounds (e.g., benzalkonium chloride) as well as defined preparations of these active ingredients have proven to be useful as chemical disinfectants. The following criteria must be observed when selecting disinfectants:

- field of application (surfaces, devices/instruments, skin/hands)
- effective spectrum (bacteria, viruses, fungi)
- concentration and residence time
- protein or soap error (deactivation by proteins/soap mixture)
- irritant effect on skin and/or mucosa (vapors)
- allergic potential
- biological degradation
- cost effectiveness.

The effective spectrum – bacteria, viruses, and so on – and the required residence time of the different chemicals and their mixtures are of particular importance.

For the purpose of the chemical deactivation of agents in HEPA (high efficiency particulate air) filters of the safety cabinets and filter cabinets of ventilation systems, more complex fumigation procedures with formaldehyde or hydrogen peroxide must be implemented as safety measure prior to performing maintenance and service work. The same applies for room disinfection in the case of disaster and emergencies in safety laboratories and areas where animals are kept.

Biological waste and consumer materials contaminated accordingly such as disposable protective globes, pipettes, cannulas, and so on, must be collected in appropriate waste receptacles or autoclave pouches and must be sterilized in an autoclave under defined physical conditions – pressure, temperature, time span – in a saturated vapor atmosphere prior to the final disposal. Work clothes must be changed and cleaned regularly. Potentially contaminated laboratory coats must be autoclaved or washed with disinfectant before being cleaned.

Generally, the use of disinfectants and procedures which were tested, permitted and listed for disinfection by the Robert-Koch Institute, are advisable for a safe and reliable operation (*disinfectant list of the Robert-Koch Institute*). However, other disinfectants and procedures that have been professionally tested for their effectiveness are to be applied safely just as well for a routine disinfection. The selection of available preparations is large. Applications, concentrations to be used, and residence times of the preparations are listed in several disinfectant lists, for instance, the Deutsche Gesellschaft für Hygiene und Mikrobiologie (DGHM) or the Deutsche veterinärmedizinischen Gesellschaft für Tierhaltung (DVG).

On the other hand, evidence of efficacy must be provided in the form of reproducible agent-related deactivation kinetics for the operation and agent-specific application of untested chemical disinfection measures.

The specific disinfection measures that are applied in the different laboratories must be summarized in the so-called hygiene plans that describe the individual procedures in note form with respect to their application, the products used, and residence times (see "sample hygiene plan").

(Annex: "sample hygiene plan")

Note: Chemicals particularly harmful to the environment which are negatively listed by official bodies/authorities, for example, hypochlorite, must no longer be used as disinfectants.

34.8.3
Personal Protective Measures

In particular cases, it is or can be necessary to supplement said hygiene measures catalog with the use of personal protective measures in addition to the generally common laboratory coats based on the outcome of the hazard assessment. Depending on the work and hazard situation, the following personal protective equipment is hereby considered:

- skin protection (ointment, crème, lotion)
- hand protection (anti-chemicals protective glove or so-called disposable gloves made from latex or nitrile)
- eye protection/face protection (protective goggles with lateral protection, face protection shield)
- particle protection filter (disposable filter mask, reusable filter mask).

In either case, a stress-specific skin protection plan is created to protect and care for the skin (see sample skin protection plan). This must be consistently observed by the employees since irreversible damage to the skin can be prevented by regularly applying the skin protection and care measures. The skin and hands must never be cleaned with organic solvents. The use of abrasive cleaning agents must be limited to the absolutely necessary extent, for example, when it is absolutely required by the degree of contamination.

(Annex: "sample skin protection plan")

When working with hazardous substances which can notably jeopardize the skin or can be absorbed through the skin, it is imperative to wear appropriate protective gloves without exception. The selection of the correct protective gloves is crucial for the protective effect. Generally, a disposable protective glove made from latex or nitrile offers enough protection in case of brief contact with sprayers. In contrast, adequate anti-chemical protective gloves must be selected in case of permanent contact. The durability specifications of the respective manufacturer are crucial and are to be observed.

If work processes and activities cause a hazard for the eyes, appropriate eye protection gear of category II is used. Depending on the type of hazard, frame glasses with lateral protection or so-called goggles are proven to be useful when handling hazardous fluids. In addition to protective glasses, to protect the face and neck, a face protection shield is worn when handling strongly corrosive acids and bases. Goggles or face protection shields, which guarantee the protection against drops and sprays, are provided just as well to prevent contamination with infectious material.

Protective glasses with corrective lenses are made available to wearers of glasses on ergonomic grounds.

The contaminations for instance caused by dust and fluids can occur when using eye and face protection and can cause skin irritation or even infections. Hence, the eye and face protection gear is cleaned, maintained and, if applicable, disinfected at regular intervals. Oculars that are excessively scratched, coated with fixed particles or are cracked must be changed in order not to impair the protective effect.

If the release of very dusty or volatile hazardous substances and/or very toxic gases or also the exposure with airborne biological agents in a hazardous concentration cannot be eliminated with certainty, despite technical and organizational safety measures, appropriate protective breathing apparatuses are properly kept at hand and used as needed.

In general, these are quarter or half-masks with P2-filter and/or particle-filtrating half-masks of the type FFP2 for the current laboratory operation. Safety areas for CMR substances, radioactive agents and airborne biological agents of risk group 3 and for enzymes are excluded from this. If the occupation exposure limit (AGW) of the chemicals mentioned last is exceeded, it is imperative to use at least partial masks with P3 filters or even half-masks with FFP3-filters if there is a significant release of corresponding biological agents.

The safety objective of supplying harmless breathing air to the wearer of the breathing protection apparatus is achieved by the filtering devices by removing the harmful substances by means of a gas, particle or combination filter. Depending on the type of filter, filtering devices can remove certain harmful substances from the ambient atmosphere within the limits of its separative and/or absorption capacity.

It must be ensured that particle filters only protect against solid and fluid aerosols and gas filters only against gases. Only combination filters are permissible for use against nitrous gases and quicksilver. Only reactor filters are effective and allowed against radioactive iodine, including radioactive iodomethane.

To provide for optimal hygienic conditions, on the one hand, and to guarantee a flawless functionality of the breathing protection apparatuses, on the other hand, these filters are purposefully stored according to the manufacturer's instructions and, in case of reusable masks, maintained and serviced regularly.

34.9 Operation of Safety Systems According to Regulations

34.9.1 Structural Barriers

34.9.1.1 Containment

For the purpose of protecting people and the environment, safety laboratories are separated among one another and from the environment in terms of space and building technology and are developed via access situations and/or access locks in line with the needs according to their hazard potential (see part I). Clear regulations and processes are to be defined and are to be observed in the operation to maintain these external structural containments. Particular attention must be given to the inward and outward transfer of people and materials in order to safely maintain this outer barrier function. This starts with the separate storage of regular and work/protective clothes in the wardrobes and dressing rooms of the different areas intended for that purpose, continues to the consistent observation of all required hygiene protection measures, and ends with the proper collection and disposal of contaminated materials and waste. The controlled access, designation and identification of black and white areas are just as important as the precise definition of logistical process sequences (see "chart people and material flow biological safety level 3").

(*Annex:* "chart of people and material flow BSL 3")

34.9.1.2 Hygiene Barriers

Personnel exclusively enters the research laboratories via gender-specific area dressing rooms in which the personal clothing is left behind and the area clothes and/or personal laboratory clothes, laboratory coat, and so on, is put on. Regular clothes and work clothes are stored separately in the available wardrobes, on coat hooks and coat lockers. The available wash basin is used for body hygiene and, in particular cases, the personal shower can be used.

Access to the safety areas of higher protection levels is made possible via controlled single-chamber or multi-chamber manlocks that serve as additional system and hygiene barriers. A marking on the floor or a sit-over that folds upwards separate the different hygiene and safety areas. The respectively required area-specific personal protective gear is changed in the lock chambers when entering and leaving. Furthermore, all necessary hygiene measures are performed according to the regulations, such as disinfection and washing of the hands, decontamination and/or disinfection of materials and objects to be deployed and, if applicable, a body shower. Last but not the least, contaminated and/or potentially contaminated waste are collected and provided for disposal.

34.9.1.3 Inactivated Lock Systems

The use of autoclaves and material locks intended to supply and dispose of materials according to the stipulations is of particular importance. Thermostable materials are sterilized in small horizontal floor-mounted appliances or in large pass-through autoclaves depending on the dimension, and thermolabile goods and cultures are chemically disinfected in material locks via fumigation with formaldehyde or H_2O_2 (fumigation locks). The latter also applies for devices and goods that do not fit in autoclaves due to their dimensions. Alternatively, chemical submerging procedures can be used for large devices. All lock systems together form the structural and/or spatial barrier function with alternatively locked openings and/or doors, which prevent the containments from breaking up in an operational sequence. Additionally, the lock and/or sterilization chambers can only be opened after successfully ending the overall disinfection and/or sterilization cycle.

Only the use of appropriate disinfection and sterilization programs in conjunction with continuous monitoring guarantees certain death and/or inactivation of pathogenic organisms in all these lock systems. Fractioned vacuum procedures are the process of choice to decontaminate the sterilized material in the autoclave. On the other hand, the fumigation with H_2O_2 is to be preferred for reasons of maintaining the protection of people and the environment when bringing thermolabile materials and devices in and out. This fumigation process takes place in a controlled manner and only releases the harmless decomposition products in the form of oxygen and water.

The regular verification of all inactivation parameters of this process, such as pressure, temperature, and duration, plays just as much of a large role as the rotational implementation of biological activity tests with appropriate indicator organisms.

Intuitive and intelligent operating tablets in conjunction with an SPS-control unit minimize the risk of a faulty operation by personnel, on the one hand, and enable the online monitoring of the program sequence. Additionally, the sequence of the inactivation program according to the regulations is monitored, evaluated, and recorded by the electronic control system itself. Any defects are recognized by the control system and, at the same time, a premature termination of the process is prevented. This is how it is ensured that the sterilization process is properly completed after eliminating the defect. As a result, an involuntary release of infectious agents is safely prevented.

Prior to commissioning the safety areas, the procedures that are used with standardized, thermostabile bioindicators and/or also with authentic reference samples are validated in relation to the agent and are afterwards revalidated accordingly at regular intervals, at least once per half a year. If any irregularities are evident, the safety system in question will be decommissioned immediately, checked by trained service personnel, and reconditioned.

34.9.2 Technical Barriers

34.9.2.1 Extraction Equipment

To avoid exposure to hazardous working materials, work processes are generally preferred that meet the following requirements, if the rule of substitution cannot be provided for:

- extensive automation
- few manual work steps with the smallest possible volume
- quick inactivation of the material
- effective possibility of decontaminating the devices used.

However, these general safety requirements frequently do not suffice.

Additionally, safety systems and barrier systems must be used in line with the needs to protect the employees depending on the nature, duration, and extent of the hazard and to minimize an effect on the employees.

Health hazardous gases, vapors, and dust must not reach the lab room. Accordingly, volatile gases and substances must be captured and diverted completely at their point of origin or source of release.

34.9.2.2 Fume Hoods

The open-air handling of these hazardous substances as part of chemical approaches and operations will only take place in the fume hoods intended for that purpose. This guarantees that no emissions in hazardous concentration and quantity affect the employees. It must be ensured that test set-ups and inserted apparatuses do not significantly impair the retention capacity of the extractor hoods. Large set-ups that can change the flow behavior in the fume hood are positioned at a distance of least 5 cm to the work surface and at least 10 cm away from the safety shield in a well-tested manner. Strong air turbulence in proximity to the fume hood as a result of personnel walking by fast or turbulent supply of fresh air from the roof also have a negative impact on the retention capacity and must be avoided during everyday laboratory operations. High thermal loads, which also do not change the air flow of normal fume hoods in predictable manner, must basically be excluded. Disintegration and vaporization extractor hoods must be used in the laboratory for such operations.

The safety shield is constantly closed in all types of fume hoods for safety reasons during the running processes and chemical reactions. Necessary brief interventions and inspections are preferably executed by the parts that can be shifted sideways (lateral shields) of the safety shield.

In case of a disaster, the volume flow clearly increases because of the manual activation of an emergency switch or automatic monitoring system. Large quantities of hazardous substances within the extractor hoods are discharged as fast as possible, and the formation of a hazardous explosive atmosphere is avoided. In either case, it must be ensured that dust emissions are prevented or kept to a harmless minimum. Failure of the ventilation function of the extractor hood is indicated by an optical and acoustic alarm system. If this is the case, all work and tests in the extractor hood are stopped immediately and the apparatuses and the test set-ups are adequately secured.

The protective effect of the extractor hood depends significantly on the overall ventilation technology and the ventilation duct in the lab room. Maintaining a minimum amount of ventilation of $12.5\,m^3\,m^{-2}\,h^{-1}$ in relation to the main effective area of the laboratory (about fourfold ventilation per hour) in the plant is just as important as having controlled airflows in the laboratory.

In principle, microbiological safety cabinets are not designed for chemical work in which large quantities of gaseous hazardous substances and vapors are released since these can divert and/or keep back these volatile substances only to a limited extent. Modifications are necessary here, such as the installation of active carbon filters during recirculating air operation or the connection to an exhaust-air plant. As for the latter, it must be ensured that the flows within the safety cabinet are not impaired. An indirect connection with interrupted pull has proven to be useful and sufficiently safe for hazardous substances being highly toxic or mutagen, for example, CMR substances. However, the direct connection to the exhaust-air plant is urgently necessary for highly effective substances and particularly hazardous substances such as radionuclides and biological agents of risk group 4. The stable exhaust air operation of the ventilation system is correspondingly significant for the protective effect of the safety cabinet. Instability and failure of the exhaust air largely neutralize the retention capacity of the safety cabinet. In this case, personnel must stop its work due to the resulting hazards immediately.

34.9.2.3 Barrier Systems

The work and the extractor hood cabins with targeted airflow or laminar flow systems are used when the development and emission of dust must be taken into account. These complex protective systems feature integrated and highly effective exhausts with filtration units and dust separators that make sure that, to a large extent, no hazardous dust particles reach the breathing zone of the employees. It is thereby essential that the exhaust systems are suitable in relation to the handled hazardous substances with respect to performance, positioning toward the source of emission, geometry of the collection orifice, and last but not the least, also with respect to the ventilation duct and the separation efficiency. The dusting conduct of the substances must be taken into account which depends on characteristics such as grain size, electrostatic charging capacity, application form (granulate, paste, powder, etc.) and fine abrasion. Additionally, the airflow must not be impaired by the experimental set-up and manual procedures.

The evidence of the effectiveness of these barrier systems must be provided through system and procedure-specific measurements as part of a hazard assessment.

34.9.2.4 Local Extraction Devices

Local exhausts are used for a reliable avoidance of emissions at devices such as gas-chromatographs, AAS-devices or HPLC-systems Thermal processes or activities with hazardous substances that are lighter than air make it necessary to have an exhaust above the point of origin (over-the-table exhausts). Hazardous substances that are heavier than air are preferably siphoned off underneath the worktables (below-the-table exhausts). It must be ensured that the vents of the source extractions, flues, and hoods are sufficiently dimensioned.

The source extractions at fixed pipes with joints or flexible hoses must be positioned as close as possible above or underneath a source of emission to guarantee its effectiveness. The throughput of the source extractions decreases significantly as the distance increases. Hence, these are not just suitable for purposes of use in which the hazardous substances are collected directly at their emersion point. Since there are generally no optical and acoustic function displays available, the effectiveness of the extraction must be checked regularly or the air must be monitored with appropriate gas-warning devices, for instance, when there are asphyxiant gases or toxic vapors. If there is an evident expansion of hazardous substances in the working environment, the emitting device must be turned off immediately and the cause of the error must be eliminated.

34.9.2.5 Insulators

Hazardous agents such as pharmacologically effective substances of hazard categories 3b and 4 for which no occupational exposure limit (AGW) is known and/or cannot be determined due to a lack of data, are handled exclusively in so-called closed systems to protect the employees. Usually, glove boxes and/or insulators are used in the laboratory. These are provided in such a way that there is no open connection, on the one hand, while hazardous substances are handled between the interior and surrounding, and low pressure is present in the interior, on the other hand. For the purpose of an inward and outward transfer of materials and substances without contamination, these containment systems feature interlockable, evacuable transfer systems, and integrated cleaning and decontamination procedures. The emission of dust and exposure of people are prevented with certainty due to this hermetic construction and function. In order to protect the product, these units that are driven by recirculating air are rinsed with inert gases such as nitrogen, argon, or helium as needed. This method of operation also enables handling substances that are, for instance, sensitive to oxygen or humidity from the air.

34.9.2.6 Aerosol-Preventing Systems

The formation of aerosols is one of the most frequent causes for the absorption of biological and other hazardous agents via the respiratory tract through inhalation. The critical droplet size is hereby $<5\,\mu m$. The following common laboratory procedures and work steps often lead to the development of bio-aerosols in case of targeted and non-targeted activities with biological agents:

- opening closed vessels such as "Eppendorf tubes"
- vortexes, mixing, and ultrasound treatment
- spillage, splashing
- centrifugation.

34.9.2.7 Microbiological Safety Cabinets

To avoid the release of bio-aerosols, the open handling as well as other activities are performed with pathogenic agents and infectious material that is transmittable with the airway essentially and exclusively in the safety cabinets at least of category II. The airborne particles are caught by the airflow in the safety cabinet, continued, and withheld via HEPA filters. When operating the safety cabinet, it must be ensured that laminar airflow that faces downward by improper covering of the vents with an excessive number of laboratory utensils, on the one hand, and by interfering airflows in the lab room as a result of an inappropriate place of set-up, on the other hand, does not neutralize the protective effect of the safety cabinet due to turbulence inside or outside. Furthermore, it is recommended to use the safety cabinets with high-performance aerosol pre-filters to increase safety. These pre-filters extend the durability of the main filters and decrease the risk of a filter breach as a result of which a contamination of the lab room can certainly be avoided.

34.9.2.8 Retaining Basins

As for procedures outside the safety cabinet, only devices and machines are used that release no aerosols. Centrifuges feature aerosol-proof, fail-safe centrifuge cups and/or sealed rotors with transparent covers. Even a breakage of glass does not remain undetected before opening the rotor. The prerequisite for using bio-reactors to cultivate organisms, pumps to extract excess cultures or to create a vacuum as well as autoclaves for waste disposal, is the fact that no contaminated process air is emitted. This is guaranteed by the appropriate process technology in the process run such as filtration or thermal inactivation, for example, through inline condensation sterilization or exhaust air heater.

The correct use and application of working materials and resources adequate for the procedure by the employees are of correspondingly great importance for the occupational health and safety protection. The most important prerequisite for the functionality of all safety systems is its flawless technical condition that can only be guaranteed by consistent testing, maintenance, and upkeep according to the regulations. The circumvention of protection and safety systems must be strongly prohibited through appropriate monitoring and control measures.

34.9.3
Storage and Disposal Systems

34.9.3.1 Solvent Waste

Solvent waste that accumulates in the laboratories on a routinely basis are collected in lockable plastic canisters with a maximum holding volume of 10 l. These containers are fail-safe and easy to handle from ergonomic point of view. Permanently aspirated safety cabinets and infrastructures of the extractor hoods serve as safe storage systems. Filled collecting tanks are placed in special transportation cars, which features sufficiently dimensioned collection basins, and are transported via the available transportation routes and elevators into the solvent waste plant. This is where the solvent waste is pumped into waste vessels or in KTC containers via an aspiration lance by means of the decanting station (see chart disposal solvent waste).

Contrary to this generally common disposal concept, the solvent waste of workplaces with devices are caught and gathered in directly in stationary, permanently aspirated and grounded waste containers with collection basins. These collecting vessels have a fire resistance of 90 min and can thus be positioned freely. These special vessels are emptied by means of an in-house disposal cart that is designed for this purpose. This cart has a pump by means of which the liquid waste is pumped into the grounded transport collecting tanks of the cart via a hose connection with quick couplings. This process conforms to the pendular principle, that is, via a parallel connection with a pump and aeration line so that the release of solvents is certainly counteracted. The further disposal takes place as previously described.

(Annex: "chart disposal solvent waste")

34.9.3.2 Hazardous Substance Waste

Solid waste is collected at designated places in the laboratory in plastic vessels with tension ring covers.

The service personnel take care of the provision of empty containers, as needed, and the disposal of the filled waste containers in the laboratory. The laboratory personnel itself is responsible to monitor the filling levels and timely request for the required service. The filled big bags and KTC containers must be picked up and properly disposed of by an expert disposal business at regular intervals.

34.9.3.3 Biological Liquid and Solid Waste

The collecting and transportation systems for biological waste – fail-safe vessels made from plastic with inserted autoclave pouch and lockable cover and/or lockable ventilated stainless steel container – are kept at hand in sufficient numbers at appropriate designated collection sites in the biological laboratories. Biologically contaminated materials and liquids in small quantities are collected directly in these transportation containers. When the fill level is reached, the transportation vessels with autoclave pouch and/or stainless steel container together with the content and the cover intended for that are sealed airtight by the laboratory personnel and placed on a transportation cart intended for that purpose to be picked up. Liquid cultures in test tubes, flasks or small fermenter, as well as collecting vessels for purely liquid waste are placed directly on the transportation cart for pickup.

The service personnel take care of the supply and disposal of the transportation containers and vessels, and the proper inactivation of the biologically contaminated material in the autoclave. The laboratory personnel are competent for the timely request of the required service.

The organization of a rotational pickup of all different waste fractions by the service personnel and the removal of hazardous waste by an expert disposal company ensures disposal according to the regulations and in an environmentally sound manner.

All the waste containers and bags are inscribed and labeled clearly and permanently for a clear allocation of the waste fractions and thus, to avoid mix-ups. The appropriate identification regulations must be taken into account.

34.10 Operational Safety in Laboratories – Conclusion

A laboratory building makes the necessary structural and technical infrastructure available for methods and activities for experimental research and use of processes and correlations of natural science. Structural and technical safety systems are a fundamental prerequisite to protect people and the environment in general. The occupational health and safety protection of the employees in particular is not insignificantly influenced by organizational and personal measures in addition to the use of safety-relevant systems according to the stipulations. The clear and obvious determination of operational rules to handle hazardous substances, on the one hand, and the consistent implementation of the required safety measures, on the other hand, contributes hereby significantly to safety. Cleanliness and hygiene take priority. Expert and hazard-conscious personnel are equally important as the availability of extensive safety information in the form of written instructions (hygiene plans, operating instructions) and registries for devices and hazardous substances (hazardous substances, biological agents, etc.). Last but not the least, the preservation of functional efficiency of systems and installations, particularly safety systems through regular maintenance and upkeep is of great importance to minimize the hazards. The establishment of an Occupational Safety and Health Administration consisting of a committee of managers and safety officers is not only very helpful for the observation of hazards and their systematic avoidance but also stipulated by law. The most important instrument to guarantee the necessary safety is the regularly implemented hazard assessment with the objective of recognizing safety deficits and eliminating these via appropriate measures. The introduction of an occupational health protection management system has proven useful for an extensive and consistent implementation of occupational health and safety protection in the laboratory.

34.11 Laboratory Rules and Regulations (Sample)

(*Status*: ....)

These general laboratory rules and regulations apply for all laboratory areas in which hazardous substances are handled. Individual operating instructions supplement these laboratory rules and regulations and must not fall short of its safety objectives.

34.11.1 General

Access to the laboratories is prohibited for unauthorized persons. This is indicated by corresponding information signs.

Pregnant and breastfeeding mothers must not come in contact with hazardous substances. Working in laboratories in which hazardous substances are used is also out of the question when the employee concerned does not even handle hazardous substances. It is prohibited to eat, drink, and smoke in all laboratories.

34.11.2 Working Hours

Arrangements must be made for the opening hours of the laboratories in relation to the work area. Prime working hours (e.g., Mon–Fri from 8AM to 6PM) and secondary working hours (e.g., Mon–Fri from 6PM to 6AM, Sat, Sun) must be differentiated. It must be possible to perform work of any nature without restriction during the prime working hours; hence, it must be guaranteed that there is always enough functional personnel (e.g., first-aid workers, etc.) present in case of emergency. As for the secondary hours, it must be ensured by agreement among the employees that there are always at least two people present in a work area who check on one another at regular intervals.

34.11.3 Work and Protective Clothes

When working with hazardous substances, it is imperative to wear

- protective glasses with lateral protection and possibly top eye cover
- a closed cotton laboratory gown
- closed and sturdy shoes.

The laboratory gown must not be worn in areas where people who do not handle hazardous substances, have access to (office, cafeteria, conference room, toilets, etc.). The use of gloves is absolutely necessary when handling certain hazardous substances (corrosive, skin irritant, sensitizing, etc.). The glove material must be selected according to the purpose of use. Gloves must not be worn outside the laboratory and must be removed when making a phone call, opening doors, using faucets, and so on.

34.11.4 Order at the Workplace

The personal workplace and all common facilities must be kept in orderly condition. The personal laboratory station must be cleaned regularly (weekly, every Friday).

Chemicals must be checked at least once a month on their need for being in the laboratory and, if applicable, must be handed over or disposed of. Furthermore, it must be ensured on a daily basis that unnecessary media (e.g., cooling water, power) are shut off at night.

34.11.5 Maintaining Safety Systems

Every person who works in a lab area must make himself or herself aware of the locations and functions of the safety systems and also about escape routes, fire detectors, and alarm plans. Escape and rescue routes must absolutely be kept clear.

Fire extinguishers are available at appropriately marked positions in all lab areas. Used fire extinguishers must be handed over to be refilled immediately upon use and must be returned promptly to its respective location.

First-aid kits are available in every laboratory and to be checked regularly for completeness.

It must be ensured that small injuries that do not require the intervention of a physician or clinic are also entered in the first-aid log pursuant to insurance law. Body and eye showers must be checked for functionality by the respective laboratory personnel on a monthly basis. This is to be documented in a journal.

34.11.6
Conduct during Hazardous Work

Hazardous work must always be performed under special protective measures (under an extractor hood, behind protective shields, in special rooms, etc.).

Any hazardous work requires that the employees performing it are informed about all possible hazards and have been briefed about the appropriate emergency measures based on the operating instructions. Essentially, it must be checked if the hazardous substances used cannot be replaced with non-hazardous substances.

34.11.7
Conduct in Hazardous Situations

In case of hazardous situations (e.g., release of gases and vapors, leakage of hazardous fluids, fire), the following applies.

> KEEP CALM!
> PROTECT YOUR OWN SAFETY WHEN PROVIDING ANY KIND OF ASSISTANCE!

If necessary,

> PERFORM FIRST-AID and/or CALL THE EMERGENCY PHYSICIAN!

Otherwise, adhere to the posted emergency plans (escape and rescue plan, alarm plant, etc.).

34.11.8
Handling Hazardous Substances and Pressurized Gases

Chemicals and hazardous substances must only be stored in containers, the form and labeling of which does not allow for a mix-up with food products. The inscription of the containers must be placed clearly and unmistakably on a proper label with substance name and hazard symbol(s). Pasting or writing over old labels is impermissible.

The hazardous substances must be recorded regularly according to type, quantity, and characteristics (hazardous substance registry). If substitutes for hazardous substances are available, these must be used (substitution ban!). Hazardous substances must be stored in such a way that only experts have access to them. Hazardous substances must only be supplied to the laboratory in quantities necessary for ordinary use and must be stored in safes.

Gas bottles must not be left behind in laboratories overnight but must be stored in a gas bottle cabinet. Gas bottles must be secured against accidents with a steel clamp or chain on location when gas cannot be supplied via a house gas line. Gas bottles must only be moved internally with special transportation carts and with screwed on valve protection caps. It is strictly prohibited to carry the bottles.

34.11.9
Correct Handling of Storage Equipment

Only vessels that are locked and provided with a content and name sign must be placed in refrigerators and cooling chambers. These must be checked at least once a month on their need for being in there and, if applicable, must be handed over or disposed of.

Flammable fluids that must be stored in cold temperatures must only be stored in refrigerators when their interior is free from ignition sources.

It is strictly prohibited to store food products and chemicals together.

34.11.10
Collection and Disposal of Hazardous Waste

The collection/disposal containers must be labeled clearly and according to the regulations and must be stored at a safe location (e.g., in the safe).

In principle, reactive, particularly hazardous waste (alkaline metals, metal hydride, cyanide, catalysts, acids and bases, etc.) and biological waste must be deactivated/inactivated before these are supplied for proper disposal.

34.11.11
People in Charge of Occupational Health and Safety Protection

Company x has appointed the following person as company physician (BGV A2):

… …

Company x has appointed the following person as safety officer (BGV A2):

… …

Place/Date ....

Management

Company:	**OPERATING INSTRUCTIONS (sample)** according to § 14 Hazardous Substances Ordinance (GefStoffV)	**Status:** *Date* *Signature:*

NAME OF THE HAZARDOUS SUBSTANCE GROUP

Corrosive hazardous substances
(present in the laboratory: hydrochloric acid, sulphuric acid, azotic acid, acid chloride, caustic soda)

HAZARDS FOR PEOPLE AND THE ENVIRONMENT

Contact leads to severe chemical burns. Risk of severe health damage even when inhaled or swallowed.
High enthalpy of solution possible if mixed with water. Consequently, risk of heat development and extrusion.
Corrosion is possible under development of corrosive gases when in contact with water or air humidity.
Substances often severely hazardous to water (WGK 3).

PROTECTIVE MEASURES AND CODE OF CONDUCT

Wear closed laboratory gowns, solid and closed shoes and frame glasses. If necessary, use acid-resistant protective gloves (e.g. type...made from nitrile).
Use goggles when working with large quantities or in case of increased risk of splashing.
Do not eat, smoke, drink, chew gum or apply cosmetics in the laboratory.
Clean hands regularly.
Protect against humidity. It is imperative to pay attention to substance characteristics when mixing with water (first provide the water then add the acid!).
Avoid contact with eyes, skin, mucosa and clothes.
Do not spill, do not allow to enter the sewage system.
Keep containers closed airtight. Store quantities in use in a cool environment and away from sunlight. Storage is only permitted in a chemicals cabinet ....in room...... designated for that purpose.
Observe small quantities for transport according to the hazardous goods regulations. Use barrows and preferably plastic encapsulated DURAN-bottles.
Use tinted bottles for UV-sensitive acids.

CONDUCT IN CASE OF HAZARD — **Call fire brigade 112**

The hazardous situation must be eliminated while protecting oneself. Please use at least goggles, protective nitrile gloves and, if gases and vapours are present, filtrating half-masks or protection level B1P2 (or higher).
Warn people in danger and, if necessary, clear and block off hazardous areas.
The laboratory manager must be informed immediately. Access by unauthorised persons must be prevented.
Leakage: Absorb the hazardous substance with the fluid binding agent Chemizorb and put in labelled, lockable containers.
Fight **fires** with carbon dioxide or powder extinguishers and notify the fire brigade in case of large fires.
Fight **personal fires** with the emergency shower or the next available fire extinguisher.

FIRST-AID — **Emergency call 19222 or 112**

Immediately remove contaminated or drenched clothes (also undergarment) and personal protective equipment.
Inhalation: Bring the people affected into the open air.
Skin: Thoroughly clean moistened skin with a lot of water and soap. Use the emergency shower in case of chemical burns across large areas.
Eyes: Constantly rinse moistened eyes while the eyes are squinted under the eye shower until medical assistance is provided.
Consult a physician or notify an emergency physician, give the safety data sheet, operating instructions and accompanying accident report to the injured, inform the physician immediately about the substance.

PROPER DISPOSAL

A

Put waste in the designated collector bottles in the fume cupboard... The disposal takes place as needed, at the latest before the weekend via the building services (competent for and to be notified as needed: Mr./Ms... tel:....). Drenched material and uncleaned common couple must be treated as the substances and collected in the designated waste receptacle in room.

Source: BG-RCI (modified)

Company:	**OPERATING INSTRUCTIONS (sample)** on handling devices, apparatuses and installations	**Status** *Date* *Signature:*

INSTALLATION –DEVICE –APPARATUS

Safety cabinet (manufacturer .... Type....)

HAZARDS FOR PEOPLE AND THE ENVIRONMENT

Risk of the release of biological agents due to improper working method.

PROTECTIVE MEASURES AND CODE OF CONDUCT

Avoid a draft in the work area, keep windows and doors closed while working in the safety cabinet.
Switch on the device with the key switch or next to the indicator light about 30 minutes before starting to work.
Use personal protective equipment: At least the laboratory gown, if necessary, additional protective glasses and disposable gloves (nitrile).
Avoid the formation of aerosol also underneath the safety cabinet as much as possible.
Avoid interferences of the laminar airflow as much as possible:

- o do not make fast or jerky movements,
- o bring in bulky devices into the safety cabinet only when absolutely necessary and remove these again after use
- o do not use the Bunsen burner permanently. only ignite briefly via sensor or foot switch as needed,
- o do not cover the core duct.

Do not store unnecessary objects in the safety cabinet. Only bring in absolutely required material and devices.
All devices that are brought into the safety cabinet must be cleaned and disinfected beforehand. Devices that are removed from the safety cabinet must be disinfected beforehand and cleaned afterwards, if necessary (observe the order).
The work area of the safety cabinet must be cleaned and disinfected after the activities are finished. Use a product according to the hygiene plan for disinfection. If flammable disinfectants are prescribed, the area must only be disinfected through wiping with quantities of alcohol of less than 20 millilitre for reasons of explosion protection.
If no work is performed at the safety cabinet, it can be switched to reduced power to save energy. As a result, the contamination of the work room by the laboratory air is avoided.
If work is performed with biological agents with a hazard potential, the device must only be switched off by an authorised person.
The sterility of the interior must be checked occasionally by openly placing Petri-dishes with food bottoms. If there is a growth of microorganisms on the bottom of the food, the laboratory manager and Mr./Ms... must be notified by the building services.

DISRUPTIONS AND HAZARDS

Call building services: .....

Safe operation is only possible if the light signal is green and the front panel is folded down.
Never ignore alarm signals.
The safety cabinet does not offer sufficient protection against gases and vapours that are hazardous to health.
Terminate work in a controlled manner in case of a complete functional breakdown during work with biological agents with a hazard potential. The superior as well as the person in charge of biological safety, Mr./Ms....tel.: ... must be notified immediately.
In case of an optical and acoustic alarm, the cause of the failure is to be determined with the aid of the operating instructions, if necessary, and to be eliminated personally if possible (e.g. by correctly positioning the front panel). If it is not possible to eliminate the failure and/or the elimination is not successful, notify Mr./Ms... tel....of building services.
If it is indicated that the pre-filter or HEPA-filter must be changed (alarm signal without warning tone): continue to work and notify the building services so that new filters can be ordered. If the indication is the same with a warning tone: terminate the work in a controlled manner and notify the building services.

ACCIDENTS AND FIRST-AID

Emergency call 19222 or 110

Rinse out open wounds, if possible, let it bleed out and spray immediately with disinfectant.
Give an additional dose of the disinfectant, if necessary and according to regulations, and let it work for at least 30 minutes.
If necessary, notify the first-aid worker, emergency services or physician. Notify the superiors.
Enter every injury, even if it is minor, in the accident log.

TESTS - MAINTENANCE - DISPOSAL

Maintenance and service work must only be performed with the written approval of the laboratory manager.
The safety cabinet must be checked annually by an expert.
Only spare parts that correspond to the original parts in material and design must be used for maintenance purposes.

34.12
Testing Equipment Registry (Sample)

Test object	Test period	Type of test	Tester	Room no.	Inventory no.	Last test	Next test	Comments
Stationary electrical systems and equipment								
Electrical systems and stationary equipment	4 years	For orderly condition	Electrically skilled person	......	......	......	......	......
Protective measures with residual current operated protective devices in non-stationary systems	Monthly	For effectiveness	Electrically skilled person or person trained in electrical engineering when using appropriate measurement and testing devices					
Non-stationary electrical equipment								
Non-stationary electrical equipment	Target value 6 months	For orderly condition	Electrically skilled person or person trained in electrical engineering when using appropriate measurement and testing devices (authorized person)	......	......	......	......	......
Extension and device connection lines with connectors	If error rate < 2%, test period can be extended accordingly							

Connecting lines with plug	At construction sites, in manufacturing plants and workshops or under similar conditions: 1 year							
Movable lines with plug and fixed connection								
Safety installations								
....	......	......	Authorized person or AUTH	......	......	......	......	......
....	.....	......	Authorized person or AUTH	......	......	......	......	......
....	.....	......	Authorized person or AUTH	......	......	......	......	......

34.13
Screening Examinations for Laboratory Activities (Selection)

Hazardous substance/activity	Work area/ area of non-targeted activities	Exposure conditions	Type of medical care
Hazardous substances			
Acryl nitrile, aromatic nitro and amino-compounds, alkyl lead compound, cadmium + compounds dimethyl formamide, carbon disulfide	Research installations/laboratories	Open handling, contamination hazard, e.g., skin irritations	Mandatory medical care
Benzol, 2-butanon, 2-hexanon, ethanol, dichloromethane, *n*-hexane, methanol, styrol, tetrachloroethene, toluol, xylol, and so on (+mixtures)	Research installations/laboratories	Open handling, contamination hazard, e.g., skin irritations	Optional medical care
Carcinogenic or mutagenic substances or preparations, category 1 or 2	Research installations/laboratories	Open handling, contamination hazard, e.g., skin irritations	Optional medical care
Biological agents			
Biological agents of risk group 2	Research installations/laboratories	Targeted and non-targeted activities with biological agents of risk group 2	Optional medical care
Bacillus anthracis, Bartonella – bacilliformis, quintana, henselae, borrelia burgdorferi sensu lato, Brucella melitensis, hepatitis-B-virus (HBV), hepatitis-C-virus (HCV), Mycobacterium – tuberculosis, bosis, and so on	Research installations/laboratories	Regular activities with possibility of contact with infected samples or suspicious samples and/or with agent-containing or contaminated objects or materials when the transmission path is given	Mandatory medical care

Biological agents of risk group 4	Research installations/laboratories	Regular activities with possibility of contact with infected samples or suspicious samples and/or with agent-containing or contaminated objects or materials.	Mandatory medical care
Activities with negative impact			
Moist work	Research installations/laboratories	For example, when wearing protective gloves daily for more than 4 h and/or moist work	Mandatory medical care – examination of the skin
Wearing latex gloves	Research installations/laboratories	Natural rubber latex gloves	Mandatory medical care – skin, lungs, respiratory tract, possibly allergology examination
Handling allergens	Research installations/laboratories	Contamination hazard caused by chemical irritant or toxic substances	Optional/mandatory medical care – lungs, respiratory tract, if applicable, X-rays, possibly allergology diagnostics
Screen work	Research installations/laboratories	At least 2 h per work day	Optional medical care-visual acuity (far, close, in relation to the workplace), visual function with both eyes, field of vision, if applicable, sense of color. Musculoskeletal system, nervous system

The table is based on the Occupational Medical Care Ordinance (ArbMedVV) of 18.12.2008, modified 23.10.2013.

Basic rules of good microbiological technique

- Windows and doors of the work areas must be closed during the activities
- It is prohibited to drink, eat or smoke in the work areas
- Laboratory gowns or other protective clothes must be worn in the work area
- Mouth pipetting is prohibited, pipette aids must be used
- Syringes and cannulas must only be used when absolutely necessary
- It must be ensured that the formation of aerosol is avoided as much as possible during all the activities
- Carefully wash hands, if applicable, disinfect and regrease them (skin protection plan) upon termination of the activity and before leaving the work area
- Work areas must be kept clean and clear. Supplies must be stored in the cabinets intended for that purpose
- The identity of the used biological agents must be checked regularly when it is necessary for the assessment of the hazard potential
- In case of activities with biological agents, the employees must be briefed before taking up the activity and afterwards at least once a year verbally and in relation to the workplace
- Employees not experienced in microbiology, virology, cell biology or parasitology must be trained particularly extensively, must be carefully guided and monitored
- If necessary, vermin must be combated regularly and properly

Hygiene plan (SAMPLE) according to the Biological Agents Ordinance Section 11

Company: ____________________ Department: ______________

Status: _Date_____________________ Signature: ______________

WHAT	WHEN	WITH WHAT	HOW	WHO
Hand disinfection	After every contamination, before leaving the laboratory	Preparation……., 1 dispenser puff = 3 ml	In dry hands and rub until dry	Everyone
Hand cleaning: (first disinfection, then cleaning!)	After contamination, and after work sections	Preparation…….., Liquid soap from dispenser	Wash hands in warm water	Everyone
Hand care	After every disinfection, and as needed	Preparation…….., Hand care lotion 1–2 dispenser puffs	Rub into the dry hands after disinfection and cleaning	Everyone

(*continued overleaf*)

(Continued)

WHAT	WHEN	WITH WHAT	HOW	WHO
Sterile workbenches	After contamination, and after individual work sections	Preparation........,	During current ventilation, disinfection of the work area through wiping	Every user
Contaminated surfaces of devices, and so on	After obvious contamination, and as needed	Preparation: ......, sprayer	Spray, leave at least 5 min., wipe	Every user
Centrifuges	After contamination	Preparation: ......, sprayer	Spray, leave at least 5 min., wipe	Every user
Contaminated glass pipettes	After use	Preparation......., ... %-solution,	Leave in the pipette rinser at least overnight	Every user
Contaminated glass devices	After use	Autoclave in the room	20 min 121 °C	
Protective clothes	1× per month or after contamination	Autoclave in the room	20 min 121 °C	
Contaminated waste, Petri-dishes disposable materials	As needed	Autoclave in the room	Autoclave pouch 20 min 121 °C Disposal pouch	
Floor	Weekly	Preparation........, concentration ... % in mixing water	With wiping mop according to the 2-bucket method	Spokesperson cleaning company Mr./Ms. .... Tel. ....

Source: BG-RCI (modified).

34.14
Skin Protection Plan (Sample)

Company: ______________________ Department: ______________

Status: _Date______________________ Signature: ______________

WHAT	WHEN	WITH WHAT	HOW	WHO
Skin protection	Before putting on gloves	Skin protection creme	Rub in	Everyone
When wearing liquid proof protective gloves (e.g., made from latex, nitrile)	Cotton undergloves are recom-mended	Preparation		
Skin protection	Before starting the work sequence	Skin protection creme	Rub in	Everyone
When handling changing hazardous substances		Preparation		
Hand disinfection (before cleaning the skin!)	After terminating the activities with biological agents and, in principle, before leaving the laboratory	Preparation	Rub in	Everyone
		Dosage Residence time		
Skin cleaning	After hand disinfection!	Liquid soap	Wash	Everyone
		Preparation Dosage Residence time		
Skin care	After hand disinfection and cleaning	Hand care lotion	Rub in	Everyone
		Preparation Dosage		

Source: BG-RCI (modified).

34.15
People and Material Flow Biological Safety Level 3

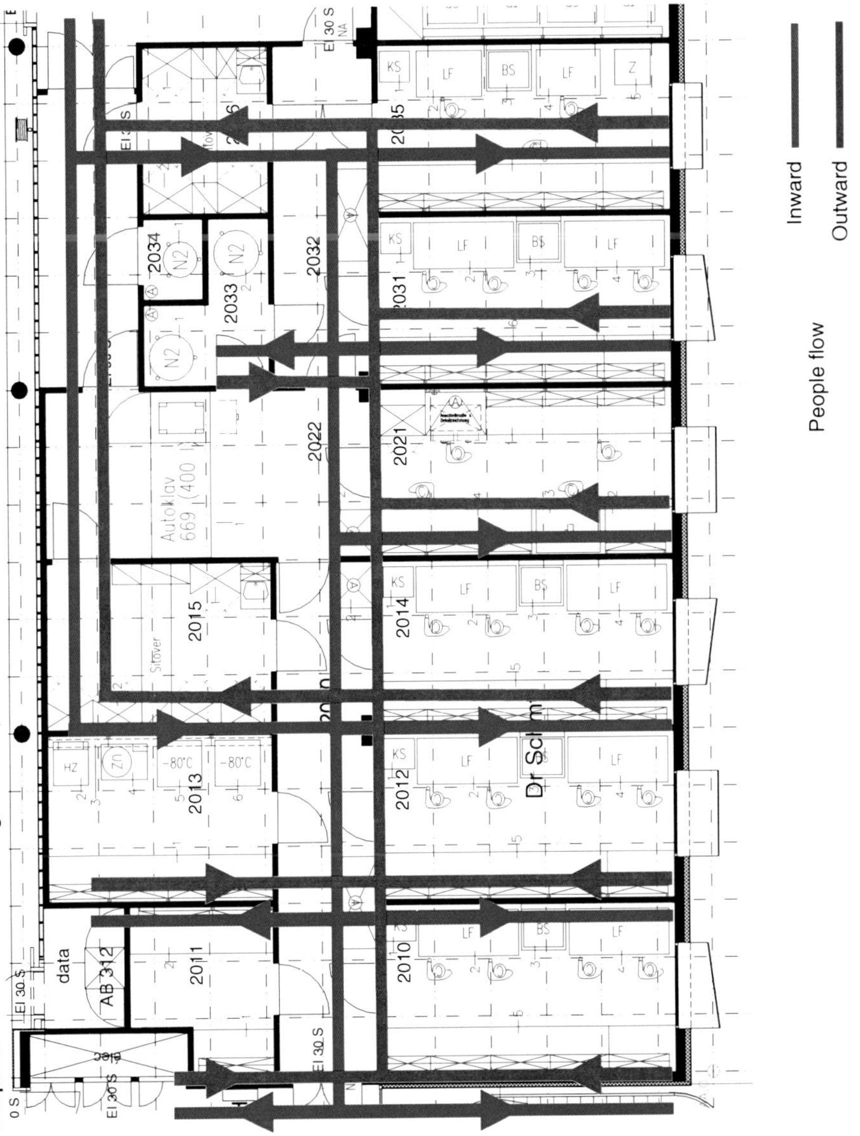

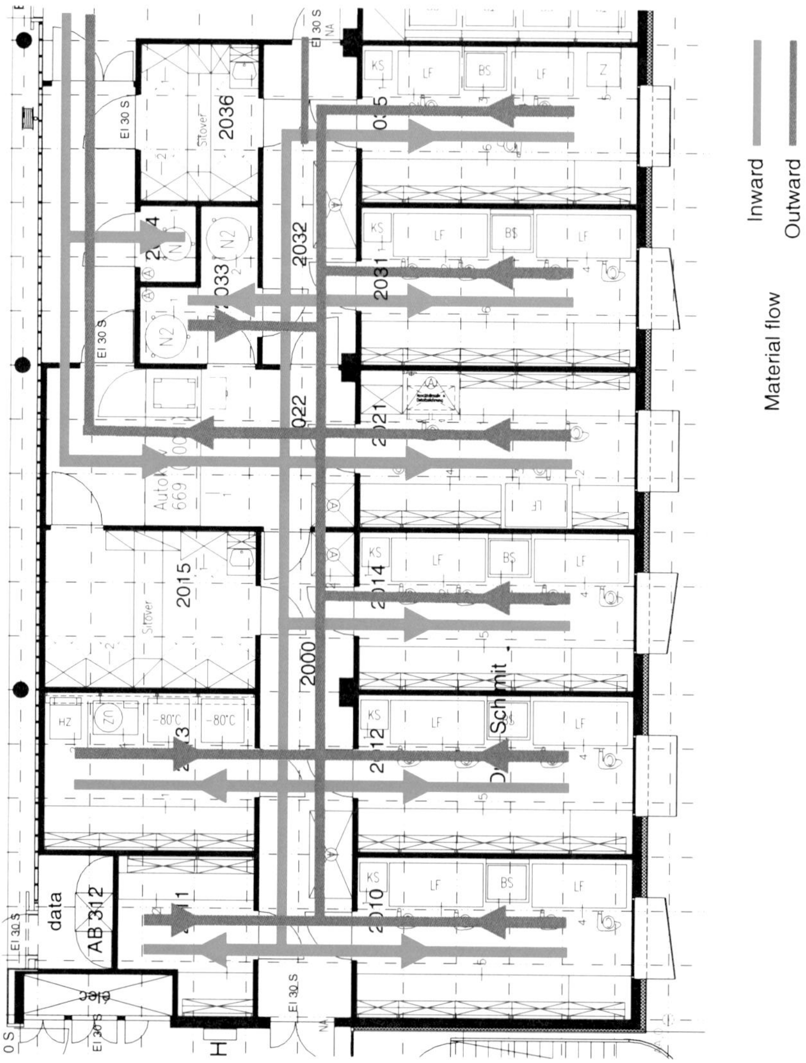

34.16
Chart Disposal Solvent Waste

References

1. ArbSchG. Labour Protection Act, 07.08.1996/19.10.2013.
2. ASiG. Occupational Safety and Health Act, 08.11.2006/20.04.2013.
3. GenTG. Genetic Engineering Act, 16.12.1993/07.08.2013.
4. MuSchG. Maternity Protection Act, 20.06.2002/23.10.2012.
5. JuSchG. Youth Protection Act, 23.07.2002/07.08.2013.
6. BetrSichV. Occupational Safety Ordinance, 03.10.2002/08.11.2011.
7. BioStoffV. Biological Agent Ordinance, 27.01.1999/15.07.2013.
8. GefStoffV. Hazardous Substances Ordinance, 26.11.2010/15.07.2013.
9. GenTSV. Genetic Engineering Safety Ordinance, 14.03.1995/18.12.2008.
10. MuSchArbV. Maternity Protection Guideline Ordinance, 15.04.1997/26.11.2010.
11. TRGS 500. Protective Measures, 01.2008/05.2008.
12. TRGS 526. Laboratories, 02/2008.
13. TRBA 100. Protective Measures for Targeted and Non-Targeted Activities with Biological Agents in Laboratories, 10/2013.
14. TRBA 500. General hygiene Measures: Minimum Requirements, 04/2012.
15. BGI/GUV-I 850-0e. Working Safely in Laboratories, 10/2011/05/2013.
16. BGI/GUV-I 850-2 (Bulletin T 032). Laboratory Extractor Hoods – Designs and Safe Operation. 08/2008.
17. BGI/GUV-I 850-2a (Bulletin T 032-1). Laboratory Extractor Hoods – Leaflet, 10/2008.
18. BGI/GUV-I 863 (Bulletin B 011). Working Safely at Microbiological Safety Cabinets, 09/2004.
19. Leaflet GW 6. Microbiological Safety Cabinets – Leaflet.
20. BGR/GUV-R 500. Operating Equipment, 04/2008.
21. BGI/GUV-I 504 ff. Operating Instructions for the Occupational Medical Care, 2009.
22. BGI/GUV-I 629 (Bulletin B002). Safe Biotechnology – Design and Organisational Measures: Laboratories. 12/2010.
23. BGR/GUV-R 190. Use of Breathing Protection Devices. 12/2011.
24. BGR/GUV-R 192. Use of Eye and Face Protection. 02/2006.
25. DIN 1946-7. (2009) *Ambient Air Technology*, July 2009.
26. DIN EN 14175-1. (2003) *Fume Hoods – Terms*, August 2003.
27. DIN EN 14175-2. (2003) *Fume Hoods – Requirements of Safety and Efficiency*, August 2003.
28. DIN EN 14470-1. (2004) *Safety Cabinets for Flammable Liquids*, July 2004.
29. DIN EN 14470-2. (2006) *Safety Cabinets for Pressurised Gas Bottles*, November 2006.
30. BG Rohstoffe und chemische Industrie (BG RCI) *www.bgchemie.de* (accessed 3 January 2015).
31. Deutsche Gesetzliche Unfallversicherung (German Legal Accident Insurance). *www.dguv.de* (accessed 3 January 2015).
32. Deutsche Gesellschaft für Hygiene und Mikrobiologie (German Hygiene and Microbiology Society). *www.dghm.de* (accessed 3 January 2015).
33. Deutsche Veterinärmedizinische Gesellschaft (German Society of Veterinary Medicine). *www.dvg.de* (accessed 3 January 2015).
34. Robert-Koch Institute *www.rki.de* (accessed 3 January 2015).

Part VII
Laboratory Operation

Helmut Martens

Not least due to the ongoing globalization and the associated transparency through the Internet, also laboratories and their services are always transparent and comparable.

Thus, the claims to the leadership of a laboratory are not only limited on the scientific aspects and "state of the art" or quality leadership, but also include all the issues such as customer satisfaction, efficiency and cost, data handling, continuous improvement, and so on.

The strategic use of indicators and the management of the laboratory as a business enterprise are always more in foreground. This requires not only higher but also other demands on laboratory management. Although the "modern" laboratory director must not do everything himself, he should be familiar with the most important areas at least so far, to assess the opportunities and risks of action and thus, can provide an opportunity for the future of the laboratory and to the staff.

The inclusion of individual skills and abilities of all employees and by the ongoing training in the most important segments of the laboratory management, modern laboratory manager enables to meet the changing challenges.

This guide is therefore a small breakdown of important topics compacted and provides information available, but cannot meet the exhaustive.

The interface between planning and operation of the building with relevance for the operator (user) and the Laboratory Management is facility management (FM). The FM has to supply the consumption figures (Benchmarks), to help the laboratory management to keep operating costs under control. In this respect, the article about the facility management was put in front of the chapter laboratory operation. This closes the circle and makes clear that the quality of planning is crucial for the functioning of the laboratory building.[1]

1) Comment of Egbert Dittrich.

The Sustainable Laboratory Handbook: Design, Equipment, Operation, First Edition.
Edited by Egbert Dittrich.

35
Facility Management in the Life Cycle of Laboratory Buildings

Andreas Kühne and Ali-Yetkin Özcan

35.1
Self-Understanding and Background

The origins of Facility Management (FM) lead back into the 1950s for the US airline Pan American World Services (PAWS). This achieved great results in increasing the productivity of operation and maintenance, and then became a service provider for the US Air Force. Thus, PAWS is considered the first company to provide FM Services.

In the 1970s, the establishment of the Facility Management Institute (FMI) in Michigan laid the key foundations for the scientific examination of the subject FM. In 1980, the establishment of the National Facility Management Association (NFMA) followed, which was later renamed the International Facility Management Association (IFMA).

From the mid 1980s, the term FM appeared in Europe for the first time, and promised quality improvement and savings in all that is not defined as core business. In this regard, the manufacturing companies took the forerunner role, which typically maintain the largest capacity to support the core business. Gradual FM was first introduced at the operational level for special achievements in the field of technology and infrastructure, and later as a strategic outsourcing issue for purchase and management. By the increased outsourcing of non-value adding tasks of a company, the market for services is growing in this area. The distinction between strategic and operational roles in FM plays a decisive part.

Along with this, the profession of the Facility Manager establishes itself in education and training. Overall, one can speak of an emerging profession whose raison d'être is indisputable, but still struggling for recognition in the public eye.

This article is based on the Fee Structure for Architects and Engineers, hereafter referred to as *HAOI*.

The Sustainable Laboratory Handbook: Design, Equipment, Operation, First Edition.
Edited by Egbert Dittrich.

35.2
Process Optimization

The benefits of FM can be measured by how well the core business (primary processes) is relieved of secondary tasks. These include technical, infrastructural, and commercial activities that are not covered in the core business of a company, but have a supporting function. This is the underlying FM service idea. Its key to success lies in the understanding of the core business and the resulting integration of primary and secondary processes.

The core business of pharmaceutical companies is the research and production of products in the health and well-being. Each of these products has an individual product life cycle and each life cycle phase has its specifics. Starting from the research of a drug product, which takes place in high-tech laboratories, through production in fully automated production facilities, to distribution in attractive office and retail space, and finally to the logistics to deliver the product to the customer.

The main task of FM is to understand the specifics of each life cycle phase and to provide suitable premises. This handbook deals with the design, construction, and operation of laboratory buildings, which is not considered by definition as the core business of a research company and therefore, falls within the scope of FM.

35.3
FM in the Life Cycle of a Laboratory Building

With both new buildings and in existing buildings, an integrated and intensive communication between the participants of the project is essential to achieve the required objectives.

Figure 35.1 illustrates the diverse and complex flow of information as it takes place between the essential parties.

Only through the consultation process between the owner's representative, architect, engineers, and FM the planning and implementation process is possible:

- avoid planning mistakes
- consider necessary requirements/customer/tenant wishes
- create synergies.

To make sure that these requirements can be taken into account, the FM requirements should be queried in each phase of the project, reviewed, and considered in the planning or execution (Figures 35.2 and 35.3).

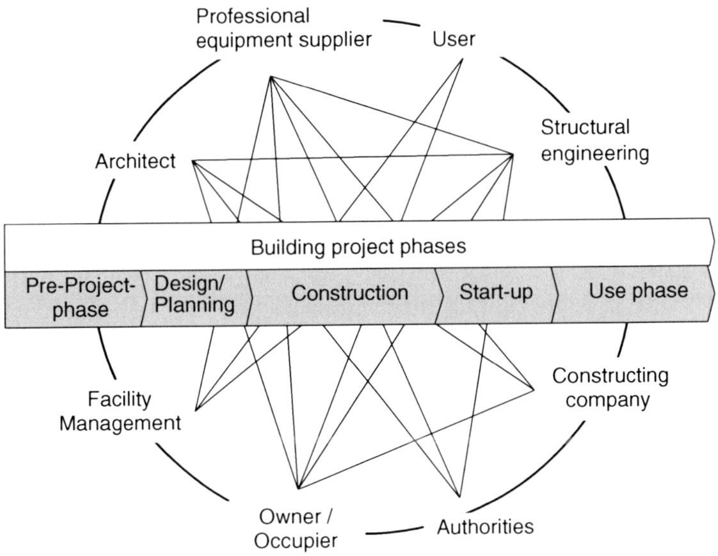

Figure 35.1 Communication in the process of building a laboratory.

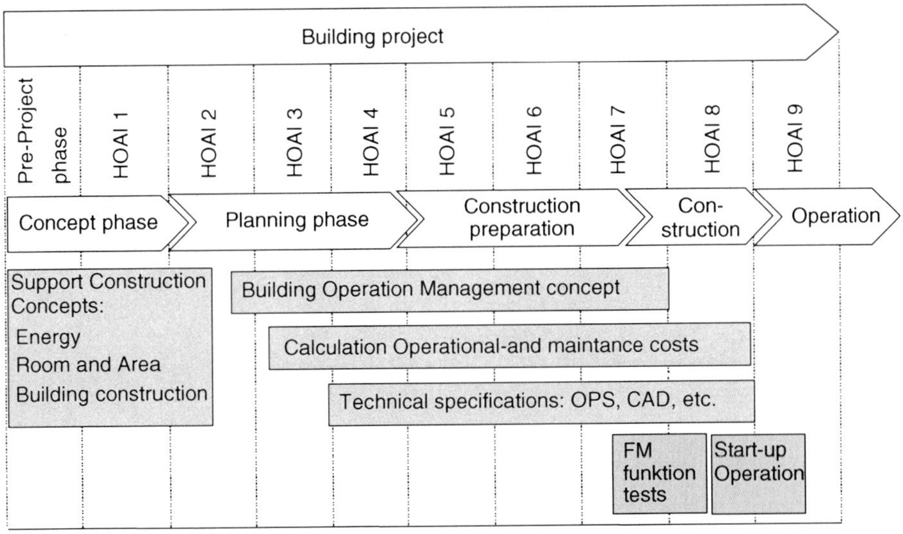

Figure 35.2 Tasks of facility management in the construction phase of building.

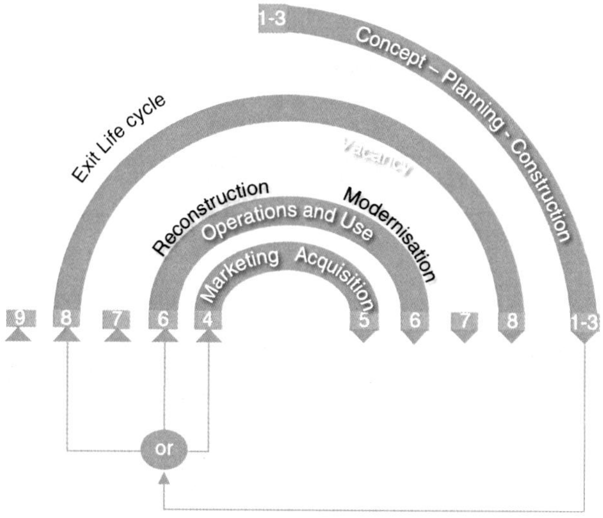

Figure 35.3 Flow-sheet concept phase (text in accordance with German HOAI LP 1).

35.4 Concept Phase Laboratory Building

35.4.1 Rough Building Concept

As a basis for all further development steps of the building with respect to essential requirements, the following aspects are to define:

- Usage of the laboratory building (e.g., chemical/physical lab building)
- Occupation (e.g., number of persons, workstations, fume cupboards)
- Functionalities (e.g., safety floor areas, usable areas, specific requirements).

35.4.2 Concept Finding

The concept finding can be characterized by several processes:

- Competition for architecture and urban planning
- Design text
- Exploratory talks.

Among other things, the essential requirements for FM are mentioned, which will be incorporated into the design of the candidates. These are quality criteria for the decision in favor of the applicant/s.

35.4.3 Project Preparation

The project list provides information about the basic integration of the Facility Manager/s in the planning and implementation process. Furthermore, the scope of duties as well as the effort required by the Facility Manager/s is defined for this purpose.

The integration of facility should be displayed in the following areas and form the following keywords:

- Project organization
- Project directory
- Project calculation FM

- Within the Project calculation the following persons must be taken into account:
  - Facility Manager (X% of eligible costs)
  - Purchasing (X% of eligible costs)
  - Operation team (overview of costs required).

35.4.4 Basic Evaluation

In this phase, the basis for the further planning process is determined. This includes the following information and should consider the influences of FM:

- Knowledge of the users and their needs and desires
- Discussion concerning the axial raster of the building
- Interplay of creative, functional, technical aspects (Construction/Technical Building)
- Formation of indicators for implementation and management phase
- Preliminary discussions with authorities on the operation (e.g., disposal).

35.4.5 Design Phase

Mandatory requirements for the operation of a building, is the compliance or the requirement of a uniform plan, documentation, and designation system. A non-compliance leads to significant additional costs in commissioning and handover.

To comply with the requirements, the below-listed technical specifications are to be handed out to the planning and realization of the process participants. Receipt and consideration of the functional specification document are to be confirmed (Table 35.1).

Table 35.1 Mandatory requirements.

1. Specification for planning documents	For example, general CAD standard for the building plan processing
2. Specification for code of room numbering	For example, object positioning system (OPS)
3. Specification for tender documents	For example, list of technical equipment for planned maintenance
4. Specification for structure of documentation	Project table of contents

OPS: object positioning system.

35.5 Construction Phase

The construction phase begins with the foundation, the ground construction up to acceptance and start-up operations. In accordance with HOAI, the construction phase includes the HOAI phases LPH 8 and LPH 9.

Since different life cycle phases (LzPh), for example, LzPh. 4, 6, and 8 during the life of a building several times, the arrangement in an arc behind each other is unsatisfactory. The view was chosen in such a way that several alternatives are always possible out from the corresponding condition:

Experience and information from operating and managing in the use phase, flow back into the planning of renovations or new construction. For this reason, building management services can be applied at the planning and construction of buildings.

Changes within the life cycle phase may have an impact on subsequent phases. These can be, for example, of qualitative or economic nature. Thus, options can be made possible by changes in the early stages or be made easier or more difficult and impossible in later stages.

In the deployment phase 1, the following phases of construction with the consequent further options are visible in the picture:

- Phases 1 to 3: Design, planning, and construction
- Options are Phase 4: Marketing,
- Phase 6: Operation and use,
- Phase 8: Vacancy (Figure 35.4).

35.6 Use Phase

In the use phase, the Phase 4 marketing, Phase 6 Operation and use of buildings, each beginning and ending with the entry is visible to the extract of a user from figure two.

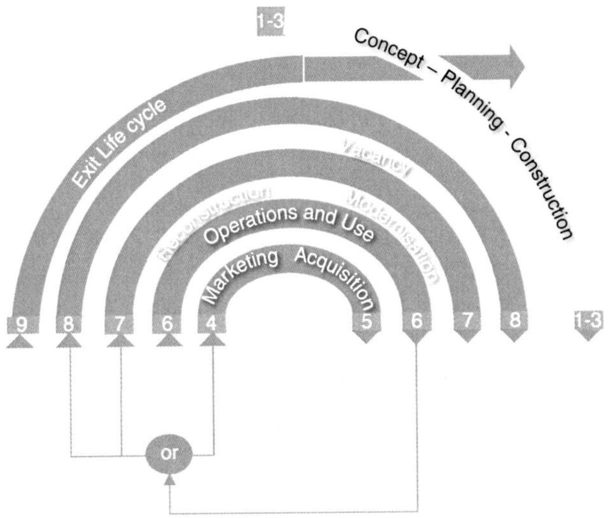

Figure 35.4 Life cycle phases after the construction in a cyclic representation (in accordance to German GEFMA 100-1).

The renting of partial floor areas, part conversions, renovations, or part vacancy while otherwise continuing use also fall into Phases 7 and 8.

By definition, the use phase is the most intensive phase of the FM, containing established real estate cost codes and floor space codes, building value conservation, optimization of use of buildings and minimizing the use of resources, taking into account environmental protection.

The optimization in the use phase increases the efficiency and quality of buildings and the associated processes.

In the use phase, there are five possible phases:

- 4/5: Marketing, such as sales or rental
- Phase 6: Operation and use options are Phase
- Phase 7: Conversion or change of use, refurbishment, or modernization
- Phase 8: Vacancy
- Phase 9: Recovery (Figure 35.5).

35.7 Revitalization Phase

The revitalization phase serves to correct structural deficiencies, which should be substantially improved or redesigned, by focusing on the uniform preparation and rapid implementation.

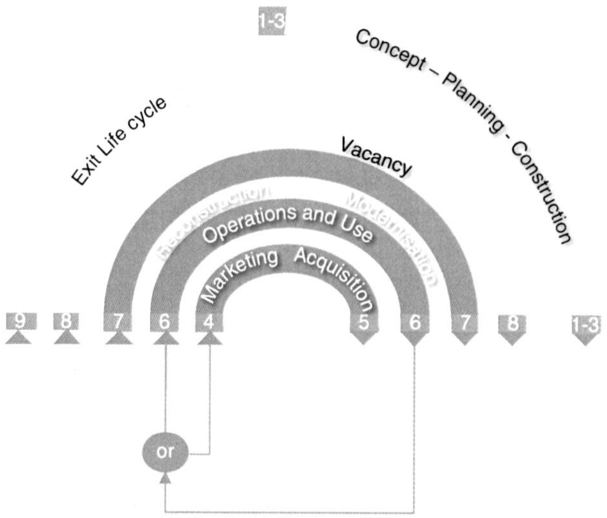

Figure 35.5 Life cycle phases in operation and use in cyclic representation (in accordance to German GEFMA 100-1).

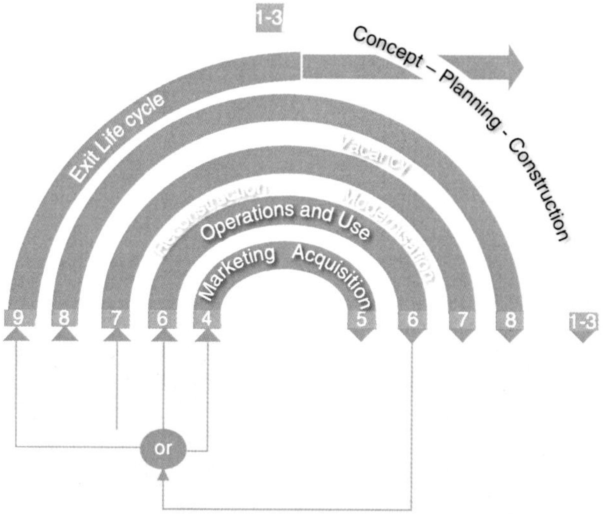

Figure 35.6 Life cycle phases after revitalization in cyclic representation (in accordance to German GEFMA 100-1).

In this phase, those extensions, total conversions, upgrades, conversions, extensions, renovations, revitalization, and so on, of objects are carried out that cannot be used and exploited throughout the total phase.

Figure 35.6 shows the following phases of revitalization with the consequent further options in the revitalization phase:

- Phases 1 to 3: design, planning, and construction
- Options are Phase 4: Marketing
- Phase 6: Operation and use.

35.8 Deconstructing Phase

With the deconstructing of buildings, the controlled demolition of whole buildings, as well as additions and renovations and floors is meant. In contrast to the earlier uncontrolled demolition, the demolition of buildings is nowadays often scheduled against a background of extensive waste separation of the different materials for the need of recycling and up-cycling.

The costs incurred during the life cycle of facilities costs, regardless of when they occur, can be made comparable by appropriate mathematical methods. For buildings, the building costs are not included in the life-cycle costs.

Furthermore, in the decommissioning phase cost, demolition and disposal costs are conceivable when recyclable building materials were used in the previous phases.

35.9 Benefits of FM

FM provides a long-term maintenance or an increase in the assets of companies in the form of buildings, equipment, and facilities. By economical and targeted use of resources, FM reduces the building and service-related cost over the entire life cycle. FM creates transparency in border areas of a company and relieves management and staff in secondary processes. With the help of an optimal workplace design, FM provides increased well-being among employees and thus, indirectly increases the productivity. Thus, FM makes an important contribution to corporate success. In concrete terms, this affects tenants and customers, owners, and the balance of sustainability.

36
Laboratory Optimization

Helmut Martens

The term *laboratory optimization* comprises a variety of claims, goals, and purposes. An individual laboratory optimization – whatever one may understand as the specific case – can be successfully completed only if it is clear as to what exactly the purpose is, so that the goal can be stated in concrete terms. The definition of the goal should always be the first step. Since laboratories are generally service providers for their customers (internal or external), and so dependent on their requirements, optimization efforts also need to always focus on the requirements and welfare of the customers in focus.

Each expected optimization – no matter what kind – must be able to be measured with reference to clearly defined indicators, because it is the only way to determine whether the target has been optimized successfully and achieved. This, in turn, means setting goals so that you can not only define objectives but also assign key figures. In case no measurable best parameter can be found, it cannot really be an optimization provable with hard facts.

This applies completely independently of the nature and scope of the optimization endeavor. Even small improvements should not leave the assessment with the feeling that it is usually a bad counselor and also an argument that is not worth the while when it comes to convincing skeptics and motivating employees in the long term. Employees develop a very keen sense of not clearly communicated changes, which are then easily understood as arbitrary, so that the successful implementation in most cases is doubtful.

Another pillar of any laboratory optimization is the determination of the initial situation, so the conscientious actual recording and description of the current state are ideally equivalent, both of which are required for measuring the achievement of the objectives. Only when the current actual condition is described and the goal defined does it make sense to be specific about the procedure envisaged, to make a calculation of profitability and assessment of sustainability, to look for suitable tools to outline the project process, to specify the resources required, and to calculate the necessary budget.

Most optimization projects in laboratories do not fail because of the lack of will but because of inadequate project planning or the lack of (or faulty) budgeting.

The Sustainable Laboratory Handbook: Design, Equipment, Operation, First Edition.
Edited by Egbert Dittrich.

36.1
The Procedure

The constant, consistent, and customer-oriented optimization of all processes and procedures in the laboratory with the aim of "state of the art" and "best practice" is in the long run the only effective strategy (see Chapter 37) to compete successfully in the market or to face the emergence of a serious competitor. This is applicable to free contract laboratories as well as public institutions and industrial laboratories because there are now virtually no laboratory services that cannot be provided otherwise and that, in the case of tenders for the services, cannot find any bidder. The comprehensive definition and development of standards by all parties as an expression of the currently best way of execution of all activities represents the highest achievable level of the continuous improvement process.

The following aspects of an optimization project should always be observed and considered applicable in each individual case:

- Determination of the need for optimization
  - Early involvement of stakeholders (employees, customers, service, etc.)
  - Definition of the optimization objective
  - Determination of the characteristic(s) to measure the achievement of objectives;
- Identification and description of the initial situation
  - Survey of the processes and procedures
  - Inclusion of the interfaces
  - Data recording;
- Determination of the optimization potential
  - Practice oriented (efficiency, quality, etc.)
  - Relative approach, for example, in a model of different stages
  - Consideration of the environment (sustainability, ecology, customers, impact on other activities and processes, interfaces);
- Planning and implementation.
  - Defining a strategy or the right tool
  - Determination of the required resources
  - Definition of responsibilities
  - Drawing up the timetable
  - Budgeting.

The essential elements will be discussed in more detail later in this chapter (Figure 36.1).

There are several ways in terms of the actual recording or taking different directions, but they lead to comparable results. The way of recording the oriented indicators actually is tangible and comprehensible, while the path by determination of the relative position in a multistage model rather follows the philosophy of a desirable ideal state.

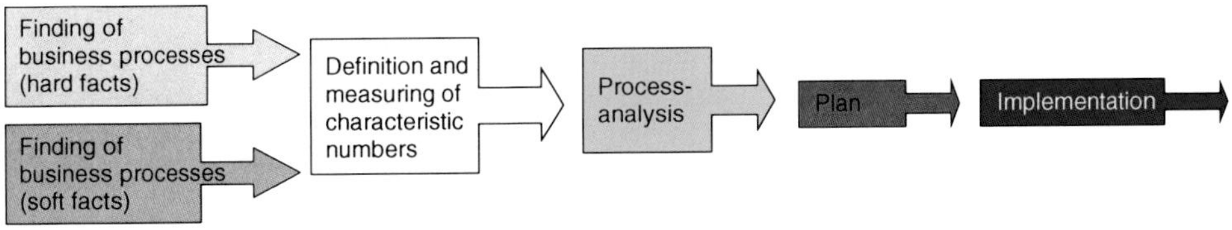

Figure 36.1 Identification and description of the initial situation.

36.2
The Actual Recording

At the beginning of each actual recording, all processes to be considered and each structure must be divided into smaller and measurable units, each in itself tangible, in order to understand the roots of potential problems and be able to draw the right conclusions from it.

The following hard facts should, where applicable, be always included in the analysis:

- process landscape, technical processes, and relationships
- equipment used
- computers and other tools used
- media used for documentation, transitions, and so on
- premises and their requirements
- interfaces
- statutory, regulatory, or other restrictions
- Data and facts that are needed for concept development and that are used as indicators in question.

The following soft facts should also be always included in the analysis as needed:

- hierarchical order
- staffing structure
- style of leadership, the leadership itself
- communication paths
- cooperation in horizontal and vertical directions
- identification of opinion makers
- identification of "pioneers."

The following example may illustrate this:

Since the processing time of the samples increased significantly in a laboratory, despite increased staffing, the processing time could not be reduced, so the whole process was divided into the following sections and the time required was recorded for each section with all abnormalities:

- Residence time in the sample input
- Residence times in the interim storage
- Processing time in the laboratory A
- Processing time in laboratory B

- Processing time in laboratory C
- Residence time to create the report
- Time for the preparation of the report
- Time for the release of the report
- Time for the submission of the report.

Two sections account for 90% of the time presented: first, the residence time in the sample input, and, second, the time for the preparation of the report, whereas the actual processing time reveals no significant problems in the laboratory. As benchmark for the internal laboratory optimization of the two identified weaknesses, the residence time in the respective area is used, while the initially defined processing time as (Key Performance Indicator) does not helpfully solve the problem, but as a distinctive unit quite impressively represents the success and therefore is reported.

The figures referred to in this example could be found by simple means in a short time. The situation is different when the task, quite succinctly is, for example, optimizing the utilization of employees. In this case, one must resort to employee retention time of each activity at each workstation in the laboratory. It requires a very high collection effort if the data is not available, in general.

Nevertheless, the utilization of the laboratory (employees as well as equipment) is one of the few interlaboratory usable parameters for evaluating the efficiency and competitiveness and, therefore, is very important to control and for flexible adaptation to changing needs.

36.3
Determination of the Optimization Potential

Whatever the objective, the optimization potential must be specifically identified and quantified if a reputable and verifiable way of a purely speculative action is preferred.

The optimization potential arises mainly from two intentions: either the customer expects an improvement or you yourself are committed to finding improvements for customers to have a long-term tie-up, or to attract new customers.

Following examples are typical for improvement:

- Media breaks in the documentation and reporting
- Unclear responsibilities
- Uncontrolled when diversification of equipment
- Lack of filling or overfilling
- Mismatch between value-sapping and value-adding activities
- Poor arrangement of the equipment, long distances
- Deficiencies in the infrastructure.

In addition to the obsolescence audit, which should be done over and over with reliable data, also the soft facts should be considered. Soft facts are evaluated that are becoming more and more customers' importance, as there would be, for example, responsiveness, speed, consulting, service, and so on. Therefore, you always must ask the question: what added value can you offer to the customer?

Besides the optimization of existing processes and services within the framework of the continuous improvement process, one may also come to the conclusion that the processes and services in its present form no longer can be optimized because too much has already been modified at too many places without a holistic, complete perspective and created such a thing that barely can be seen through nor can be systematically improved.

In such cases, it can be much more useful to utilize the tool "business process reengineering" (BPR). Here, what you can improve on what exists is no longer asked, but here the process is practically reinvented by asking the question, "How would the process (or performance) look like if we would have to invent it again now?" You must quite consistently disengage from everything you previously thought was normal to face and start all over again. This very exciting and challenging approach generally leads to leaner processes, which are then also supported by modern relief and means of communication and are oriented primarily to customers. Particularly successful can be such a project if all affected employees are involved and if you possibly can utilize external expertise, especially when it comes to the integration of current relief and means of communication.

36.4 Planning and Implementation

Depending on the size and the affected interfaces of the project, the concept is developed.

For this, managers have to be involved as well as performers and possibly also special experts (Figure 36.2).

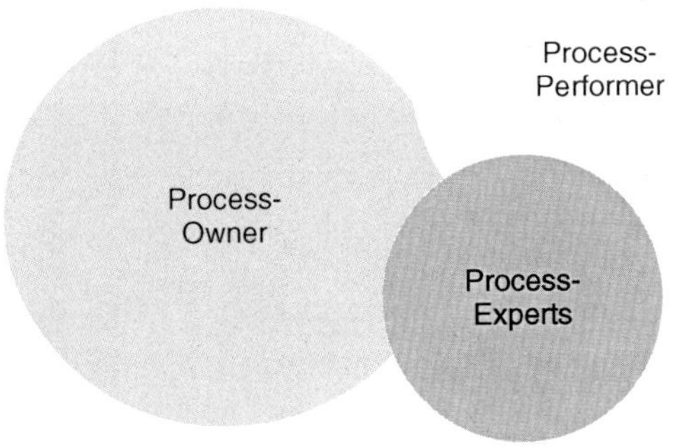

Figure 36.2 Process.

The team determination must necessarily be given to enable successful project implementation at all appropriate times.

If, for example, several or all areas are affected and if it is about a fundamental change, then the concept development proceeds hierarchically.

- At the managing level, the framework and overall objectives are defined.
- At the process level, the potentials are identified and found to be in conformity with the overall objectives of solutions.

For example, if only one area is affected and the measures have no influence on other areas, the concept must be developed by managing and process levels together.

Complex processes are split into subprocesses ("milestones") of meaningful size and a project plan in which the relations and conditions of the subprocesses are set for successful integration.

A milestone marks not performance, but the achievement of a defined state, which is caused by tasks or must be still achieved. Milestones serve the content and temporal control of projects.

In addition to the exclusive implementation of defined measures, which involve highly qualified personnel and do not present an insurmountable hurdle, optimization processes are often – also for those involved at all levels – represent a major challenge.

Besides changing one's own past behavior, it can also lead to lasting changes in the relationship between process participants. Also, one must say goodbye to cherished habits and a long-held view of things.

To face such a deep change, the assistance of a consultant and/or coaches should definitely be included in the budget planning.

36.5 Permanent Need for Optimization

Optimization needs must be achieved in the lab again and again throughout the process, as well as the logistics chain from the ordering or receipt of the samples, the (interim) storage, handling, preparation, and audit reports to their shipping and the whereabouts of the documents, records, samples, patterns of leadership, and employee involvement, the ergonomics in the workplace, dealing with suggestions for improvement, and so on, by industrial safety, occupational safety, and cooperation in the laboratory, with customers and with other departments.

In addition to the generic and subject-specific quality requirements, which usually have a high priority and always had, and inherent with the potential for natural science-oriented self-realization in the self-understanding of scientists, the economic and logistical requirements does not always have the ranking in regard to their importance adequate to existing laboratories. It is obvious here as a rule – in addition to the operation of soft facts – by far the most effective improvement in most laboratories.

In the laboratory – as with most service providers – the human resource plays an important role. Staff cost forms by far the biggest proportion of the total cost (about 60–75%, depending on the type of laboratory and structure) as also most know-how support without which it would be impossible to operate the laboratory. Therefore, this resource must be employed carefully and usefully, and handled economically and cost consciously.

Second, the customers, their expectations, and their desires play a crucial role. A laboratory optimization that serves only the needs of the employees of the laboratory but leaves out the needs of the customer would be a really poor approach.

Therefore, it is vitally important to consider all aspects and all potentially affected parties at an early stage.

36.6 An Example

In a chromatography lab, a gas chromatograph has been purchased that is equipped with a sampler for 100 samples. Per day, 10–15 samples are delivered to the lab. Employees collect the samples until the sampler is filled, that is, about once a week, to allow the device to run samples.

So far, the predecessor device was charged daily, manually with the incoming day samples; the next day they were all evaluated, at the latest. The customer got his data always the next day. He was happy for years.

The one-sided optimization of processes in the laboratory according to the wishes and needs of the employees would mean a wait for a week for the customer for his or her data with all process-relevant consequences. In the laboratory, in turn, under this condition, the unit in general would (application development, validation, etc.) be available on the remaining days of the week for other tasks, regardless of the actual need.

To achieve the optimum condition under the given circumstances, therefore, only means finding the best compromise between the different positions and not to enforce a maximum position. It would have been better in any case to take into account these considerations before purchasing the equipment, to assess and possibly to purchase two not quite as large-scale and correspondingly lower cost devices depending on the actual demand. So, you would have been able to provide the customer with a redundancy of systems with correspondingly higher reliability to his advantage.

As this example clearly shows, the goal of optimization should be defined in time and precisely. Each upcoming procurement action, each planned refurbishment or new construction, in short any change taking place in the laboratory environment quite basically presents an opportunity for the realization of the potential for optimization. Often it offers itself precisely in such situations over a narrow time window the chance to achieve much with as little effort as possible, while you have to run a much greater effort for improvements in hindsight.

36.6.1 Another Example

A prosperous company is planning to build a laboratory because the old building is bursting at the seams. The people responsible are aware that the existing building infrastructure is unfavorable and in no case should serve as a template for the new building. Hopes are raised for a better infrastructure that will also be able to process more orders with the same occupants.

The management of the laboratory feels that, due to lack of experience, it is not specialized in such things to be able to detect all possible optimization potential and to develop sustainable guidelines for the design of the new building. Therefore, to take on this task, an experienced consultant is to be engaged.

Example of economic efficiency: The fee of the consultant for the entire project amounts to €100 000. In the laboratory, nearly 50 people are employed, 5 of them in management positions. The total staff costs amounts to €2.5 million, and the number of samples processed is approximately 50 000 per annum. If the consultant assumes, as is quite common with experienced consultants, that 30% of optimization potential can be realized in the context of planning at no extra cost, they could process in future instead of the 50 000 samples/a 65 000 samples/a with the same personnel expenses, with a cost advantage of 750 000 €/a. Thus, the fee of the consultant would have paid for itself in less than 2 months.

If you were planning to take the existing structure into the new building and only later try to utilize a similar potential for optimization, you would have to make a refurbishment of the new building, which would easily eat up the calculated savings in the first 2 years and still would not make any improvement. Also, this would be profitable over time, but it would then be net € 1.4 million more expensive. Should the building be only rented, you could forget all optimization potentials, because changes would be upon the request of the landlord – if at all – paid dearly, as each specialization reduces the chance of a subsequent alternative use.

After many years of experience, unfortunately, it is far too often the case that such plans are operated for too long without the direct participation of those affected; and only then, when plans and financing are in the bag and perhaps even the equipment is ordered, the persons concerned are informed and their opinion is sought.

36.7 Utilization of Staff

The "utilization of employees" is an outstanding key performance indicator (KPI), which nevertheless in most laboratories cannot be supported with concrete data.

The capacity utilization of the laboratory is one of the few parameters for efficiency and competitiveness and, therefore, very important to control and adapt to changing needs.

The actual utilization of the staff is usually specified as a percentage in relation to the net working time, that is, the sum of the planned hours of all employees minus off days and other absences. The exact number can thus always be determined afterwards, what should absolutely be done for a very good reason. If the planning underlying absenteeism at the end of the survey period will no longer be corrected, and is – as it happens in individual companies as well – calculated across all departments with a fixed failure time base, and this openly communicated also to the entire staff as a quasi-fixed size, for sure a claim of ownership is developed because these cost even appear as a fixed calculator size.

In order to plan for the future, experience of past years is mostly used.

Determination of the staff utilization makes the following statements:

- The actual utilization is a factor for determining productive and contingency cost, and, thus, is also an important aspect of pricing.
- Disparities between different areas and over or underutilization are identified; it can be taken as targeted measures.
- With current numbers, it can be determined immediately whether new inquiries and orders with the existing staff in the expected time can be processed.

36.8 Utilization of Equipment

In parallel with the acquisition of employee retention times, measurement of the utilization of devices is of great importance for the planning the laboratory operation.

The device running times depend on the application in the first place, and with the different degrees of diversity (e.g., titration depending on consumption) provided. Thus, one should use the average figures and not try to capture the accuracy in seconds.

The actual utilization and the available capacity of the unit are recorded analogous to the retention times.

Similarly, the numbers of capacity planning and the current needs assessment are used.

In the optimization in the area of equipment, the following should be noted:

- The more heterogeneous the equipment fleet, that is, the more manufacturers and types for an inspection exist, the more protracted will be the optimization if you cannot make a major investment in a short time.
- The more heterogeneous, for example, the use of columns, gases, eluants, and so on, on the systems, the more is needed to invest in time and know-how.
- The possible availability of equipment, however, often (autosampler, automatic control, automated sample preparation, different working time models) increases efficiency with relatively little effort.

36.9
Employee Retention, Employee Retention Time, Device Runtime

To get to the characteristic of employee loyalty to truly resilient material data, you have to capture all the activities of all employees and then break them down into defined and measurable steps. The time spent by the employee for the performance of each substep is then determined as the retention time. Great care must be taken that no device running times but only the times in which the employee actively participates are recorded in the process and he or she has no time for other activities.

Each time recording, whether for personnel or equipment, should always be performed on the basis of a single measurement and not on the basis of samples. The number of measurements is, in principle, variable, and therefore serves only as a multiplier. The same applies for fixed, predetermined calibration, comparison standard, and blank measurements.

36.10
An Example

The sample preparation for a chromatographic test procedure is 8 min, during which the employee is fully involved. Then, he starts the device, lets it run in 10 min, then, subsequently, enters in 2 min the sample data, positions the vial, and starts the run. The measurement lasts for 30 min, during which time the unit is running unattended. After the measurement, the employee needs 10 min to evaluate the chromatogram, and enter the data into the laboratory information management system (LIMS). The total processing time in this case is exactly 60 min and the device runtime is 50 min, while just 20 min is directly taken by the staff in conducting the examination.

This 20 min is the determined retention time for this process.

As can be seen already, the effort taken to capture all times can be immense depending on the number of test methods and other activities, and with large units the cost sometimes is a mid-six-figure sum. Since this parameter depends very much on the individual, the individual residence, individual paths between the stations, and the in-between required tasks make a precise acquisition with a stopwatch in practice absolutely no sense, because the thing to accept fault tolerance is simply too large.

36.11
Cost

Unlike in the manufacturing sector and in the industry, in the laboratory the cost structure is a special case. The staff cost is about 3–5% of the total cost in the industry, but in the laboratory it is about 60–75% of the cost. This also makes clear at a glance where lies the most potential to reduce cost. Certainly, the other costs, such as material costs, rent, depreciation, maintenance, and so on, are also worth watching, but big jumps here are not to be expected, especially since it is often only through additional investments that the payoff occurs, that too only in the medium term.

Cost should ideally apply only where value to the customer takes place. Cost to the consumer has thus to be minimized, and avoided, in the ideal case. Such cost is generated in the laboratory wherever there exists unnecessary workplace or where there is idle time.

Contingency costs, vacancy costs, or deployment costs are, for example, as follows:

- Fixed cost of service provision (energy, space, depreciation, training, maintenance, quality management measures, etc.)
- Personnel cost for service provision
- Personnel costs for substitution (vacation, illness, etc.). That is, expenses are incurred even without service output, since in most laboratories employees are made permanent.
- Contingency cost or cost of idle capacity incurred. This can take place if the forecast or agreed performance increase is not observed or it is only temporary, so that more personnel should be employed than would be required on average.
- Contingency costs can also arise.
- Uncertain or unregulated service (lack of knowledge of the market, a lot of "walk-ins," etc.)
- Safety aspects (measurement only when needed, in case of failure)
- Preservation of know-how in the laboratory, without concrete service demand (on its own or on customer request)
- Excess capacity (inefficient and expensive equipment, only a few samples)
- Lack of information (missing data acquisition and evaluation, no or poor customer contact, etc.)
- Unreflective procurement of equipment without specific identification of needs
- "Hoarding" (storage of no longer used methods, reference materials, equipment, etc.).

36.12 Logistics

Both the logistic processes within the laboratory and those between the laboratory and customer leave a lot of room for improvement.

Within the laboratories, it is in most cases because, though they are well-equipped initially, subsequent tasks have to be performed with this prescribed equipment. Long distances by suboptimal arrangement of the individual instruments and equipment, interim storage, and workstations between each other are then also often found, as well as a lack of information about what was on, which sample already made, for example, where it is currently located, what time is left for the still outstanding work steps, and what capacities are available. True time wasters and information killers are such related activities, which are carried out in widely separated areas or even floors.

Especially in the area of logistics, LIMS would be required. But, even without expensive computers, laboratory logistics can be much improved often with simple tools and the good will of all employees, for example, through the use spaces differently.

The same applies to the integration of the customer in the laboratory logistics. With modern computers and the Internet, it is possible without much effort to move all the sample application from status tracking to statistical analysis to the customer so that he is completely independent of the presence and availability of laboratory personnel at all times to take a look at his samples and its data, generate any statistical analyses and graphical representations of himself, and use them in reports, and thus is constantly informed.

36.13 Quality

By "quality" (see Chapter 37) in the laboratory, very often the quality of the results of all services for customers is understood. This should now be self-evident, at least it will be provided to the customer.

Meanwhile, the quality of a laboratory by customers is also defined by other factors.

So a quality laboratory is understood as a laboratory with competent and customer-oriented staff who provide the clients with the information in an understandable, open, and honest manner in order to do something with it and the results can realistically be assessed. By quality of a laboratory, it is also understood that the customer is constantly advised and his problems are solved step by step. With the Internet and the increasingly connected comparability of services, and the inexorable exchange of information on the subject, the key competitive advantages will be increasingly found outside the "classical" terms of quality. This is what laboratories should face up to.

36.14 Customer Satisfaction and Customer Loyalty

As mentioned in the sections earlier, the customers are less a fixed calculable size that one owns quite naturally. Laboratory services can be performed around the globe, and big, global logistics companies make available samples and materials within a short time (within 24 h) under defined conditions to almost any place in the world, so that almost no client is bound to a specific laboratory. Thus competition is not only about the price but also about an extremely high level of customer satisfaction and loyalty, in the medium and long term.

Active feedback from the customers plays an increasingly important role. When customers give willingly a constructive feedback, it is often already a good sign. Poor suppliers are rebuked for failures or are left out. To develop from this customer feedback improvements, the service provider should join hands with the client in an ongoing shared culture of improvement of mutual relations that binds the customer, which in the medium and long term helps to overcome any temporary disturbance of the climate relatively unscathed.

36.15 Laboratory Indicators

Depending on the objectives of the optimization, the following exemplary figures, with which success can be measured, are listed (no claim to be complete):

Aim	KPI	Definition
Improvement of efficiency	Utilization of employees	% of nominal time
	Utilization of equipment	% of nominal time
	Decrease of unit cost	€ per unit
	Share of hold-up cost	Ratio of hold-up cost and total cost
	Share of set-up cost	Ratio set-up cost and total cost
	Share of QA cost	Ratio of QA and total cost
	Flexibility of employees	Ratio of controlled working methods and total methods
	Decrease in cost of complains	Ratio of cost and total cost
	Decrease in cost of repetitive tests	Ratio of repetitive test and total tests
	Flexibility of devices	Number of processes per device
	Decrease of maintenance	Ratio maintenance cost and total cost
Improve quality	Progress in customer satisfaction	Ratio of unsatisfied and satisfied customers
	Reduction of complaints	Ratio of complaints and orders
	Reduction of inaccurate tests	Ratio of inaccurate tests and orders
	Participation on round robin tests	Ratio of the processes confirmed by round robin tests and total tests
	Reduction of processing time of complains	Time to accomplish
	Quality of test processes	Time between customer request to KPIs

Aim	KPI	Definition
Qualification and motivation of staff	Training	Ratio of participants and total staff
	Number of staff away sick	Ratio of sick days and nominal days
	Suggestions for improvement	Number of suggestions per 100 employees
	Fluctuation	Labor fluctuation rate
Staffing policy	Age pattern	Comparison with average in business
	Labor costs	Comparison with average in the business
Test methods	Employee loyalty	Time of direct loyalty per test method
	Calibration effort	Ratio of calibration and test
Improving logistics	Cycle time	Time between order of resp. sample entry and delivery of KPIs
	Time to produce reports	Ratio of report time and cycle time

The detection and tracking of indicators must not be an end in itself, but also should be subject to strict economic constraints. Possible optimization potentials are not to be eaten up by additional work elsewhere or even reversed. Ideally, laboratories have, therefore, to ensure on computerized systems the collection of relevant laboratory characteristics with minimal effort and in a short time. But even in these cases, the initial collection and entry of data is associated with a high cost, as it would always arise recurrently at a purely manual collection and tracking.

Here, the calculation of costs and benefits to be expected is not easy. Maybe that is why so far only a minority of laboratories consistently work with KPIs, although there is actually no alternative in the long run.

37
Quality Management

Helmut Martens

Quality control, quality assurance, quality management: are these all the same? What is behind the terms and what requirements to laboratories also arise from the different standards and laws?

37.1
Quality Control

Quality control checks at the end of production in compliance with the specifications, in order to remove bad parts. Thus, the quality control ensures that only specification compliant products reach the customer. Due to lack of evaluation and control elements of quality control, it lacks any ability to improve processes and thus, to reduce a possible error rate systematically. Rejects, no matter how extensive, is accepted as part of the whole and is usually recorded in the pricing as a fixed figure.

37.2
Quality Assurance

In quality assurance, there is a systematic monitoring and evaluation (validation) of all production processes in such a way to avoid on the way of preventing possible errors and so to get there, only to produce as specified.

37.3
Quality Management

The quality management system coordinates and controls the most comprehensive tool, linking and integrating all quality-related processes and procedures throughout the organization, with the aim to improve them permanently and thus, strengthen the company's position in the market and ultimately to secure the jobs.

The Sustainable Laboratory Handbook: Design, Equipment, Operation, First Edition.
Edited by Egbert Dittrich.

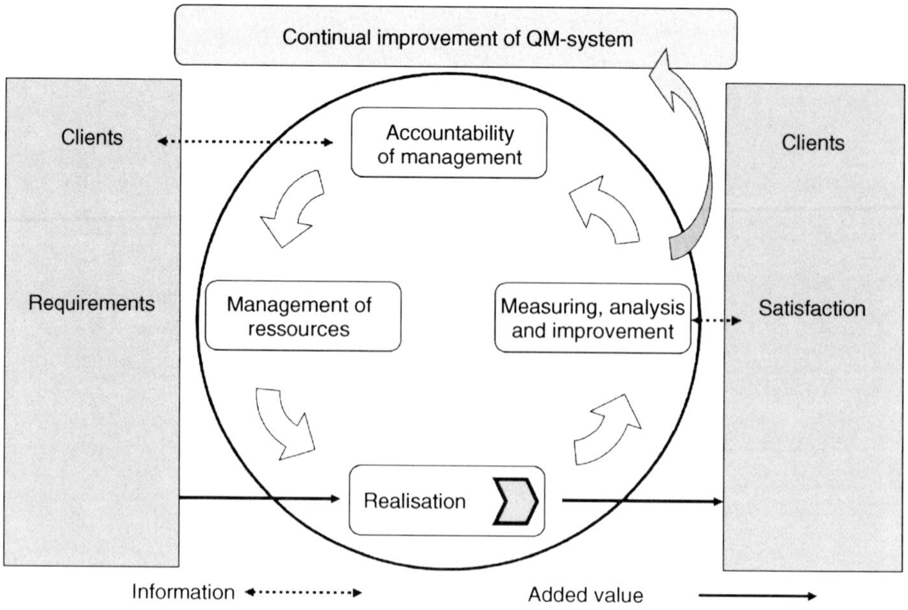

Figure 37.1 QM system.

In the legally regulated areas (GMP, GLP, GCP, etc.), the focus is on safety of patients (e.g., pharmaceutical industry) or more generally, on human beings (e.g., environmental monitoring, approval of new substances). Here, we find strong elements of quality assurance and in production, as a kind of last resort before the release of products, a highly regulated quality control. U.S Food and Drug Administration (FDA) in 1997 introduced ISO 9001 and ISO 13485 for medical devices as the basis of a generally binding quality management system in the pharmaceutical industry. The introduction of the continuous improvement process (CIP) fails in the laboratories in the majority of cases of misunderstanding and cost arguments, since continuous improvement also means a constant update of the documentation and thus, also represents a latent risk of having to update recurring filings with the approval authority, which in turn significantly is associated with cost. But, this is also due to the erroneous belief that all possible or to-be-aimed improvements relate only to the pure inspection process and so must inevitably lead to the follow-up cost (Figure 37.1).

In areas not regulated by bylaws, these include virtually all other industries and institutions, various quality standards are applied, having one thing in common, in essence:

37.4 Creation and Maintenance of a Quality Management System

On closer examination, however, more closely with the material and the underlying objectives, it must be said that the whole purpose of a management system is too often not understood nor implemented accordingly.

37.5
The Purpose of Systematic Quality Management

All known current quality standards can be very significantly reduced to works of the American, W. Edward Deming, who was taught by Walter A. Shewhart, the founder of statistical process control, who from the 1940s, developed the process-oriented view of the activities of the company, which later reflected in various quality standards and quality regulations.

The ideas of Deming found in 1950, was adopted only in Japan, where they were from some engineers, who quickly and very successful transferred into corporate philosophies such as Kaizen or TPS (Toyota Production System) and perfected over many years. It was not until about 1980, where in the Western Hemisphere for the first time the ideas of Deming and their effects consciously perceived the production in the consumer goods industry to the Far East without it being properly understood and it was hardly put into practice meaningfully. Only after Porsche rose for most profitable car factory in Europe from impending insolvency case by the former CEO, Wendelin Wiedeking took Japanese engineers in the company and successfully commissioned with the implementation of a similar system with TPS, and slowly changed the perception. The current ISO 9001, the " mother" of all other standards for quality management, is largely based on the ideas and work of Edward Deming and includes the CIP (Deming Cycle PDCA cycle) as an essential core element of all action.

To be described and to be coordinated, process steps cover all the contributions that have discernible impact on the quality of services and products. The improvement as a permanent job to all employees must be tested for their effectiveness; this is only possible by means of suitable indicators. With Key Performance Indicators, you can not only measure the instantaneous success but also define targets and monitor their achievement, continuous or discontinuous. A functioning quality management system is thus at the heart of good corporate activity and source of measurable business success.

37.6
Integrated Management Systems

The systematic approach of ISO 9001 on the tasks of a management system has now also found imitation in related systems such as ISO 14001, which describes an environmental management system. This system makes it much easier for combined and integrated management systems (IMSs) to introduce, for example, quality and the environment in a variety of forms and so to keep the administrative and documentation requirements manageable. In addition to the commonly held and, in principle, for any organization usable ISO 9001, there are also derived rules (e.g., in clinical area), which set specifically different priorities, not in spite of but just are more combinable with other standards in the laboratory.

In principle, any combination of different standards and also the involvement of statutory regulations such as GMP, GLP, and so on, in an IMS are conceivable. The important thing is always the consistent observance of the interfaces and overlaps,

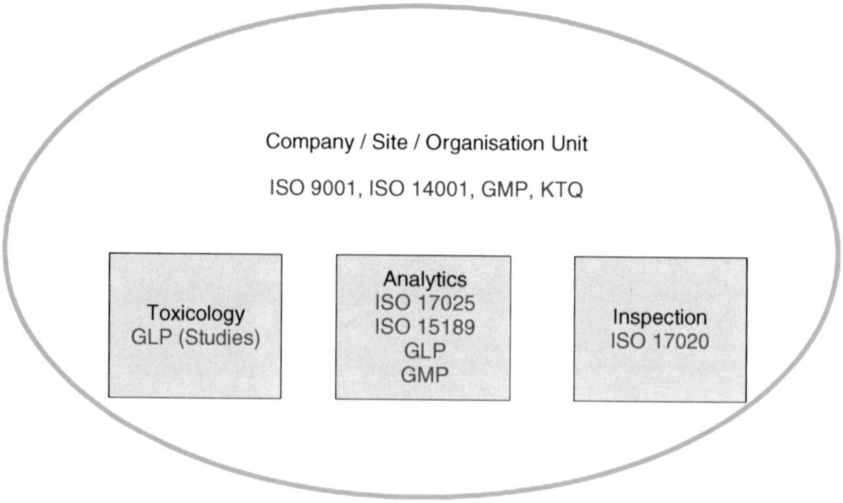

Figure 37.2 Management system certified.

and in particular, the consideration of special requirements of individual standards in the description of the host system in order to realize this fully and in all elements. For example, a superior system should not only allow certain types of documents, if subordinate systems have needs beyond for other types of documents. Here, one would violate own rules of the organization (Figure 37.2).

37.7 Certification or Accreditation

Certification and accreditation are two central concepts of quality management in laboratories that again and again create discussions, misunderstanding, and misuse. The distinction is quite simple.

Certification is provided by an independent third party verification of conformity to a standard (compliance audit). There, the proof is thus furnished that all elements and requirements of a standard or a standard-like regulation are understood and applied. A certification may refer to people, products, or management; it is the certificate of a property. The certification of a quality management system, for example, says nothing about the quality of services rendered and the expertise of the staff, but merely that all requirements of the standard have been met.

Accreditation, in turn, is the formal recognition of the professional and technical competence and independence of an authority to carry out certain tasks or to perform services on the basis of the aspects to demonstrate the competence advanced management system. This is to certify ability to precisely defined activities that are explicitly named.

Bodies wishing to carry out internationally recognized certifications must be accredited by an internationally recognized body.

Laboratories that want to show their expertise and their skills to potential customers or authorities are, therefore, well advised to apply for accreditation, insofar as this is possible for their services. Interesting and somewhat surprising here is that the FDA, for example, for pharmaceutical laboratories only have the ISO 9001 expected to be based on a quality management system and does not require proof of competence in accordance with ISO 17025. Also, in the accreditation, it can give similar as in the example above, combinations of different standards into a single, integrated quality management system. Maybe mentioned as typical here would be the combination of the standards ISO 17020 and ISO 17025 in the forensic science, or the combination of ISO 17025 and ISO 17043 in an application laboratory, which also offers round-robin tests.

Accreditation standards compared

- Testing and calibration laboratories (ISO/IEC 17025)
- Human Medical and diagnostic laboratories (EN ISO 15189)
- Inspection bodies (ISO/IEC 17020)
- Certification bodies for products (DIN EN 45011/ISO/IEC
- Guide 65/Draft ISO 17065)
- Certification bodies for management systems (DIN EN ISO/IEC 17021)
- Certification bodies for persons (DIN EN ISO/IEC 17024)
- Producers of reference materials (DIN EN ISO/IEC 17025 and ISO Guide 34)
- Provider of round-robin laboratory tests/DIN EN ISO/IEC 17020/DIN EN ISO/IEC 17025/ISO 17043.

37.8
International Recognition of Accreditation

The international comparability of the quality of all accreditation bodies is ensured by international cooperation through multilateral agreements and through the regular evaluation of accreditation bodies by the respective umbrella organization (Figure 37.3).

Following measures are essential in helping:

- Recognized national organization (e.g., German Accreditation Body, DAkkS),
- Multilateral Agreement (EA, ILAC),
- Involvement of stakeholders (advisory boards),
- Regular monitoring of compliance with the requirements of ISO 17011 (EA evaluations).

37.9
Central Functions of Quality Management

According to ISO 9001:2008, "top management" must appoint a board member to the tasks that are necessary to achieve and maintain a quality management

National: DAkkS Deutsche Akkreditierungsstelle — e.g. German Accreditation Body
www.dakks.de

International: EA European co-operation for Accreditation — European co-operation for accreditation
www.european-accreditation.org

ilac-MRA — International Laboratory Accreditation Cooperation
www.ilac.org

IAF International Accreditation Forum — International Accreditation Forum
www.iaf.nu

Figure 37.3 International accreditation bodies.

system. Here, the "Quality Manager" is thus always a member of the governing body. In the accreditation standards, there are slightly different definitions and also an additional.

First of all, here is the "Quality Manager," as it is usually in the majority of quality systems, an employee of the company or just a person who needs to have this dedicated function access to the highest governance body. Unlike from the ISO 9001, thus, allows you to derive any mandatory membership in the top management.

In all cases, the quality manager, however, is responsible for establishing and maintaining the quality management system, that is, it is equipping its force function also with the necessary powers. Otherwise, it would remain a toothless tiger, fighting a losing battle and be constantly burned in danger.

In addition to ISO 9001, in the accreditation standards that are based on technical competence, there is the function of the technical management. This is responsible for the technical implementation of the provision of services and their results, and providing the necessary resources (people, devices, materials, etc.). This responsibility of the technical management in turn yields, notwithstanding the provisions in ISO 9001, also a necessary task distribution between him and the Quality Manager.

If the Quality Manager draws ISO 9001 responsible for the maintenance of the quality system and for the release of all documents of the quality management system, we have in the world of standards for accreditation a division of tasks and responsibilities, which the QM officer cannot meet alone.

If the technical management is responsible for providing the technical expertise and – liability law – is responsible for all results, then the technical management needs, with their competence also, to gain the power to release the relevant documents (sampling and testing instructions including auxiliary processes) responsible for the content.

The Quality Manager, in accordance with standard indeed can be an employee who is under certain circumstances, perfectly legitimated even, but has not the technical expertise to assess each of these technical documents adequately and find possible errors.

Correctly, the Quality Manager in the accredited laboratory is therefore responsible for reviewing all documents of the quality system as they are released by the Chief Executive. The Quality Manager, unlike in the ISO 9001, required, usually does not belong to this governing body.

The technical documents, in turn, are accordingly approved by the technical management. This is reasonable and the persons in charge of considering solution must be allowed in accordance with an IMS (integrated management system).

37.10
Responsibilities of the Quality Manager in Practice

Besides the basic functions already described in detail, the QM manager is confronted in daily practice by tasks which he has to face and also requires a qualification that goes beyond the described facts, for example:

- Regulate and implement all requirements of the standard.
- Updates to the standard and rules of the accreditation are taken in a timely manner.
- All of the laboratory is covered by the regulations.
- The arrangements are appropriate and economically feasible.
- The rules must be given and explained to all employees in a timely manner.
- Compliance with these regulations is monitored regularly.
- Detected errors shall be eliminated.
- The QM system shall be constantly improved.
- The communication with the accreditor must be maintained.

In order to meet this, the following properties are required:

- Human skills, such as natural authority or enthusiasm.
- Psychological expertise, for example, in the fields of communication, motivation, or conflict management.
- Organizational competence; QM manager must plan steering concepts and enforce validity beyond the big day, reliable and appropriate for the laboratory.
- Expertise, such as technical intelligence and comprehensive expertise.
- Pedagogical competence, such as the knowledge of teaching methods and learning techniques.

37.11
Implementation of a Quality Management System in the Laboratory

The introduction of a quality management system is a project, which temporarily binds considerable resources and which should therefore be carefully planned and prepared.

Professional support by an external consultant can be very helpful and speed things up greatly if he is actually familiar with the explicit underlying standards of laboratories, also knows from long personal practical experience and has experience in introducing quality management systems. Especially, it should be considered when choosing the big difference between the aiming for certification and the pursuit of accreditation.

Otherwise, it is important to identify the individual and specific laboratory-related priorities and systematically tackle the project. Example is given here to the introduction of an accreditation as is the rule in the laboratory world. Here it is, first of all, to clarify the following questions:

- Who is named as Quality Manager?
- Is our project so ever accreditation efficient? If in doubt, contact with the national accreditation body should be taken as early as possible in order to clarify the basics.
- How should the registration systems for documents, facilities, reference, and reference standards, and so on, look like?
- What are the different types of documents, in addition to the quality manual, and which mandatory is actually still needed?
- How do we distinguish these types of documents so that they can be sorted any time or can be selected?
- How should the documents be usefully created and controlled?
- Are there standards or normative procedures for the application applicable during processing for the customer order or mainly house methods? These results provide the still open validation effort, which can be very time consuming.

Only when these issues have been resolved and brought to the way of solution, it is time to devote to the other requirements of the respective standard. These further requirements claim resources either, but they are to meet generally faster and with small effort.

37.12
Documents

The standards differ between documents and records. Documents within the meaning of the standards contain guidelines on how something should be done concretely. Records in accordance with the standards contain all the information about how something was done. If one keeps in full to the specification documents, you have to this end, in contrast, for example, work according to GMP regulations, not yet even record in detail (Figure 37.4).

This image shows that the scope of the specification documents and then the resulting records can be adapted to the needs of the laboratory and also should. Other aspects such as turnover rate of staff, level of education, employment of interns, trainees, and so on, can make it beyond useful and necessary to make the

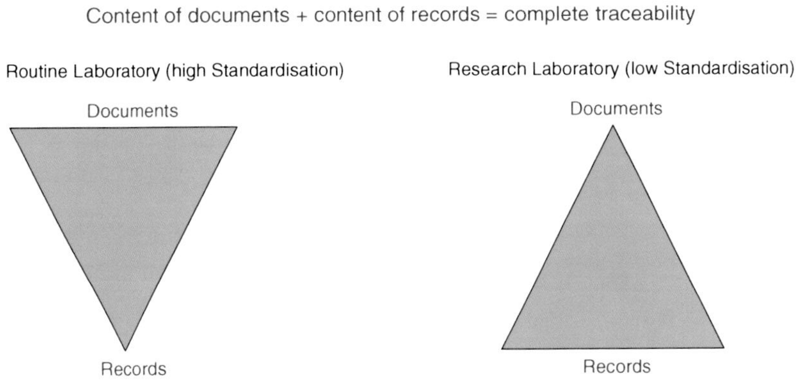

Figure 37.4 Specification documents.

specification documents extensive than it actually be necessary for the qualified personnel.

The specification documents then in turn also differ in two groups: the documents for regulating and maintaining the quality management system, and the documents to ensure the competence of service delivery. The two groups clearly differ in the content, in the structure, and the responsibility.

Main document regulating and maintaining the quality management system is the quality manual, which must always be present. Depending on the size of the laboratory or the number of individual regulations in certain areas, or to the relatively high update requirements, it may be useful to all the rules to make some of the procedures outside the QM manual and refer to it only in the QM manual. This document has the more or less popular term *method statement*, the term is, however, arbitrary.

The designation of documents for service provision is arbitrary. Encountered most often are work instructions, test instructions, inspection statement, sampling instructions, method of analysis, and so on. Actually, coming only from legal requirements of the GLP – range, the term *SOP* which is available in a standard controlled quality management system is used more frequently for all kinds of instructions. The problem here is, first, that there can be no, for all document types, equally well-suitable uniform structure and, second, that the documents are no longer distinguishable and not selectable, which usually increases the workload for the QM officer unnecessarily.

The use of model documents (templates), whether acquired by purchase, as "give away" somewhere added free at a training or obtained from another laboratory is usually not a problem and must therefore be carefully considered. No lab is like another; the specific procedures, responsibilities, and meaningful level of regulation can or should not be assumed by third parties. In addition, such patterns can simply be totally out of date or incorrect, or even nonsensical and contain hazardous business rules or simply too complex and expensive for its own needs. Sample documents can provide an important input when it comes to what one might regulate. But nothing more!

37.13
Expiration of Accreditation Project

After fundamental issues are clarified and the QM officers have been appointed, it is recommended to create an actual recording of the laboratory in order to meet the requirements of the standards. Suitable for this purpose, in addition to the relevant standard itself, are checklists of accreditors for each standard, as it is used for external audits. Only when the actual recording is completed and has been evaluated, the actual action can be identified and quantified. In addition, it is also clear where any outside support is useful because too little information is available about the meaning and purpose of the normative requirements or because it is simply the "knowhow" lack. Then, the individual tasks are identified, addressed, and placed in the correct order.

Roughly outlined, the order of good manner would be like follows:

- If necessary, information of operating or staff council about the meaning and purpose of the project
- Preparation of documents (quality manual, instructions, etc.)
- Training for all employees affected
- Validation of tests and possibly software
- Registration and identification of facilities
- Registration and identification of reference materials and reference standards
- if necessary, concept for computer (data protection and security, audit trail, etc.)
- Permanent implementation control.

It applies to the QM representatives in the laboratory to be as much as possible present to explain the new rules again and again and to monitor whether and how the reaction proceeds, or actively promote as needed. The CIP is a basis of standards. Therefore, it makes no sense, for example, to delay the release of the documents to end. The right thing is to drive forward as soon as possible and then to test in everyday life, whether the determined regulations prove themselves in any arrangements. Otherwise, they are just improved and system certified to comply with the provisions taken thereto. The same applies, for example, for necessary procurement actions, which under any circumstances should be postponed but should be pulled through. Finally, you can show only for current events that the quality management system also works as described. If not, there is still the CIP. In all the arrangements, it is always to be observed that:

- the balance between quality and cost-effectiveness is highly respected,
- customer orientation is advance formalizing and,
- it is refrained to exceed the regulatory requirements of the standard.

The last point is to watch over and over again when a system is newly introduced. Here, the variety of motifs may play a role. But, all have in common the danger that it is hard to get rid of the once exaggerated arrangements. To explain internally why they had overdone it in the heat of the moment is not easy for everyone. To explain to the accreditation body that a previously established regulation is senseless can lead to discussions and problems with auditors. Therefore, the case should be limited to the essentials and then rework rather useful when needed. Importantly alone is that any requirement of the standard is disregarded and nothing has been regulated and described additional. At which time the contact for accreditors is wanted depends on how specific the services of the laboratory are to be classified. In very special services, it may make sense to look first for the contact. Otherwise, the time is reached when all documents have been submitted, are effected, and an internal audit showed no unregulated requirements or serious flaws. Cooperation with the accreditation body is organized as follows:

- Application procedure
  Request, application, application verification, accreditation agreement, if necessary preliminary interview
- Review process
  Selection of assessors, confirmation by the applicant, assignment of assessors, examination of the documents, on-site assessment, assessment report
- Accreditation
  Decision by the accreditation committee, issuing the document recording the directory, extension/change of accreditation
- Monitoring procedures
  Regular surveillance audits (max. every 15 months) re-accreditation after 5 years.
  From the position of the application up to the grant of the certificate, a period of about 6–9 months must be calculated in general.

  Cost drivers of an accreditation project can be:

- Laboratory size (number of employees, various locations)
- Scope of accreditation (testing field number, number of test methods, etc.)
- Starting level,

  but also:

- Lack of experience of the QM manager and/or the technical management
- Frequent change of managerial staff
- Misinterpretation of the standards or any further requirements of the accreditation body
- Lack of planning, several new beginning
- Missing concept.

Although the vast majority of cost is never recognized as a rule, because "only" labor internally is generated, such a project requires a major effort from all involved, which could be alleviated effectively and efficiently through targeted external support and kept bearable.

38
Data

Helmut Martens

Data comprise logically grouped information units that are transmitted between systems or stored on a system. That simply sounds a possible definition. But data and knowledge in the form of various pieces of information forms the backbone of a laboratory operation. Data in the above sense practically represents the special asset of any laboratory (brain capital), which is the basis of customer relationship, and should be maintained accordingly, protected, and kept confidential.

Data can appear in any form in the laboratory and on every conceivable medium. In their systematic collection and administration, corresponding demands are made.

Laboratory data is collected and stored, like it has always been, very carefully and accurately on paper ("lab notebooks"). Bundled by topic or otherwise in strict chronological order, they are understandable with some effort, even after many years. The mere linking of various pieces of information that are stored in different locations can take a large amount of time. The accumulated data per time unit was manageable until modern working methods and equipment came into being, which increased the performance of a laboratory unit and the amount of data collected per unit of time substantially. Even in small laboratory units with manageable amounts of data, it has therefore become the rule to capture the data in computer systems to be processed and stored. The former practice of dealing with extreme care these data on paper has been done away with, and the computer, with its advantages but also with its various quirks, has become the popular tool and is dealt with so often that carelessness can sometimes lead to anxiety and fear.

By now, most laboratories depend to such a high degree on electronic data systems that the laboratory operation can be jeopardized in case of loss of data or information, affecting the customer relationship or even the basis of contract.

On the other hand, however, the power of modern computer systems is not optimally used, as the person responsible is seldom known, by which the efficiency of the process suffers unnecessarily.

The Sustainable Laboratory Handbook: Design, Equipment, Operation, First Edition.
Edited by Egbert Dittrich.

Errors committed before the procurement (e.g., no thought-out concept, no specification, lack of involvement of those affected directly, lack of resources, etc.) do the rest, turn the blessing of the art too often into a curse for the employees.

38.1
Data Systems

In each lab, there are various types of data that are often kept and maintained in different data systems, under different persons in charge. While referring to data, we distinguish roughly between personal data, monetary data, and laboratory data, and their ranges overlap and have common interfaces.

In addition, specialized knowledge is an important part of the capital of a laboratory. In order to keep this knowledge searchable, we need to use modern database systems.

As is usual in all other organizations, the data in different areas or departments are handled in the laboratories as a rule, which all use their own systems to do so. The introduction of quality management systems on the one hand and increased cost and performance pressure on the other entail integrated concepts and solutions necessary to enable the laboratory manager to lead his laboratory according to demand increase and meet his increasing responsibility with reasonable effort.

38.2
Data Systems at the Corporate Management Level

Stepping onto the variety of different systems and to gain an overview is beyond the scope of this chapter. Therefore, here we give only a few exemplified procedures.

At the corporate management level, personnel data, orders, release data, data of performance, and cost accounting are typically handled. For these tasks, systems such as SAP or other techniques are applicable. But, some of the data requested that are held in some form in the laboratory are also needed there. We shall think of personnel data, such as the entire issue of training and qualifications or calculations for services rendered to occupancy rates, and so on. Often, there are no viable interfaces, or the data controllers deny access, and so on. So, a lot of data is stored redundantly in multiple systems, which does not make economic sense and results in additional risk of inconsistency.

Where the companies are too small to afford such systems, personnel data, cost data, or performance data can often be transferred to the accounts department, for example. From there, the day's business gets done. On request, data on economic development will be provided from there. However, these data are rarely processed in a way that they are really lab intern used for recognition, and minimization of vulnerabilities. Whoever is interested in this shall install an internal system. Thus, we are facing a redundancy of incompatible systems also.

38.3 LIMS

Laboratory Information Management System (LIMS) is the general name for IT solutions of various kinds that are developed and marketed to be intersectoral but also very industry-specific. Recent market surveys are found at regular intervals in various professional journals and other publications.

Following changing customer needs, systems that were once exclusively developed for sample and sample-data management have been increasingly expanded in their functionality. Thus, all LIMS vendors have migrated from them to more or less foreign territory, which itself appears in different quality and characteristics of the enhancements. Even though in the last few years much has happened, there are still significant differences.

This is not to say here that you basically have to use everything that modern LIMS offers. This applies, first of all, because of the premise that the maintenance of the system and the data should not cost more effort than it delivers in savings and advantages. Less can sometimes be more!

In addition to these enhancements, some other fundamental aspects also play a role that you should know in order to avoid serious and costly errors in procurement. Some are mentioned here, with no guarantee of completeness:

- To what extent are the needs of the laboratory of standard applications covered?
- Can the necessary adjustments all be made with reasonable effort?
- Which interfaces are included in the standard?
- How user friendly is the licensing policy?
- What adjustments are included in the purchase price?
- Who guarantees and what does it cost to maintain the adaptation software releases and updates?
- How complex is the maintenance of master data?
- Can changes to layout and reports, for example, by the provision of appropriate tools, by own staff be accomplished?
- Is the software compatible with existing programs?
- Are the software running on the existing operating systems and on more recent versions the same (servers and clients)?
- Does it conform to the quantity structure of the data streams and print jobs with the performance of the desired system configuration?

38.4 LIMS Selection and Procurement

If you wish for potatoes you better grab a hole.

Such mottos one should heed, because before entering into a purchase, one should have made a thorough comparison of eligible systems based on concrete requirements.

This can be done only when you, on one hand, develop a coherent and comprehensive approach and, on the other hand, collect a quantity of structured data streams and interfaces present in the laboratory.

A prepared concept should address the following:

Description of the actual object as the starting point:

1) General requirements
   a. Access to the system and its features
   b. Technical framework
   c. Maintenance of master data
   d. System support
2) Functions and processes
   a. General functions
   b. Master data management
   c. Carrying out the test order
   d. Status dependencies
   e. Area or customer-specific functions
3) Involved areas, and so on
   a. Software
   b. Hardware
   c. Interfaces
   d. Premises
   e. Scope
   f. Procurement period
4) Cost-benefit analysis
5) Quality aspects
6) Ongoing costs
   a. Hardware
   b. Software
   c. Master data maintenance
7) Feasibility.

At the same time, the requirements are worked out, which are regarded as essential and these are clearly distinguished from those that can be defined as optional and negotiable (nice to have).

The to-be-determined quantity structure should include the following:

- Number of potential users
- Number of systems concurrently accessing users
- Sample throughput per day/per year
- Availability of the system (24 h operation? Weekends?)
- Sample types
- Percentage distribution between the sample types
- Transmission of data in a higher level system
- Integration into a company network/intranet/Internet
- Number of different test methods

- Number of different testing scopes
- Kinds of transmission of the results
- Number and type of output units
- Type and number of interfaces to analyzers
- Number and spatial distribution of the sites
- Form of long-term documentation.

Based on these findings, the detailed specifications will be created.

The final specification is then passed to the prospective provider, who in turn makes a comparison with their software and describe in a specification sheet how they intend to solve that problem, including an indication of the cost. Here, users should definitely ask for possible follow-up costs caused by exchange in operating systems, releases, and so on.

The specification is then the basis for the creation of a concrete and comprehensive offer.

In addition to this rather basic illustration, there are in practice two very different approaches that you should know.

The approaches here depend on the personnel involved, that is, whether the whole is planned at the theoretical basic management level, or by practitioners who subsequently make use of it.

In the first case, you will create specifications based on a systematic analysis that will become known or scrutinized completely, and the service is rendered completely on this basis. Additions and changes are difficult, and definitely costly. So, you commit yourself quite early. Experiences of such a procedure have shown that the larger the number of participants and the farther away they are from direct use, the greater the risk of new additional claims. This explodes the cost, and the budget goes completely wrong, or when a fixed price has been agreed upon, the supplier drops out. In both cases, the investments are lost.

After bids have been solicited on the basis of the concept, two to three providers that appear to be most suitable could be determined. Alternatively and purposefully, a delta analysis can be made on a test installation in concrete terms, with all functions based on the concept and all recorded amendments, which can then be converted into a specification. This approach has its advantages but is suitable primarily only for people who can deal with a time-intensive system and have good knowledge and routine in dealing with LIMS.

How many test installations of this are played out in the first place is a question of resources and also of the evaluation. If exclusion criteria encountered, one must willingly or otherwise install an additional test candidate. If it is more a question of taste, you should pragmatically weigh it. Of course, the possibilities of testing are also dependent on how interesting the project is for the provider, and one could even find several vendors who are willing to test the installation without significant functional limitations over a period of several weeks. It is not uncommon here for a considerable fee to be charged that will only be credited, in part, in case of a later complete or part purchase.

The second option has some advantages. First, the users also gain timely experience with it, how the software can be used, and how intuitive the user interface presents itself as a whole. On the other hand, also in due course meaningful features may be added, to which one might not otherwise have access in the classical form of performance specifications, or otherwise may not have been in focus. No one benefits from a system that is on paper and costly when it is not accepted because of serious deficiencies; it may even be boycotted. A significant portion of installations fails exactly because of this.

In the field of good manufacturing practices (GMP), the general procedure for the procurement of LIMS also depends on how it can be represented in the validation master plan, which comprises company-wide rules. The specification determines the design and is later qualified and validated.

38.5 Requirements for a Specification

38.5.1 Content and Classification

This introduction gives a brief overview of the realizable system in terms of

- product scope (integration into business processes, the enumeration of the main and sub-functions of the system);
- definitions, terms, abbreviations (In this chapter, all concepts, terms, and abbreviations are defined that are necessary for understanding the specifications. This can also be done by reference to the appendices or other documents.);
- restrictions for system development (It refers to, for example, organizational matters, hardware, software, network, interfaces to other systems, operating conditions, data protection requirements, audit or other requirements, integration of equipment, etc.);
- standards to be complied with (showing all requirements derived from higher level standards and instructions, such as data names (balance concept), billing regulations, regulatory requirements).

38.5.2 Conditions

- Access rules/user administration (password assignment, assigning the access rights to functions, deputy-control, logging (audit trail) to data and user);
- Data security, cryptographic methods, logging (audit trail), backup data, programs (frequency), formation of checksums for data transfers;
- System reliability and assurance measures (availability, reliability, replacement system, redundancy, manual continue working, restart);
- User interface and layout (function key assignment, general screen layout, general list layout, barcode format, backup system);
- Description of the dialog sequence (menu);

- Requirement on product quality (usability, changeability, universality, portability, efficiency, robustness).

38.5.3 Descriptions of Function

- Main and sub-function name;
- Conditions for the execution of the function;
- Assigning access permissions to functions;
- Input variables (quantitative and timing of keyboard, files, barcodes, scanners, online devices);
- Processing (representation of processes, algorithms, controls, alarms, error handling, logging (audit trail), plausibility checks, load behavior);
- Output variables (quantitative and timing, to screen, file, print lists, barcode, follow-up processing, receiver/location);
- Interfaces to other systems (quantitative and temporal occurrence or calling internal/by other systems to over-bearing data, communication channels).

Regarding the licenses actually required, there are also significant variations in the licensing policies between providers. Therefore, the part procured must be clarified with respect to how many users actually need a full license, for example, to maintain master data, make corrections, or perceive other tasks that require this. In most LIMS, the normal data is inputted, and usually a limited research meanwhile can take place on access via web browsers, which in any case are much cheaper (usually free of charge or requiring a full license).

In this way, you can also enable your customers to have a cost-effective, individual access to his data without spatial or temporal restriction, so this research at any time and in almost any location is independent and can make statistical analysis without burdening the laboratory staff. This added value for the customer is rarely actively promoted and used.

38.6 Selection of Suitable Suppliers

In addition to meeting the requirements of the product, some additional points should be made, with the focus on the product to be able to gain medium- and long-term satisfaction for the customer.

Therefore, the following questions should be answered:

- Is the manufacturer well established on the market?
- Does the manufacturer offer adequate services (e.g., hotline)?
- Has the manufacturer implemented a quality system?
- Is the dealer (if there is no direct reference by the manufacturer) authorized by the manufacturer?
- Is the warranty valid for a reasonable period of time?

- Is the product sophisticated?
- Are there a number of active installations and users?
- Are there any good references in the organization or external users with high reputation?
- Does the product meet the requirements of relevant audits?

In some cases, it may also be also useful to clarify the following questions on the site, for example, by an audit:

- Does the company have an economically sound basis (avoid bad investments)?
- Is there an established quality assurance system (normative basis, documentation, validation, etc.)?
- How is the product development carried out (tasks, tests, release periods, user involvement, etc.)?
- How is the quality of the product (platform, complaints, etc.)?

38.7
Data Privacy and Data Security

While data and information on paper can be protected against unauthorized access and against loss with relatively simple and inexpensive measures, in the case of electronic data it is quite different. If all computers and devices are networked with each other and with the Intranet or Internet, the risk increases dramatically. Not only the responsible parties and the staff in the laboratory but also the customers have the fundamental and comprehensive right that the data do not fall into someone else's hands or get lost. The simple claim can be satisfied if the risks are explicitly known and the appropriate measures to minimize or eliminate them are established.

From this, obligation cannot be freed, shifting the responsibility to other departments. A realistic assessment of the risks and the importance of the different data and information for laboratory and customers can, in fact, be carried out only by the person responsible for the data.

Practical examples show again and again that it can otherwise lead to serious misunderstandings and unnecessary risks. For example, it is of little use if the laboratory determines in its quality management system that the data will be kept for at least three years, while the IT department, whose responsibilities are to provide backup, already discards the monthly backup files.

The people in charge have to capture all data streams and then carry out targeted risk assessments, and define the intended use at an adequate level of safety.

The risks of ensuring data security are, for example,

1) Technical risks
   a. Error of computers and peripherals
   b. Error of the disk
   c. Unsuitable environmental conditions

   d. Error or inadequate operator safety of the software used;
2) Organizational risks
   a. Insufficient or unclear division of responsibilities at, for example, support, data entry, checking, grant the right to write
   b. Operator error by inexperienced users
   c. Operator error by rush in the workflow
   d. The risk of falsification of datasets and programs.

38.8 Risk Assessment

The nature and extent of the arrangements to be made for the use of computer systems depend on the determination of the necessary level of safety, which is to be made by the risk assessment.

The risk of an event may jeopardize the availability, integrity, or confidentiality of the data, and its processing involves a pair of values of the two components.

- Frequency of occurrence
- Damage potential.

The aim of risk analysis is to identify the risks and assess and reduce the consequence of this risk to the extent that the residual risk is quantifiable and acceptable. The expenditure for the implementation of the method should be appropriate to the value of the computer applications and the values of the laboratory generally.

In a risk or safety needs analysis, for example, the following points are highlighted and evaluated:

- Is quality-related data processed?
- Is quality-relevant data programs that were not developed according to a GMP guideline used?
- Is there any program established by the appropriate licenses?
- Is there any program that uses guaranteed nonchangeable, write-protected original version?
- Is data shared by multiple users?
- Are adequate manuals available?
- Are there records for the configuration?
- Has the hardware been involved in testing?
- Are ambient conditions in the hardware on which the programs are installed adequate to the manufacturer's specifications?
- Are there any add-ons in these programs?
- Are there any interfaces to other programs?
- Is there a backup plan?
- Is there an authorization concept?
- Is there an audit trail that covers all quality-related data comprehensively?

38.9 Safety Management

The appropriate arrangements should include the following organizational measures:

- Responsibility for regular data backup
- Responsibility for the reconstruction of damaged or lost data
- Management of backup data
- Responsibility for determining the user profiles
- Updating of hardware and software and the associated liability of usability of older datasets
- Maintenance of shared data
- Advice and support to users
- Care and maintenance of equipment
- Training of users
- Training of administrators
- Training of those responsible for the master data
- Training of data protection aspects
- Contact for problems
- Access logging
- Terminate applications during breaks or at the end of work
- Hardware protection
- User ID
- Plans for emergency.

In addition, the following technical measures should be regulated:

- Safety procedures for data
- Backup plans for data
- Backup media for data
- Storage or treatment of the backup files
- Check the backup files
- Backup of configuration files
- Securing the masks and reports
- Backup of add-ons, macros, modules, and so on
- Backup of data before changes
- Frequency of the periodic data backup
- Regular backup of the total files (physical backup)
- Determination of the backup storage
- Identification of the backup carrier
- Storage of the backup carrier
- Avoiding damage to the backup storage
- Consideration of the ageing of backup carrier
- Versioning or overwrite protection for quality-relevant data
- Versioning of the master data

- Access to the source code
- Traceability of data.

On the subject of antivirus and firewall, the following stipulations should be taken:

- Virus scanning and protection programs used
- Viruses occurring during logging
- Procedure during the occurrence of viruses
- Use of a firewall
- Ensuring continuous update.

Finally, arrangements for maintenance should be taken into account:

- Sufficient dimensioning of the hardware
- Checking for changes to the hardware
- Ensuring confidentiality and integrity of data during maintenance, in particular by external forces
- Commitment from external contractors to maintain confidentiality of all received information
- The content and format of the logbook.

In the rule system, administrators are responsible for the organization of data security and the implementation of data backup. The back storage of quality-related data should be performed by giving the reasons. The initiator is responsible for the conformity of the data back storage with the regulations. (The back storage of data-relevant quality work could be subject to documentation of the manipulation of raw data.)

Back storage should be performed and documented with tested procedures. In the case of manual implementation, the individual commands should be recorded.

The actual implementation of information security measures should be regulated clearly and unambiguously in a statement. The following points should be noted:

- Data backup type (full backup, incremental backup, data export, system security)
- Interval (daily backup, weekly backup, monthly backups, annual backup)
- Media type (HDD, DDS tape, DLT tape, CD-RW, etc.)
- Number of generations (number of consecutive identical performed backups)
- Identification of the disk (each disk must be identified in a way such that it can be assigned to one or more backup runs)
- Storage (disk must be stored either in a fireproof storage safe or in a separate fire zone room from the computer)
- Secure storage of the disk against unauthorized access
- Behavior in case of error (here, actions are described that are carried out after a failure, such as repetition of the backup. Mistakes are always documented)
- Utilized backup tool and associated procedures
- Backup format

- Control of data backup
- Management of failed disk.

The controlled operation of a computer system for creating and processing data-relevant quality also requires systematic planning and documentation of all changes. In order to ensure that in complex systems changes in all aspects have been taken into consideration, arrangements should be determined to make a statement.

Changes to be understood are as follows:

- Changes of hardware components of a system
- Changes to software programs and modules with new functionality
- Changes in the mask layout
- Changes to configuration tables
- Any others.

If the modification plan is designed in the form of a checklist, modifications can be recorded simultaneously. The following aspects should, if necessary, be taken into account through a statement for a systematic approach:

- Aim and description of the change
- Risk analysis
- Clarification with respect to other systems (interfaces)
- Documentation of the change
- Deliverables change measures
- Estimated cost/time effort if possible
- If necessary, qualification and validation activities
- Approval for implementation/installation
- Entries in the logbook
- Examination after installation
- Acceptance of changes.

In case of any changes, attention should be paid to recheck also in the appropriate documents. With such update, all system documents may be affected.

38.10
System Documentation

To be able to use permanently, reliably, and safely a LIMS or similar software, the necessary documents must be kept safe, updated, and producible on demand. In particular, these include:

- Implementation and results of the risk analysis
- Specifications
- Source code or parameter descriptions
- Test plans
- Installation plans

- Configuration plans
- User information
- Manual for managers
- Acceptance reports
- Qualification and validation plans and manifestations
- Changes in hardware and software
- Other documents created during the development and those that describe, for example, workstation, network, and server.

38.11 Emergency Plan

Any laboratory work is time-critical, so a temporary system failure is difficult or impossible to cope with. If any operation is to be maintained without interruption, a corresponding emergency plan must be worked out and activated when necessary. Under the failure of a computer system are all malfunctions to understand that keep the system over a period to be determined unproductive. The period depends essentially on how close the system is involved in the production or which QA aspects of the productivity of the system are necessarily required.

The type of failure (hardware, software, network disruption, planned intervention, for example, for release change) is irrelevant.

The period and the measures to be taken are fixed by the user responsible for the system, possibly in cooperation with competent managers.

The organizational measures should be examined at regular intervals for their efficiency and effectiveness, and also the staff trained systematically. The training measures to be considered in particular are:

- Behavior in case of system failure
- Behavior in case of an incident
- Meaning of error messages
- Dealing with error messages
- Measures to protect systems.

An effective emergency plan at least describes the following items:

- List of individuals/service providers who are to be involved mandatorily in the implementation of the measures and the time in which to make this happen;
- Responsible persons who must be informed on the implementation of the emergency concept. It is also advisable to fix the way this should be done;
- The actions to be taken to restore the productivity of the system, which must be described as rules or to be referred to as related documents. Here, a possible necessary data adjustment or recovery of data integrity and completeness is included;
- The organizational measures to bridge the failure or to maintain the productivity must be described and, if necessary, assisted with tools.

Index

The Sustainable Laboratory Handbook: Design, Equipment, Operation, First Edition.
Edited by Egbert Dittrich.

g

n

o

p

Printed and bound by CPI Group (UK) Ltd, Croydon, CR0 4YY